Verilog Digital
Computer Design

Algorithms into
Hardware

ISBN 0-13-639253-9

9 780136 392538

90000

Verilog Digital Computer Design

Algorithms into Hardware

Mark Gordon Arnold
University of Wyoming

Prentice Hall PTR
Upper Saddle River, NJ 07458
http://www.phptr.com

Library of Congress Cataloging-in-Publication Data

Editorial/Production Supervision: *Craig Little*
Acquisitions Editor: *Bernard Goodwin*
Manufacturing Manager: *Alan Fischer*
Marketing Manager: *Miles Williams*
Cover Design Director: *Jerry Votta*
Cover Design: *Talar Agasyan*

© 1999 by Prentice Hall PTR
Prentice-Hall Inc.
A Pearson Education Company
Upper Saddle River, NJ 07458

All product names mentioned herein are the trademarks of their respective owners.

Prentice Hall books are widely used by corporations and government agencies for training, marketing, and resale.

The publisher offers discounts on this book when ordered in bulk quantities.
For more information, contact: the Corporate Sales Department at 800-382-3419, fax: 2012367141,
email: corpsales@prenhall.com or write
        Corporate Sales Department
        Prentice Hall PTR
        One Lake Street
        Upper Saddle River, NJ 07458

Printed in the United States of America
10 9 8 7 6 5 4

ISBN 0-13-639253-9

Prentice-Hall International (UK) Limited,London
Prentice-Hall of Australia Pty. Limited, Sydney
Prentice-Hall Canada Inc., Toronto
Prentice-Hall Hispanoamericana, S.A., Mexico
Prentice-Hall of India Private Limited, New Delhi
Prentice-Hall of Japan, Inc., Tokyo
Pearson Education Asia Pte. Ltd., Singapore
Editora Prentice-Hall do Brasil, Ltda., Rio de Janeiro

Table of Contents

Appendices

List of Figures

3. VERILOG HARDWARE DESCRIPTION LANGUAGE

4. THREE STAGES FOR VERILOG DESIGN

5. ADVANCED ASM TECHNIQUES

6. DESIGNING FOR SPEED AND COST

7. ONE HOT DESIGNS

8. GENERAL-PURPOSE COMPUTERS

9. PIPELINED GENERAL-PURPOSE PROCESSOR

10. RISC PROCESSORS

11. SYNTHESIS

APPENDIX C COMBINATIONAL LOGIC BUILDING BLOCKS

APPENDIX D.SEQUENTIAL LOGIC BUILDING BLOCKS

APPENDIX E.TRI-STATE DEVICES

Preface

When I started teaching Verilog to electrical engineering and computer science seniors at the University of Wyoming, there were only two books and a handful of papers on the subject, in contrast to the overwhelming body of academic literature written about VHDL. Previously, VHDL had been unsuccessful in this course. For all its linguistic merits, VHDL is too complex for the first-time user. Verilog, on the other hand, is much more straightforward and allows the first-time user to focus on the design rather than on language details. Yet Verilog is powerful enough to describe very exotic designs, as illustrated in chapters 8-11.

As its subtitle indicates, this book emphasizes the algorithmic nature of digital computer design. This book uses the manual notation of Algorithmic State Machine (ASM) charts (chapter 2) as the master plan for designs. This book uses a top-down approach, which is based on the designer's faith that details can be ignored at the beginning of the design process, so that the designer's total effort can be to develop a correct algorithm.

Chapters 2-11 use the same elementary algorithm, referred to as the *childish division algorithm*, for many hardware and software examples. Because this algorithm is so simple, it allows the reader to focus on the Verilog and computer design topics being covered by each chapter. This book is unique in showing the correspondence of ASM charts to *implicit style* Verilog (chapters 3, 5 and 7). All chapters emphasize a feature of Verilog, known as *non-blocking assignment* or *Register Transfer Notation (RTN)*, which is the main distinction between software and synchronous hardware. Except for chapter 6, this book ignores (abstracts away) propagation delay. Instead, the emphasis here is toward designs that are accurate on a clock cycle by clock cycle basis with non-blocking assignment. (Many existing Verilog books either provide too much propagation delay information or are so abstract as to be inaccurate on a clock cycle basis. Appendices C and D motivate the abstraction level used here.)

Chapter 4 gives a novel three-stage design process (behavioral, mixed, structural), which exercises the reader's understanding of many elementary features of Verilog. Chapter 7 explains an automated one hot preprocessor, known as VITO, that eliminates the need to go though this manual three-stage process.

This book defers the introduction of Mealy machines until chapter 5 because my experience has been that the complex interactions of decisions and non-blocking assignments in a Mealy machine are confusing to the first-time designer. Understanding chapter 5 is only necessary to understand chapters 9 and 10, appendix J and sections 7.4 and 11.6.

The goal is to emphasize a few enduring concepts of computer design, such as pipelined (chapters 6 and 9) and superscalar (chapter 10) approaches, and show that these concepts are a natural outgrowth of the non-blocking assignment. Chapter 6 uses ASM charts and implicit Verilog to describe pipelining of a special-purpose machine with only the material of chapter 4. Chapters 8, 9 and 11 use the classic PDP-8 as an illustration of the basic principles of a stored program computer and cache memory. Chapter 8 depends only on the ASM material of chapter 2. Chapter 9 requires an understanding of all preceding chapters, except chapter 7. The capstone of this book, chapter 10 (which depends on chapter 9), uses the elegant ARM instruction set to explore the RISC approach, again with the unique combination of ASMs, implicit Verilog and non-blocking assignment.

Chapters 3-6, 9 and 10 emphasize Verilog *simulation* as a tool for uncovering bugs in a design prior to fabrication. Test code (sometimes called a testbench) that simulates the operating environment for the machine is given with most designs. Chapter 10 introduces the concept of Verilog *code coverage*. Chapters 7 and 11, which are partially accessible to a reader who understands chapter 3, uses specific *synthesis* tools for programmable logic to illustrate general techniques that apply to most vendors' tools. Even in synthesis, simulation is an important part of the design flow. Chapter 11 will be much more meaningful after the reader has grasped chapters 1-9. The designs in chapter 11 have been tested and downloaded (www.phptr.com) into Vantis CPLDs using a tool available to readers of this book (appendix F), but these designs should also be usable with minor modifications for other chips, such as FPGAs.

Appendices A, B and G give background on the machine language examples used in chapters 8-11. Appendices C and D give the block diagram notation used in all chapters for combinational logic and sequential logic, respectively. Chapters 1-11 do not use tri-state bidirectional buses, but appendix E explains the Verilog coding of such buses.

This book touches upon several different areas, such as "computer design," "state machine design," "assembly language programming," "computer organization," "computer arithmetic," "computer architecture," "register transfer logic design," "hardware/software trade-offs," "VLSI design," "parallel processing" and "programmable logic design." I would ask the reader not to try to place this book into the pigeon hole of some narrow academic category. Rather, I would hope the reader will appreciate in all these digital and computer design topics the common thread which the ASM and Verilog notations highlight. This book just scratches the surface of computer design and of Verilog. Space limitations prevented inclusion of material on interfacing (other than section 11.6) and on multiprocessing. The examples of childish division, PDP-8 and ARM algorithms were chosen for their simplicity. Sections labeled "Further reading" at the end of most chapters indicate where an interested reader can find more advanced concepts and algorithms, as well as more sophisticated features of Verilog. Appendix F indicates postal and Web addresses for obtaining additional tools and resources. It is hoped that the simple examples of Verilog and ASMs in this book will enable the reader to proceed to these more advanced computer design concepts.

In places, this book states my opinions rather boldly. I respect readers who have differing interpretations and methodologies, but I would ask such readers to look past these distinctions to the unique and valuable approaches in this book that are not found elsewhere. I have sprinkled (somewhat biased) historical tidbits, primarily from the first quarter century of electronic computer design, to illustrate how enduring algorithms are, and how transient technology is. Languages are more algorithmic than they are technological. Just look at the endurance of the COBOL language for business software. Hardware description languages will no doubt change as the twenty-first century unfolds, but I suspect whatever they become, they will include something very much like contemporary implicit style Verilog.

Acknowledgments

The author would like to thank the following people who have contributed to the content of the computer design laboratory at the University of Wyoming: Tighe Fagan, Rick Joslin, Bob Lynn, Susan Taylor McClendon, Philip Schlump, Elaine Sinclair, Tony Wallace and Cao Xu.

The author would also like to thank his colleagues at the University of Wyoming, especially Tom Bailey whose editorial advice is appreciated very much, John Cowles who cheerfully endured a semester of teaching the computer design laboratory, Jerry Cupal who convinced the author to try Verilog and Richard Sandige who has engaged the author in stimulating discussions about digital design.

The author wishes to acknowledge the tremendous contribution of James Shuler of SUNY (Brockport, NY) in the development of the VITO preprocessor described in chapter 7 and appendix F. The author gratefully acknowledges Elliot Mednick of Wellspring Solutions (Salem, NH) for making a limited version of its VeriWell™ Verilog simulator available to everyone. The author is extremely grateful to Kevin Bush of MINC, Inc. (Colorado Springs, CO) for making a limited version of its VerilogEASY synthesis tool available to readers of this book.

The author sincerely thanks Freddy Engineer of MINC, Inc. and Neal Sample of Stanford for proofreading the manuscript. The author also thanks Karolyn Durer, Peggy Hopkins and Phyllis Ranz for their assistance in the manuscript preparation. The author wishes to acknowledge the contribution of his editors at Prentice Hall: Russ Hall, Camille Trentacoste, Diane Spina, Bart Blanken, Craig Little and Bernard Goodwin.

Finally, the author wishes to express deep appreciation to Frank Prosser and David Winkel of Indiana University for the inspiration they provided with *The Art of Digital Design*. It is hoped that the material presented here will contribute a fraction of what *The Art of Digital Design* has.

This book is dedicated to the memory of my father,
Gordon William Arnold,
whose encouragement and sense of humor
during the last year of his life stimulated
the writing of this book.
He was a wonderful dad,
and I cherished him.

1.   WHY VERILOG COMPUTER DESIGN?

1.1   What is computer design?

A computer is a *machine* that *processes information*. A machine, of course, is some tangible device (i.e., hardware) built by hooking together physical components, such as transistors, in an appropriate arrangement. Processing occurs when the machine follows the steps of a mathematical *algorithm*. Information is represented in the machine by *bits*, each of which is either 0 or 1. This book only considers digital information (i.e., bits) and does not consider analog information. Analog information can be approximated by digital information by using a sufficient number of bits.

Computer design is the thought process that arrives at how to construct the tangible hardware so that it implements the desired algorithm. The goal is to turn an algorithm into hardware. Computer designers have two ways to look at the machines they build: the way they act (known as the *behavioral* viewpoint, which is closely related to algorithms), and the way they are built (known as the *structural* viewpoint, which is like a "blueprint" for building the machine).

1.1.1   General-purpose computers

When you say the word computer today, it brings to mind what we refer to as a *general-purpose computer*, which you can program with *software* to implement any algorithm. With a general-purpose computer it is not necessary to build a new machine to implement each new algorithm. Programming such a general-purpose machine is often done with a conventional high-level language, such as C, C++, Java or Pascal.

1.1.2   Special-purpose computers

If you accept the definition of a computer given in section 1.1, there are many kinds of machines that fit this description in addition to general-purpose computers. We will refer to these other kinds of machines as *special-purpose computers*, which are non-programmable machines that implement one specific algorithm. A general-purpose computer is actually like a special-purpose computer that implements one algorithm, known as *fetch/execute*, that interprets a software program. The fetch/execute algorithm is fairly complex, so it is easier to study computer design by first looking at how simpler algorithms (than fetch/execute) can be transformed into hardware. For example, a traffic light is controlled by a machine that indicates when different colored lights are to be turned on. This machine is not programmable. Once it is designed, it always does the same boring thing: green, yellow, red, green, yellow, red, ... Nevertheless, by the above definition, it is a computer. It follows a particular (although boring) algorithm.

Special-purpose computers are ubiquitous because, in large volumes, it is more economical to manufacture a special-purpose computer that implements one boring algorithm than to purchase a general-purpose computer and waste most of its capabilities on that boring algorithm. However, for small volumes, or for problems where the specifications change frequently (such as tax accounting), the software approach is more economical.

There are only a handful of general-purpose computers on the market, and so there are not many jobs for designers for these popular machines. On the other hand, many non-computer industries use special-purpose computers as parts for the products they manufacture, and so job opportunities exist for designers of special-purpose machines. Also, special-purpose computers play a role in the peripheral devices, such as modems, that attach to general-purpose computers.

1.2   A brief history of computer/digital technology

The history of computer design highlights two things: changing technologies and lasting concepts. It is important to make a distinction between a concept and a technology. Information and algorithms are mathematical concepts that exist regardless of the physical details of their implementation with a particular technology. Many of the algorithms used in computers today were discovered by the great minds of mathematics decades or centuries ago.

Almost four centuries ago, Blaise Pascal (for whom the language is named) built one of the first mechanical calculators (which required a great deal of human intervention to operate). Pascal is remembered today however because he discovered several interesting algorithms, such as "Pascal's triangle," which are still in common use. A century and a half ago, Charles Babbage succeeded in using the technology of his day (precision cams and gears) to build the first fully automatic special-purpose computer for tabulating mathematical functions. Babbage also envisioned a general-purpose machine (with its fetch/execute algorithm) but was unable to complete it due to financial difficulties.

The invention of the vacuum tube was the technological advance that made building computers affordable. For a fraction of the cost of a machine built with cams and gears, a vacuum tube computer could automatically carry out hundreds of algorithm steps in a second. During the 1930s, C. Wynn-Williams in Great Britain built the first binary counter with vacuum tubes and the team of John Atanasoff and Charles Berry at Iowa State University built the first vacuum tube special-purpose computer for solving simultaneous equations. During World War II, several computers were built, including Colossus (in Great Britain), which was used to break coded German messages. In 1945, the mathematician John von Neumann popularized the idea of a general-purpose computer, and his name is often synonymous with a machine that implements the fetch/execute algorithm. The first operational general-purpose computer was the Manchester Mark I, which was a vacuum tube machine built in England that ran its first program in 1948.

In the 1950s, general-purpose vacuum tube computers cost millions of dollars, and only large corporations and governments owned them. The next major technological advance came with the invention of the transistor, which can do the same thing that a vacuum tube can do faster and more economically. Transistors also have the advantages that they run cooler and have a longer life than vacuum tubes.

This, of course, lowered the cost of general-purpose computers so that smaller corporations could own them, but it also made the application of *digital design* practical. Digital designs are special-purpose computers built using electronic circuits that process bits. Devices like digital watches, digital microwave oven timers, digital thermostats, hand-held calculators, etc. are all controlled by special-purpose computers that became economical with the invention of the transistor and related digital electronics.

In the 1960s, it became possible to manufacture hundreds or thousands of transistors on a chip of semiconductor material, known as an integrated circuit, at very low cost. Integrated circuits made it possible to mass-produce general-purpose computers, as well as digital electronic chips. Special- and general-purpose computers are now so powerful and affordable that they are part of almost every complex device built, from children's toys to the space shuttle.

Since the 1960s, there have been continual improvements in semiconductor technologies. It is now possible to get millions of transistors on a single chip. Of course, today's chips cost a fraction of the price of, and run faster and cooler than, their predecessors. But the algorithms that these chips implement are similar to the algorithms implemented with earlier technologies.

1.3   Translating algorithms into hardware

In the beginning, hardware designers were programmers and vice versa. The world of hardware design and software design fragmented into separate camps during the 1950s and 1960s as advancing technology made software programming easier. The industry needs many more programmers than hardware designers and programmers require far less knowledge of the physical machine than hardware designers. Despite this, the role of software designers and hardware designers is essentially the same: solve a problem.

Although many hardware designers realized in the 1960s and 1970s that their primary job was to develop an algorithm that solves a problem and translate that algorithm into hardware, some hardware designers lost sight of this essential truth.

An early notation for describing digital hardware that provides tremendous clarity in this regard is the Algorithmic State Machine (ASM), which was invented in the early 1960s by T.E. Osborne. As the name suggests, the ASM notation emphasizes the algorithmic nature of the machines being designed. Chapter 2 explains the ASM notation, and how it can be used manually to translate an algorithm into hardware. This notation is used throughout the rest of this book.

1.4 Hardware description languages

Unfortunately, hardware designers were inundated with the overwhelming technological changes that occurred with semiconductor electronics. Many hardware designers lost track of the advances in design methodology that occurred in software. Around 1980, as semiconductor technology advanced, it got more and more difficult to design hardware. Up to that time, most hardware design was done manually. Designers realized that the ever-increasing power of general-purpose computers could be harnessed to aid them in designing the next generation of chips. The goal of using the current generation of general-purpose computers to help design the next generation of special- and general-purpose computers required bringing the worlds of hardware and software back together again.

Out of this union was born the concept of the Hardware Description Language (HDL). Being a computer language, an HDL allows use of many of the timesaving software methodologies that hardware designers had been lacking. But as a hardware language, the HDL allows the expression of concepts that previously could only be expressed by manual notations, such as the ASM notation and circuit diagrams.

As technology advances, the details about HDLs will undoubtedly change in the future, but studying an HDL instills fundamental concepts that will endure. These ideas, originally thought of as hardware concepts, are becoming more important in software due to the increased importance of software parallel processing and object-oriented programming. There is a deep theoretic similarity between the concepts in software fields (such as operating systems and data structures) and the concepts in computer design. The growing popularity of HDLs attest to this fact: hardware is becoming more like software, and vice versa.

Chapter 3 discusses a popular HDL, known as Verilog, which is easy to learn because it has a syntax similar to C and Pascal. Verilog was developed in the early 1980s by Philip Moorby as a proprietary HDL for a company that was later accquired by Cadence Design Systems, which put the Verilog standard into the public domain. It is now known as IEEE1364. Verilog is used together with ASM charts in the rest of this book. This book is not just about Verilog or ASM charts for their own sake; this book also describes how these notations illuminate the thought processes of a computer designer. Ultimately, computation takes place on hardware. As children, all of us were inquisitive about everything: "How does this work?" Even if you do not plan on becoming a computer designer, it seems reasonable that you should be able to answer that question about the machines that are at the heart of your chosen career. The power of the Verilog and ASM notations give us insight for answering this question.

1.5 Typography

Fonts are used in this book to distinguish between different kinds of text, as explained below:

Times	is used in the bulk of the text for discussion.
Bold Times	is used to emphasize important or surprising concepts.
Italic	is used for the definition of an important term or phrase.
`Courier`	is used for Verilog text, exactly as it is typed into the file, and for similar notations taken from ASM charts, hardware diagrams, and simulation results. This font is also used for parts of other high level languages, such as C.
`Bold Courier`	is used for parts of Verilog text that are important in the discussion that precedes or follows them.
`Italic Courier`	is used to describe parts of Verilog syntax, such as a *`statement`*, which can be replaced with some particular symbol, such as `while`. Also, it is used to highlight complex simulation results.

1.6 Assumed background

It is assumed that the reader has a reasonable amount of experience programming in a conventional high-level language, such as C, C++, Java or Pascal. Programming experience in assembly language (appendices A, B and G) is very helpful. It is assumed that the reader can understand binary, octal and hexadecimal notations, can convert these to and from decimal and can perform arithmetic in these bases. It is also assumed that the reader is familiar with the common combinational logic gates (AND, OR, NOT, etc.), and that the reader knows about the common digital building blocks used in digital design (appendices C, D and E).

1.7 Conclusion

The few computers built in the nineteenth century were based on classical mechanics (cams and gears visible to the naked eye). Almost all the computers built in the twentieth century have been based on electronics. It is hard to say what technologies will be prevalent for computers in the twenty-first century.

Conventional semiconductor technology will someday reach its limit (based on the minimum size of a transistor and the speed of light). Technologies based on recombinate DNA, photonics, quantum mechanics, superconductivity and nanomechanics (cams and gears built of individual atoms) are all contenders to be the computer technology of the twenty-first century. The point is that it does not matter: technology changes every day, but concepts endure. The intellectual journey you travel by turning an algorithm into hardware illustrates these enduring concepts. I hope you enjoy the journey!

2. DESIGNING ASMs

This chapter explains the graphical notation used throughout the rest of this book. This graphical notation helps a hardware designer working only with pencil and paper. Chapters 4 and above require that the reader understand the notation explained in this chapter. Chapter 3 describes an alternative textual notation, known as Verilog, more suitable for automation (computer-aided design), where software tools help the designer produce a correct machine. The reader of this book will acquire a thorough understanding of both the notation in chapter 3 and the notation in this chapter, but we begin at the beginning with the most important question a hardware designer can ask: "How do I write down the particular algorithm that the hardware is supposed to follow?"

2.1. What is an ASM chart?

An *Algorithmic State Machine* (ASM) chart is a flowchart-like notation that describes the step-by-step operations of an algorithm in **precise** units of time. ASM charts are useful when you want to design hardware that implements some particular algorithm. The ASM chart can describe the behavior of the hardware without having to specify particular hardware devices to implement that algorithm. This allows you to make sure that the algorithm is correct before choosing an interconnection of particular hardware ("a structure") that implements the behavior described by the ASM. The most serious errors in hardware design do not result from connecting wires to the wrong place ("bad structure") but instead are the fault of designers not thinking through their algorithms completely ("bad behavior"). Designing a hardware structure is much more expensive than describing its behavior, and so it is sensible to spend extra time on the behavioral ASM chart before considering how to implement it with a hardware structure.

Although an ASM chart looks similar to a conventional software flowchart, the interpretation of an ASM chart differs from a conventional software flowchart with regard to how the passage of time relates to the operation of the algorithm. In software, the exact amount of time from one algorithm step to the next is not explicitly described by a flowchart. In the ASM chart, each step of the algorithm takes an exact amount of time, known as the *clock period* or *clock cycle*. There are also other time-related distinctions in the ASM chart notation which are described later.

An ASM chart is composed of rectangles, diamonds (or equivalently diamonds can be drawn as hexagons for notational convenience), ovals and arrows that interconnect the rectangles, diamonds and ovals. ASM charts composed only of rectangles and diamonds are said to describe *Moore* machines. ASM charts that also include ovals are said to describe *Mealy* machines. Mealy machines are described in chapter 5, and some of the more advanced concepts in chapter 7 and chapters 9 and above require the reader to understand Mealy notation. At this time, we will ignore the use of ovals and concentrate only on ASM charts for Moore machines.

Each rectangle is said to describe a *state*. A label, such as a number or preferably a meaningful name, can be written on the outside of the rectangle. The term *present state* refers to which rectangle of the ASM chart is active during a particular clock period. The term *next state* indicates which rectangle of the ASM chart will be active during the next clock period. The ASM chart indicates how to determine the next state (given the present state) by an arrow that points from the rectangle of the present state to the rectangle of the next state. Each arrow eventually arrives at one of the rectangles in the ASM chart. Since it has a finite number of rectangles, there is at least one loop in an ASM chart. An ASM chart is said to describe a particular *finite state machine*. Unlike software, there is no way to stop or halt a finite state machine (unless you pull the plug).

There is a relationship between the ASM chart and its behavior. For example, consider the following ASM chart with three states:

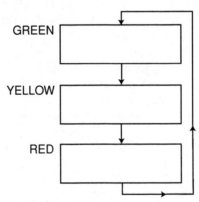

Figure 2-1. ASM with three states.

Assuming that we start in state GREEN, and that the clock has a period of 0.5 seconds, the ASM chart will make the following state transitions forever:

time	present state	next state
0.0	GREEN	YELLOW
0.5	YELLOW	RED
1.0	RED	GREEN
1.5	GREEN	YELLOW
2.0	YELLOW	RED
2.5	RED	GREEN
. . .	. . .	. . .

For example, between 0.5 and 1.0, the ASM is in state YELLOW. It is again in state YELLOW between 2.0 and 2.5.

The following sections explain the commands that can occur in rectangles of an ASM chart, the decisions that can occur in diamonds of an ASM chart, the input and output connections to a machine described by an ASM chart and issues of ASM chart style. Although the examples in the following sections vaguely resemble a traffic light controller, they are not intended to solve such a practical problem. They are instead intended solely to illustrate ASM chart notation and style.

2.1.1 ASM chart commands

Normally, the rectangle for a state is not empty. There are three *command* notations that a designer can choose to put inside the rectangle, which are described in the following sections.

2.1.1.1 Asserting a one-bit signal

A *signal* is a bit (or as explained in section 2.1.1.2, a group of bits) that conveys information. A signal is transmitted via a wire (or similar physical medium). The designer gives a signal (and its corresponding wire) a name to document its purpose. When the name of a signal occurs inside a rectangle of an ASM chart, that signal is asserted when the machine is in the state corresponding to the rectangle in question. In other state rectangles, where that signal is not mentioned, that signal takes on its default value. As an example, assume the default value for the signal STOP is 0. In the following:

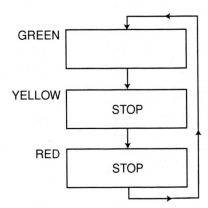

Figure 2-2. ASM with command outputs.

STOP will be 1 when the ASM is in state RED or state YELLOW. STOP will be 0 when the ASM is in state GREEN. The following illustrates this situation:

	time	present state	next state	
	0.0	GREEN	YELLOW	STOP=0
	0.5	YELLOW	RED	STOP=1
	1.0	RED	GREEN	STOP=1
	1.5	GREEN	YELLOW	STOP=0
	2.0	YELLOW	RED	STOP=1
	2.5	RED	GREEN	STOP=1
	. . .	. . .	. . .	. . .

2.1.1.2 Outputting a multi-bit command

When the name of a signal is on the left of an equal sign (=) inside a rectangle, that signal takes on the value specified on the right of the equal sign during the state corresponding to the rectangle in question. In other state rectangles, where that signal is not mentioned, that signal takes on its default value.

The following two diagrams show ASM charts that use =. The first of these ASMs is equivalent to the ASM given in section 2.1.1. The second example introduces a two-bit bus SPEED whose default value is 00.

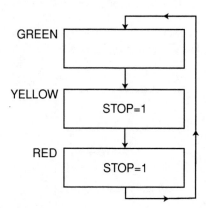

Figure 2-3. Equivalent to figure 2-2.

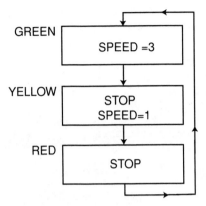

GREEN

SPEED =3

YELLOW

STOP
SPEED=1

RED

STOP

Figure 2-4. ASM with multi-bit output.

In the above, SPEED is 0 in state RED, 1 in state YELLOW and 3 in state GREEN.

time	present state	next state	output
0.0	GREEN	YELLOW	STOP=0 SPEED=11
0.5	YELLOW	RED	STOP=1 SPEED=01
1.0	RED	GREEN	STOP=1 SPEED=00
1.5	GREEN	YELLOW	STOP=0 SPEED=11
2.0	YELLOW	RED	STOP=1 SPEED=01
2.5	RED	GREEN	STOP=1 SPEED=00
...	...	...	... ...

2.1.1.3 Register transfer

The last two notations are simply a way of indicating how state names translate into physical signals, such as STOP and SPEED. Although we will eventually find many uses for these two notations, they are by themselves not the most convenient way to describe an algorithm.

Most algorithms manipulate variables that change their values during the course of the computation. It is necessary to have a place to store such values. Eventually, the designer will choose some kind of synchronous hardware register (appendix D) to hold such temporary values. In ASM chart notation, it is not necessary to make this design decision in order to describe an algorithm. Register Transfer Notation (RTN) (denoted by an arrow inside a rectangle) tells what happens to the register on the left of the arrow at the beginning of the next clock cycle. If a particular register is not mentioned on the left of an arrow in a state, the value of that register will remain the same in the next clock cycle. For example,

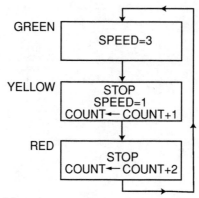

Figure 2-5. ASM with register output.

in the above assume that a three-bit COUNT register is 000 at time 0:

```
        present   next
 time   state     state
 0.0    GREEN     YELLOW    STOP=0  SPEED=11  COUNT=000
 0.5    YELLOW    RED       STOP=1  SPEED=01  COUNT=000
 1.0    RED       GREEN     STOP=1  SPEED=00  COUNT=001
 1.5    GREEN     YELLOW    STOP=0  SPEED=11  COUNT=011
 2.0    YELLOW    RED       STOP=1  SPEED=01  COUNT=011
 2.5    RED       GREEN     STOP=1  SPEED=00  COUNT=100
 ...    ...       ...       ...     ...       ...
```

Unlike STOP and SPEED, COUNT is not a function of the present state. For example, whenever the ASM is in state GREEN, STOP is 0. The first time in state GREEN, COUNT is 000, but the second time in state GREEN, COUNT is 011.

Notice that the ← causes a delayed assignment, which is different than assignment with conventional software programming languages, such as C. This is an important distinction to keep in mind when designing ASM charts. **One of the central topics that reoccurs throughout this book is the consequence of designing algorithms that use this kind of delayed assignment.** Although at first a novice designer may find ← unnatural and may make mistakes because of a misunderstanding of ←, once the reader masters the concept of this delayed assignment, all of the advanced concepts in later chapters will become much more understandable.

2.1.2 Decisions in ASM charts

One or more diamonds (or hexagons) following a rectangle indicate a decision in an ASM chart. The decision inside the diamond occurs at the **same time** as the operations described in the rectangle. There are two kinds of conditions that a designer can put inside a diamond in an ASM chart. These decisions are described in the following sections.

2.1.2.1 Relations

Relational operators (==, <, >, <=, >=, !=) as well as logical operators (&&, ||,!) can occur inside a diamond. It is also permissible to use the shorter bitwise operators (&,|,^,~) inside a diamond when all of the operands are only one-bit wide. When the relation in the diamond involves registers also used in the rectangle pointing to that diamond, the action taken is often different than would occur in software. Because the decision in the diamond occurs at the same time as the operations described in the rectangle, you ignore whatever register transfer occurs inside the rectangle to decide what the next state will be. The register transfer is an independent issue, which will only take effect at the beginning of the next clock cycle. As a illustration of such a decision, consider:

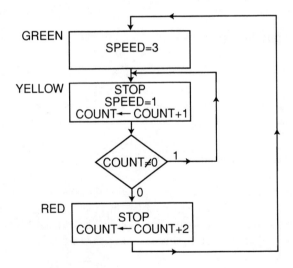

Figure 2-6. ASM with decision.

At first glance, it might appear that the ASM will get stuck in the loop the first time the machine enters state YELLOW because COUNT+1 is 001. However the decision COUNT != 0 is based on the current value of COUNT, which remains 000 until the beginning of the next clock cycle. Therefore the ASM exits from state YELLOW and proceeds to RED. On the second time the machine enters state YELLOW, COUNT is 011, and so it stays in state YELLOW for six clock periods. The only reason the ASM ever leaves state YELLOW is because the three-bit COUNT wraps around from 7 to 0.

```
           present   next
    time   state     state
    0.0    GREEN     YELLOW   STOP=0  SPEED=11  COUNT=000
    0.5    YELLOW    RED      STOP=1  SPEED=01  COUNT=000
    1.0    RED       GREEN    STOP=1  SPEED=00  COUNT=001
    1.5    GREEN     YELLOW   STOP=0  SPEED=11  COUNT=011
    2.0    YELLOW    YELLOW   STOP=1  SPEED=01  COUNT=011
    2.5    YELLOW    YELLOW   STOP=1  SPEED=01  COUNT=100
    3.0    YELLOW    YELLOW   STOP=1  SPEED=01  COUNT=101
    3.5    YELLOW    YELLOW   STOP=1  SPEED=01  COUNT=110
    4.0    YELLOW    YELLOW   STOP=1  SPEED=01  COUNT=111
    4.5    YELLOW    RED      STOP=1  SPEED=01  COUNT=000
    5.0    RED       GREEN    STOP=1  SPEED=00  COUNT=001
    ...    ...       ...      ...     ...       ...
```

The highlighted line shows the last time the ASM is in state YELLOW. The next state is RED because COUNT is 000.

2.1.2.2 External status

Many hardware systems are composed of independent actors working cooperatively but in parallel to each other. We use *actor* as an ambiguous term that incorporates other digital hardware (i.e., special- and general-purpose computers) as well as non-digital hardware and people who communicate with the machine described by an ASM chart. From a designer's standpoint, the details of the other actors are normally unimportant.

These actors need to send information to the machine described by the ASM chart. When such external information can be represented in only one bit, it is known as *external status*. (Multi-bit signals can be broken down into several single-bit status signals if desired.) External status signals have names that are simply labels for physical wires connecting the machine that implements the ASM chart to the outside world. By convention, the name of a status signal can occur by itself inside a diamond. The meaning of such a diamond is the same as testing if the status signal is equal to one. For example,

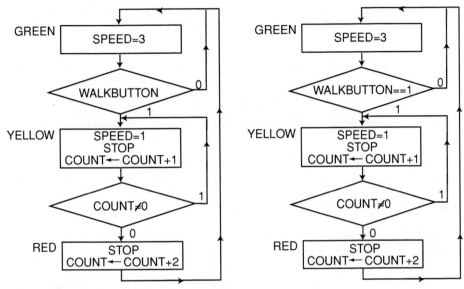

Figure 2-7. ASMs with external status.

The above ASMs are equivalent to each other. They will stay in state GREEN until WALKBUTTON is 1 during the last clock period that the ASM is in state GREEN. When WALKBUTTON is 1 in state GREEN, the next state will be state YELLOW. The machine ignores WALKBUTTON when it is in state YELLOW or in state RED.

2.1.3 Inputs and outputs of an ASM

An ASM chart describes the behavior of a piece of digital hardware without specifying its internal structure. The machine described by an ASM chart is often part of a larger structure. The machine receives inputs from other actors in that larger structure and provides outputs to other actors in that larger structure. In order to be part of such a larger structure, the machine described by an ASM chart must have a limited set of *input ports* (to receive information) and *output ports* (to send information). These ports are either single wires or buses having names that correspond to names used in the ASM chart.

In addition to its ASM chart, we can draw a *block diagram* of a machine. Such a diagram is sometimes called *a black box* because it hides the internal structure that implements the machine. The only thing we know is that the behavior of the machine is described by the corresponding ASM, and that the machine has inputs as specified by arrows pointing into the black box, and outputs as specified by arrows pointing out of the black box. As is standard notation in all hardware structure diagrams, when the input or output ports are more than one bit wide, the width is specified by a slash.

The following is a block diagram of the machine described by the ASM chart in 2.1.2.2:

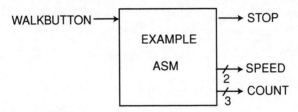

Figure 2-8. Block diagram.

2.1.3.1 ASM inputs

There are two kinds of inputs to a machine described by an ASM chart and black box diagram. It is somewhat arbitrary which of these two approaches a designer uses. The designer is free to choose the way that seems most appropriate for the problem at hand. It is permissible for a designer to mix these two approaches in a particular ASM. Since it plays a role later in the design process, the distinction between these two kinds of inputs is important to note.

2.1.3.1.1 External status inputs

A designer may consider a one-bit input port as an external status input when it is mentioned only by itself in diamond(s). Such status inputs are usually interpreted as providing an answer from the outside world to some yes/no question, such as "has the button been pressed?" As an example, in the block diagram above in section 2.1.3, WALKBUTTON is the only external status input.

2.1.3.1.2 External data inputs

When an input of any width is used only on the right-hand side of register transfers in rectangles and/or only in relational decisions, it is considered an external data input. Such data inputs usually play the same role that input variables play in conventional programming languages.

It is arbitrary whether the designer wishes to consider a one-bit input as a status or data input, as was illustrated in section 2.1.2.2. When a multi-bit input is used only with relational operations in diamonds (and not on the right of register transfers in rectangles), a designer may consider such a multi-bit input as being composed of several status input bits. For example, consider a machine with an external three-bit data input A. At some point in the behavior of the machine, the ASM needs to test if the value being input to the machine from the outside world on the bus A is equal to two. Figure 2-9 shows the block diagram and two equivalent ASM charts. The ASM chart on the left treats A as a single data input, but the ASM on the right treats A as being composed of three status inputs (A[2], A[1] and A[0]).

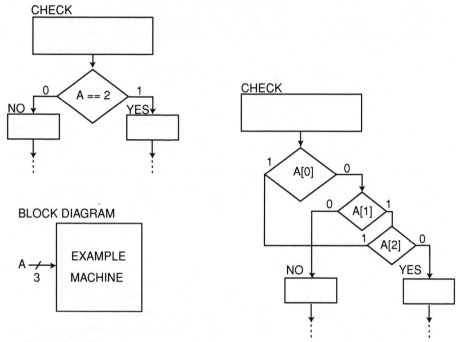

Figure 2-9. Two ways to test multi-bit input.

It is natural for designers to treat yes/no information as status. In most other cases, it is easier for the designer to consider something as a data input than to consider it as a status input, as the above ASMs illustrate. Inputs used on the right of register transfers must be treated as data inputs.

In the block diagram above in figure 2-8, there are no external data inputs.

2.1.3.2 ASM outputs
There are two kinds of outputs from a machine described by an ASM chart and black box diagram. Unlike the two kinds of inputs, with outputs the designer makes a choice based on how the output is generated in the ASM. From that point on, the designer is committed to that kind of output. Since it plays a role later in the design process, the distinction between these two kinds of outputs is important to note.

2.1.3.2.1 External command outputs

External command outputs are generated as described in sections 2.1.1.1 and 2.1.1.2. They are a function only of the present state of the ASM (assuming, as we have been so far, that ovals are not present in the ASM so that it represents a Moore machine). Command outputs do not retain their value when changing from one state to the next. If a particular command output is not mentioned in the next state, it reverts to its default value. In the block diagram given in section 2.1.3, the external command outputs are STOP and SPEED.

2.1.3.2.2 External data outputs

External data outputs are register names mentioned on the left of a register transfer in at least one rectangle. Unlike command outputs, data outputs can retain their value when changing from one state to the next. If a particular data output is not mentioned in the present state, it continues to hold its current value during the next state. Not all registers used in an ASM are necessarily output by the ASM. If a register is not specified as an output from the block diagram, it is an internal register, which the outside world is not allowed to examine (because no bus has been provided to connect that register to the outside world.) The only external data output of the block diagram in section 2.1.3 is COUNT.

2.1.4 Goto-less style

ASM charts allow a designer to specify an arbitrarily complex set of decisions to determine the next state. In theory, the possible next states for a particular state could be any of the rectangles in the ASM chart. Of course, in every particular case (based on register values and status inputs during that clock period), the ASM deterministically describes a particular next state to which the ASM goes in the next clock period. The problem for a careless designer is that the number of possible next states to consider could be quite large if the ASM is large. The flaw here is not in the technical capability of ASM charts (and corresponding hardware) to correctly implement such complex decisions, but rather in the capability of designers to comprehend their designs.

Similar problems were encountered decades ago in software design. At that time, the high-level language `goto` statement was quite popular. Psychological studies have shown that people are only able to keep a few details in their short-term memory at any time. Using `goto` statements correctly requires that a designer remember too many details when the program gets to be of any size. Dijkstra popularized the idea in 1968 that software programmers should avoid the use of `goto` statements in order to make software more readable, and more likely to be correct.

Arbitrary next states in a flowchart are just like `goto`s in software. Therefore, we will mostly use "`goto`-less" style ASM charts. Such ASMs are limited to decisions that are analogous to the high-level language style that is nowadays standard practice in software. In other words, we will try to make our decisions act like high-level language `if` statements, and our loops act like high-level language `while` statements. Although it is technically possible to make an ASM chart look like a plate of spaghetti, the goal of the `goto`-less style is to avoid such a mess. On rare occasions, there may be a compelling need to use an ASM chart which violates the `goto`-less style.

2.1.5 Top-down design

There are two basic approaches to solving problems: bottom up and top down. With the bottom-up approach, the designer begins with some tiny detail of the problem and solves that detail. The designer then goes on to some unrelated tiny detail, and solves it. The designer will eventually have to "glue" the details together to form the complete solution. On large problems, bottom-up designed pieces usually do not "fit" together perfectly. This problem happens because the designer did not view the separate details as fitting into some master plan. An unreasonably large percentage of the designer's time at the end of the project is wasted on integrating the details into a complete system.

The opposite of the bottom-up approach is top-down design. In the top down approach, the designer starts with a master plan. The details come later, and because the designer has a master plan, the details will fit perfectly into the final solution. Top- down design is not natural for novice designers. If you have never built hardware before, it is natural to worry about how the details will work (voltages, wires, gates, etc.). Learning something new is a mostly bottom-up process, but experienced designers use what they have already learned in a top-down fashion. Top-down design is based on **faith**: you have solved details similar to those in your current problem before, and so you can ignore the details when you begin the solution to the current problem. An ASM chart is a useful notation for describing the overall actions of a hardware system without getting into the hardware details. Therefore, the ASM chart makes a good starting point for the top-down design process.

We will take the top-down design process through three stages, briefly described in the following subsections (2.1.5.1 through 2.1.5.3).

2.1.5.1 *Pure behavioral*

This is the most important stage. It is the stage that most of the examples in later chapters concentrate on. In this stage of the top-down design process, the machine is described with a single ASM chart using primarily RTN and relational decisions. The only differences between the pure behavioral solution and a software solution is that the ASM chart describes the passage of time relating to the hardware system clock (as explained in sections 2.1.1 and 2.1.2) and the machine described by the ASM chart connects to the external world via hardware ports (section 2.1.3). A practical example of taking a simple problem and exploring various solutions using pure behavioral ASM charts is given in section 2.2. The only kind of structure that exists in the pure behavioral stage consists of the input and output ports, as illustrated by the following:

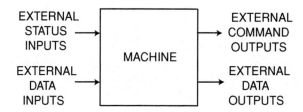

Figure 2-10. Pure behavioral block diagram.

2.1.5.2 Mixed

A pure behavioral ASM chart is merely the statement of an algorithm with precise timing information and includes an indication of which operations occur in parallel. It does not describe precisely what hardware components implement the computation. The goal of computer design is to arrive at a "blueprint" of a physical machine. The pure behavioral ASM chart is merely a description of what the designer wants the machine to do. It does not tell how to connect the physical components together. Software people wonder why the problem is not done upon completing the behavior ASM. After all, we **do** have a solution (an algorithm). Hardware people wonder why we spend so much time with ASM charts. After all, we **do not** yet have a solution (physical hardware). The answer to both groups is: have patience. The pure behavioral stage is important because it enhances the likelihood the designer will produce a **correct** solution. The next stage, which is known as the *mixed* stage, accomplishes part of the transformation from the algorithm into a physical **hardware** structure.

The mixed stage of the top-down design process partitions the problem into two separate but interdependent actors: the *controller* and the *architecture*. The architecture (sometimes called the *datapath*) is the place where physical hardware registers will implement the register transfers originally conceived in the pure behavioral stage. The architecture also contains combinational logic circuits that perform computations required by the algorithm. What the architecture cannot do by itself is sequence events according to the master plan given in the behavioral ASM. This is why the controller exists as an independent actor. The controller tells the architecture what to do during each clock cycle so that the master plan is carried out. Although it may seem the concepts of controller and architecture make things more complicated, in fact working in this fashion simplifies the thought process. In theory, it is possible to design a machine in an extreme way that either has no architecture or has no controller. Such extreme designs are as unnatural to think about as software without variable declarations.

The controller issues commands (as explained in sections 2.1.1.1 and 2.1.1.2) instead of RTN. The architecture receives and acts upon those commands and responds by outputting status. The controller makes decisions based on such status signals received from the architecture (as explained in section 2.1.2.2) instead of relational decisions. It is still possible to draw an ASM chart at this stage of the design, but the ASM chart only describes the independent action of the controller (in terms of commands and status), rather than the complete behavior of the system. This is what top-down design is all about: moving from one master plan (the behavioral ASM) to greater detail on how to carry out the master plan (the mixed ASM). The hardware structure in the mixed stage now has more detail. From the standpoint of the outside world, the mixed stage is identical to the pure behavioral stage, but internally we now see the interconnection of the controller and the architecture.

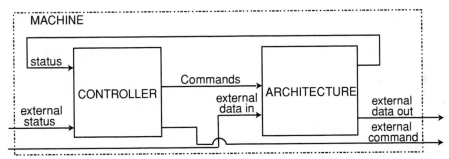

Figure 2-11. Mixed block diagram.

Although, in theory, the architecture could be described by ASM chart(s), it is usually more effective to use a hardware structure diagram. This is because a single ASM chart for the architecture could easily have billions of states (corresponding to all the combinations of values that all the registers in the system could have). Therefore, at the mixed level of abstraction, we use an ASM chart to describe the controller (which still has the same number of states) but use a hardware block diagram to describe the architecture. This stage of the design is known as mixed because it is a mixture of behavior and structure. Examples of translating some of the pure behavioral solutions of section 2.2 into mixed behavioral controller/structural architecture solutions are given in section 2.3.

2.1.5.3 Pure structure

The final stage of the design process is to implement the ASM chart for the controller as a hardware structure. This translation from the mixed stage to the pure structure stage is quite mechanical, and in fact software tools exist that create controller hardware automatically. One can simply describe the controller as a table that says, given the present state and status inputs, what the next state and command outputs will be. Various techniques exist to turn such a table into a hardware structure. For example, such a table can be burned into a Read Only Memory (ROM). The only other hardware required for the controller besides the ROM is a register to hold the present state. Examples of translating some of the mixed ASM charts of section 2.3 into pure structural solutions are given in section 2.4.

2.1.6 Design automation

Synthesis tools exist (chapter 11) that automate much of the final stages of the design process explained above. When using such tools, the designer's job is essentially complete at the end of the pure behavioral stage. Many designers skip over the mixed stage and go straight to the pure structural stage, from which the synthesis tool can automatically create the netlist of gates needed to fabricate an integrated circuit.

Nevertheless, it is important for a designer to understand how all these stages can be carried out manually in order for the designer to know how to create an efficient and correct design. The remainder of this chapter gives manual examples of the three stages described above (pure behavioral, mixed, pure structural). The most important of these stages is the pure behavioral stage because, unless that stage is correct, the rest of the design process (either manual or automated) is pointless.

2.2   Pure behavioral example

To illustrate the design process for a pure behavioral ASM, we need a simple algorithm to implement in hardware. One such algorithm comes from the definition of unsigned integer division. This definition is probably the first thing you ever learned about division when you were a child, and so we will refer to the algorithm that derives from this definition as the *childish division algorithm*.

By way of illustration, suppose you give a child the following problem: "You have seven friends and twenty-one cookies. How can you divide your cookies equally among your friends?" One solution is to give each friend one cookie, and note that each friend received a new cookie. Check to see if there are enough cookies left to give another one to each person. Since there are, repeat this process. When you are done, you will have noted that each person has received three cookies.

The name "childish" may seem pejorative, but this algorithm has a very honored place in the history of computer design (section 8.1). Like a child, this algorithm is simple and unpretentious, yet it raises important issues that also apply to much more complicated algorithms. Variations on the childish division algorithm are used throughout the rest of this book. Even Snoopy can tell whether or not this algorithm has been implemented correctly:

PEANUTS   **by Charles Schulz**

PEANUTS © United Feature Syndicate. Reprinted by Permission.

Of course, high-level software languages are sophisticated enough to have integer division built in. If the variable x is the number of cookies, and y is the number of friends, x/y is the solution to this problem. In hardware, division is seldom implemented as a combinational logic building block (although for small bus sizes this is certainly feasible). This means we need to use an ASM chart to describe a division algorithm. There are much more efficient algorithms than this childish algorithm that are normally implemented in hardware, but the childish division algorithm will allow us to emphasize the properties of ASM charts without having to get into obscure mathematical detail to justify the algorithm. Why this childish algorithm works is obvious.

Before considering the hardware implementation of this childish algorithm, let's consider how to code it in software, such as in the C programming language:

```
r1 = x;
r2 = 0;
while (r1 >= y)
  {
     r1 = r1 - y;
     r2 = r2 + 1;
  }
```

Upon exiting from the loop, $r2$ will be x/y. This is a slow algorithm when the answer $r2$ is of any appreciable size because the loop executes $r2$ times.

This software algorithm would still work when the statements inside the loop are interchanged. These two statements are independent of each other (the new value of $r1$ does not depend on the old value of $r2$ and vice versa):

```
r1 = x;
r2 = 0;
while (r1 >= y)
    {
        r2 = r2 + 1;
        r1 = r1 - y;
    }
```

2.2.1 A push button interface

Before we can put the childish division algorithm into an ASM chart, we need to consider how the eventual hardware will communicate with other actors. In this case, we will assume the only other actor is a person, referred to as the user, who will give our ASM data inputs for x and y, and who will wait for the machine to answer back with x/y. Software designers have sophisticated user interfaces to choose from (using keyboard and mouse software drivers) that make it easy for people to interact with software. Such software solves two fundamental problems that always arise when two actors try to communicate: what data is being communicated and when should the data transfer occur? These problems arise because the actors operate independently in parallel to each other. In hardware, these same problems also arise, but it is the responsibility of the designer of the ASM chart to solve them.

A simple user interface scheme that will work here is to assume that we have a *friendly user* working with a push button that produces a status signal, pb, that our ASM can use. When the user pushes the button, pb will be 1 for exactly a single clock cycle (the design of such a push button is actually another exercise in working with ASM charts which we will ignore for the moment).

Also, we will assume that the division machine produces a READY signal. When READY is 1, the machine is waiting for the user to push the button. When READY is 0, the machine is busy computing the quotient. When READY becomes 1 again, the machine will output the valid quotient. We will assume that the user is patient enough to wait at least two clock cycles before pushing the button again. The user sets the value of the buses x and y in binary on switches before pushing the button. The user leaves x and y alone during the computation of the quotient. We will not consider how the machine would malfunction if an unfriendly user who does not obey these assumptions tries to use the machine.

The overall appearance of the ASM chart and hardware structure block diagram will be:

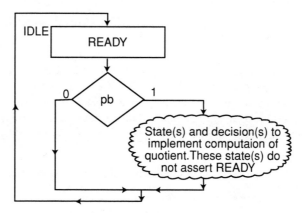

Figure 2-12. ASM for friendly user interface.

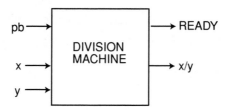

Figure 2-13. Block diagram.

The cloud will be replaced by one or more rectangles and one or more diamonds required to implement the childish division algorithm. None of the rectangles that replace the cloud will assert the READY signal. This means that on the next clock cycle after the user pushes pb, READY will become 0. READY only becomes 1 when the states in the cloud have finished and the ASM loops back to state IDLE.

In the following sections, we will examine several different ways to implement the childish division algorithm hidden in the cloud. An experienced hardware designer would not need to go through so many alternatives to arrive at our final solution. The reason we will look at so many different ways to do the same thing is to illustrate important properties of ASM charts that are somewhat different than conventional software. In this discussion, we will see that certain ASM charts that look reasonable to one familiar with software are actually incorrect and that some ASM charts that look somewhat strange are actually correct. Later, in section 2.3, we will use some of the pure behavioral ASMs we develop in this section as the starting point for the mixed stage of the top-down design process. These examples will also be used in later chapters.

2.2.2 An ASM incorporating software dependencies

The software paradigm used by conventional programming languages, such as C, can be described as each statement completes whatever action it is meant to accomplish before the software proceeds to execute the next statement in the program. Ultimately, all such software programs execute on some kind of hardware, therefore, it must be possible to describe this software paradigm using the ASM chart notation. Although it is often **inefficient**, a software algorithm can always be translated correctly into an ASM with the following rules:

1. Each assignment statement is written by itself in RTN in a unique rectangle that is not followed by a diamond.[1]

2. Each `if` or `while` is translated into an empty rectangle with a diamond to implement the decision.

With this approach, either of the following ASMs correctly implements the software algorithm for division given earlier:

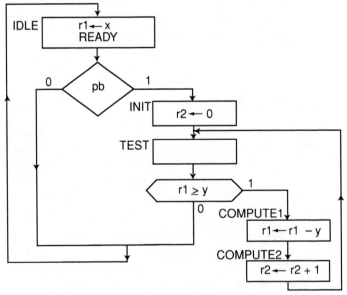

Figure 2-14. ASM for software paradigm (COMPUTE1 at top).

---

[1] Although in the following example there is a diamond in state IDLE involving an external status signal, the original software algorithm does not mention this status signal (pb), and so the software paradigm is preserved in this example.

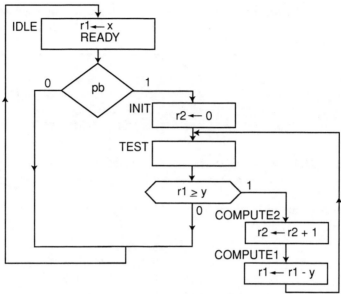

Figure 2-15. ASM for software paradigm (COMPUTE1 at bottom).

The only difference between these two ASMs is whether state COMPUTE1 is at the top of the loop or at the bottom of the loop. Since these ASMs exactly model the way software executes one statement at a time (one software statement per ASM rectangle), whether r1 or r2 gets a value assigned first is irrelevant, because this was also irrelevant in software.

The value of x is assigned to the register r1 in state IDLE. Although this could have been done in an additional state, since we have assumed (see section 2.2.1) that the user waits at least two clock cycles when READY is 1 before pushing pb, the initialization of r1 can occur here. The value of x will not be loaded into r1 until the second of these two clock cycles. If pb is true, the ASM proceeds to state INIT, which will eventually cause r2 to change. If pb is false, as would be the case most of the time, state IDLE simply loops to itself. Since state IDLE leaves r2 alone and r2 typically contains the last quotient, this user interface allows the user as much time as required to view the quotient. The user interface, not the division algorithm, requires that r2 be assigned after the pb test.

State INIT makes sure that r2 is 0 at the time the ASM enters state TEST. State TEST checks if r1>=y, just as the `while` statement does in software. States COMPUTE1 and COMPUTE2 implement each software assignment statement as RTN commands in separate clock cycles.

Both of these ASMs work when x < y. For example, the following shows how the ASMs proceed when x is 5 and y is 7 (all values are shown in decimal for ease of understanding):

```
IDLE      r1=   ? r2=   ? pb=0 ready=1
IDLE      r1=   5 r2=   ? pb=1 ready=1
INIT      r1=   5 r2=   ? pb=0 ready=0
TEST      r1=   5 r2=   0 pb=0 ready=0
IDLE      r1=   5 r2=   0 pb=0 ready=1
```

The way each of the above ASMs operates is slightly different when x >= y. The following shows how the ASM with COMPUTE1 at the top of the loop proceeds when x is 14 and y is 7:

```
IDLE      r1=    ? r2=   ? pb=0 ready=1
IDLE      r1=   14 r2=   ? pb=1 ready=1
INIT      r1=   14 r2=   ? pb=0 ready=0
TEST      r1=   14 r2=   0 pb=0 ready=0
COMPUTE1  r1=   14 r2=   0 pb=0 ready=0
COMPUTE2  r1=    7 r2=   0 pb=0 ready=0
TEST      r1=    7 r2=   1 pb=0 ready=0
COMPUTE1  r1=    7 r2=   1 pb=0 ready=0
COMPUTE2  r1=    0 r2=   1 pb=0 ready=0
TEST      r1=    0 r2=   2 pb=0 ready=0
IDLE      r1=    0 r2=   2 pb=0 ready=1
IDLE      r1=    ? r2=   2 pb=0 ready=1
```

The time to compute the quotient with this ASM includes at least two clock periods in state IDLE, a clock period in state INIT, and the time for the loop. The number of times through the loop is the same as the final quotient (r2). Since there are three states in the loop, the total time to compute the quotient is at least 3+3*quotient.

Here is what happens with the ASM that has COMPUTE2 at the top of the loop:

```
IDLE      r1=    ? r2=   ? pb=0 ready=1
IDLE      r1=   14 r2=   ? pb=1 ready=1
INIT      r1=   14 r2=   ? pb=0 ready=0
TEST      r1=   14 r2=   0 pb=0 ready=0
COMPUTE2  r1=   14 r2=   0 pb=0 ready=0
COMPUTE1  r1=   14 r2=   1 pb=0 ready=0
TEST      r1=    7 r2=   1 pb=0 ready=0
COMPUTE2  r1=    7 r2=   1 pb=0 ready=0
COMPUTE1  r1=    7 r2=   2 pb=0 ready=0
TEST      r1=    0 r2=   2 pb=0 ready=0
IDLE      r1=    0 r2=   2 pb=0 ready=1
IDLE      r1=    ? r2=   2 pb=0 ready=1
```

The latter ASM illustrates the need for an empty rectangle in state TEST. State COM-PUTE1 schedules a change in register r1 (for example, from 7 to 0), but the change does not take effect until the beginning of the next clock cycle. Therefore the decision cannot be part of state COMPUTE1 but instead needs the empty rectangle (state TEST).

2.2.3   Eliminating state TEST

The empty rectangle for state TEST was introduced only to allow a mechanical trans-lation from software to an ASM. In many instances, a decision like this can be merged in with other states. Remember with an ASM that a non-empty rectangle having a diamond following it means the computation in the rectangle and the decision in the diamond take place in parallel. It is inappropriate to merge a decision onto states doing computation when the outcome of the decision (in the software paradigm) could de-pend on the computation. Consider the following modified version of the ASM (figure 2-15) that has COMPUTE2 at the top of the loop:

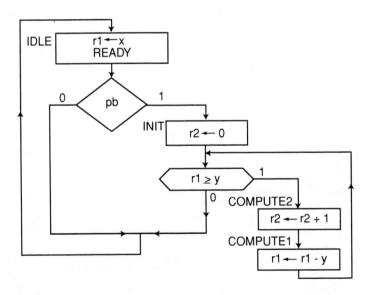

Figure 2-16. Incorrect four-state division machine.

The only difference here is that state TEST has been eliminated. Although this works for x<y, it fails to compute the correct quotient for x>=y. As an illustration of this error, assume that r1 is twelve bits and consider when x is 14 and y is 7:

```
IDLE          r1=    ? r2=    ? pb=0 ready=1
IDLE          r1=   14 r2=    ? pb=1 ready=1
INIT          r1=   14 r2=    ? pb=0 ready=0
COMPUTE2      r1=   14 r2=    0 pb=0 ready=0
COMPUTE1      r1=   14 r2=    1 pb=0 ready=0
COMPUTE2      r1=    7 r2=    1 pb=0 ready=0
COMPUTE1      r1=    7 r2=    2 pb=0 ready=0
COMPUTE2      r1=    0 r2=    2 pb=0 ready=0
COMPUTE1      r1=    0 r2=    3 pb=0 ready=0
IDLE          r1=4089 r2=    3 pb=0 ready=1
IDLE          r1=    ? r2=    3 pb=0 ready=1
```

The decision $r1>=y$ actually occurs separately in two states: INIT and COMPUTE1.
In state INIT, the only computation involves $r2$, and so the decision (14 is $>= 7$) proceeds correctly. The problem exists in state COMPUTE1 because the computation
changes $r1$, and the decision is based on $r1$. The second time in state COMPUTE1,
$r1$ is still 7, although it is scheduled to become 0 at the beginning of the next clock
cycle. The decision is based on the current value (7), and so the loop executes one more
time than it should and the incorrect value of $r2$ (3) results. The mysterious decimal
4089 is the side effect of 12-bit underflow ($4089+7=2^{12}$).

Although it is incorrect to remove state TEST in the last example, what about removing
state TEST in the other ASM (figure 2-14, with COMPUTE1 at the **top** of the loop)?

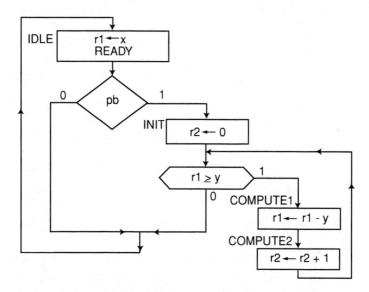

Figure 2-17. Correct four-state divison machine.

This ASM has the decision `r1>=y` happening in two different states: INIT and COMPUTE2. The difference here is that the decision is not dependent on the result of the computation in state COMPUTE2. Therefore, this ASM is correct. As an illustration, consider when `x` is 14 and `y` is 7:

```
IDLE        r1=   ? r2=   ? pb=0 ready=1
IDLE        r1=  14 r2=   ? pb=1 ready=1
INIT        r1=  14 r2=   ? pb=0 ready=0
COMPUTE1    r1=  14 r2=   0 pb=0 ready=0
COMPUTE2    r1=   7 r2=   0 pb=0 ready=0
COMPUTE1    r1=   7 r2=   1 pb=0 ready=0
COMPUTE2    r1=   0 r2=   1 pb=0 ready=0
IDLE        r1=   0 r2=   2 pb=0 ready=1
IDLE        r1=   ? r2=   2 pb=0 ready=1
```

The second time in state COMPUTE1 schedules the assignment that changes `r1` from 7 to 0. This takes effect at the beginning of the clock cycle when the ASM enters state COMPUTE2 for the second time. The decision, which is now part of COMPUTE2, is based on the correct value (0). This means the loop goes through the correct number of times and the quotient in `r2` is correct. As was the case with the earlier ASMs, `r2` will remain unchanged until `pb` is pushed again.

Although the ASMs in section 2.2.2 are also correct, this ASM has the advantage that it executes faster as it requires only `3+2*quotient` clock cycles.

2.2.4 Eliminating state INIT

In addition to being able to describe a decision and a computation that occur in parallel, the ASM chart notation can describe multiple computations that occur in parallel. Consider eliminating state INIT by merging the assignment of zero to `r2` into the rectangle for state IDLE:

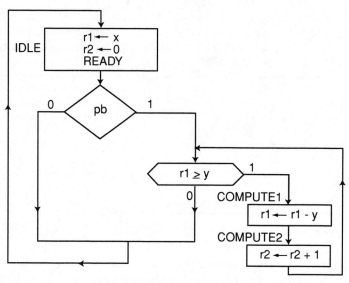

Figure 2-18. Incorrect user interface (throws quotient away).

You may have as many RTN assignments occurring in parallel within a state as you want as long as each left-hand side within that state is unique. In this instance, `r1` and `r2` are scheduled to have new values assigned at the beginning of the next clock cycle. Since we have assumed that the user will ensure that the ASM stays in state IDLE while x remains constant for at least **two** clock cycles, `r1` and `r2` will be properly initialized before entering the loop. This ASM will correctly compute the quotient and leave the loop after the proper number of times for the same reason. To illustrate what this ASM does, consider the same example as the other ASMs (when x is 14 and y is 7):

```
IDLE       r1=   ? r2=   0 pb=0 ready=1
IDLE       r1=  14 r2=   0 pb=1 ready=1
COMPUTE1   r1=  14 r2=   0 pb=0 ready=0
COMPUTE2   r1=   7 r2=   0 pb=0 ready=0
COMPUTE1   r1=   7 r2=   1 pb=0 ready=0
COMPUTE2   r1=   0 r2=   1 pb=0 ready=0
IDLE       r1=   0 r2=   2 pb=0 ready=1
IDLE       r1=   ? r2=   0 pb=0 ready=1
```

There is a new problem with this ASM that we have not seen before: the quotient (2) exists in `r2` for only one clock cycle. This ASM throws it away because the assignment of 0 to `r2` is in state IDLE. From a mathematical standpoint, this ASM is correct, but from a user interface standpoint, it is unacceptable.

2.2.5 Saving the quotient

One way to overcome the user interface problem in section 2.2.4 is to introduce an extra register, $r3$, that saves the quotient in a new state COMPUTE3:

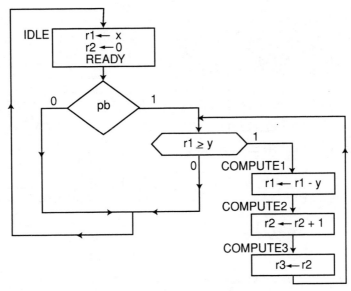

Figure 2-19. Saving quotient in $r3$.

This ASM works for $x>=y$ (the quotient is now in $r3$, not $r2$). For example, when x is 14 and y is 7:

IDLE	r1= ?	r2= 0	r3= ?	pb=0	ready=1
IDLE	r1= 14	r2= 0	r3= ?	pb=1	ready=1
COMPUTE1	r1= 14	r2= 0	r3= ?	pb=0	ready=0
COMPUTE2	r1= 7	r2= 0	r3= ?	pb=0	ready=0
COMPUTE3	r1= 7	r2= 1	r3= ?	pb=0	ready=0
COMPUTE1	r1= 7	r2= 1	r3= 1	pb=0	ready=0
COMPUTE2	r1= 0	r2= 1	r3= 1	pb=0	ready=0
COMPUTE3	r1= 0	r2= 2	r3= 1	pb=0	ready=0
IDLE	r1= 0	r2= 2	r3= 2	pb=0	ready=1
IDLE	r1= ?	r2= 0	**r3=** **2**	pb=0	ready=1

Unfortunately, there is a subtle error in the above ASM: when the answer is supposed to be zero, $r3$ is left unchanged instead of being cleared. This occurs because the only assignment to $r3$ is inside the loop, but the loop never executes when the quotient is zero. One way to overcome this problem is to include an extra decision in the ASM to test for the special case that $x<y$ (which can be done by testing if $r1>=y$ is false):

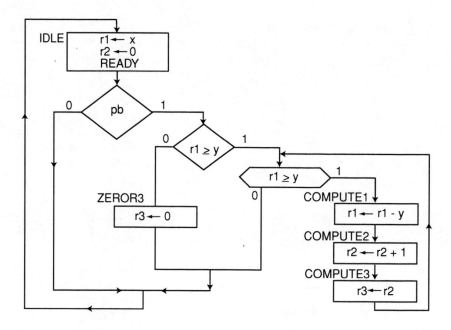

Figure 2-20. Handling quotient of zero.

Of course, this has the disadvantage of taking longer (2+3*`quotient` clock cycles), but sometimes a designer must consider a slower solution to eventually discover a faster solution.

2.2.6 Variations within the loop

Let's take the final ASM of section 2.2.5 and consider some variations of it inside the loop that will make it incorrect. Our eventual goal is to find a faster solution that is **correct**, but for the moment, let's just play around and see how we can break this ASM.

One incorrect thing to do would be to assign to r3 before incrementing r2:

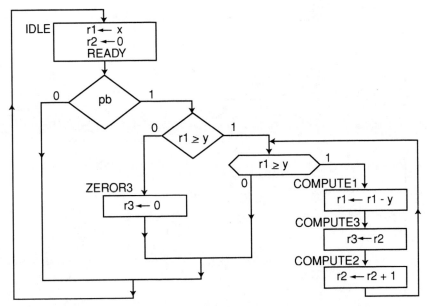

Figure 2-21. Incorrect rearrangement of states.

Here is an example when x is 14 and y is 7 of what kind of error occurs:

```
    IDLE       r1=    ? r2=    0 r3=    ? pb=0 ready=1
    IDLE       r1=   14 r2=    0 r3=    ? pb=1 ready=1
    COMPUTE1   r1=   14 r2=    0 r3=    ? pb=0 ready=0
    COMPUTE3   r1=    7 r2=    0 r3=    ? pb=0 ready=0
    COMPUTE2   r1=    7 r2=    0 r3=    0 pb=0 ready=0
    COMPUTE1   r1=    7 r2=    1 r3=    0 pb=0 ready=0
    COMPUTE3   r1=    0 r2=    1 r3=    0 pb=0 ready=0
    COMPUTE2   r1=    0 r2=    1 r3=    1 pb=0 ready=0
    IDLE       r1=    0 r2=    2 r3=    1 pb=0 ready=1
    IDLE       r1=    ? r2=    0 r3=    1 pb=0 ready=1
```

The value in r3 is one less than it should be since it was assigned too early.

Another thing to try (which unfortunately will also fail for similar reasons) is to merge states COMPUTE2 and COMPUTE3 into a single state COMPUTE23:

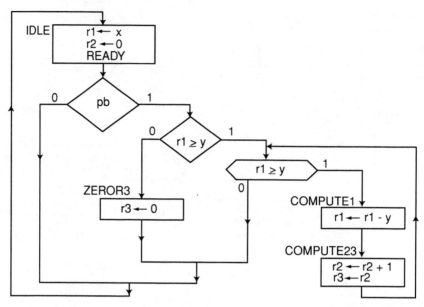

Figure 2-22. Incorrect parallelization attempt.

Here is an example when x is 14 and y is 7 of what kind of error occurs:

```
IDLE          r1=   ? r2=    0 r3=    0 pb=0 ready=1
IDLE          r1=  14 r2=    0 r3=    0 pb=1 ready=1
COMPUTE1      r1=  14 r2=    0 r3=    0 pb=0 ready=0
COMPUTE23     r1=   7 r2=    0 r3=    0 pb=0 ready=0
COMPUTE1      r1=   7 r2=    1 r3=    0 pb=0 ready=0
COMPUTE23     r1=   0 r2=    1 r3=    0 pb=0 ready=0
IDLE          r1=   0 r2=    2 r3=    1 pb=0 ready=1
IDLE          r1=   ? r2=    0 r3=    1 pb=0 ready=1
```

Even though inside the rectangle the assignment to r2 is written above the assignment to r3, they happen in parallel. **The meaning of a state in an ASM is not affected by the order in which a designer writes the commands inside the rectangle.** Since there is a dependency between the commands in state COMPUTE23, this ASM is not equivalent to the correct solution of section 2.2.5 but is instead equivalent to the incorrect solution given a moment ago. After the second time in state COMPUTE23, r2 is incremented (from 1 to 2), but r3 changes to the old value of r2 (1), which is not what we want.

Although all of the above variations may seem hopeless, there is in fact a correct and faster solution if we press on with this kind of variation. Let's merge all three commands into a single state COMPUTE:

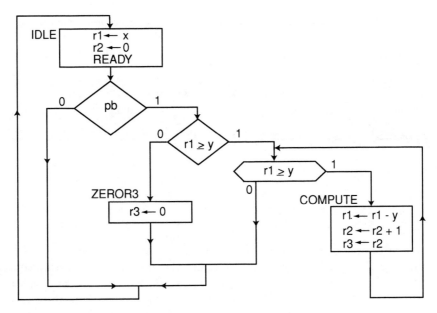

Figure 2-23. Correct parallelization.

This ASM is correct, as illustrated by the example used before (when x is 14 and y is 7):

```
IDLE      r1=      ? r2=  0 r3=   ? pb=0 ready=1
IDLE      r1=     14 r2=  0 r3=   ? pb=1 ready=1
COMPUTE   r1=     14 r2=  0 r3=   ? pb=0 ready=0
COMPUTE   r1=      7 r2=  1 r3=   0 pb=0 ready=0
COMPUTE   r1=      0 r2=  2 r3=   1 pb=0 ready=0
IDLE      r1=   4089 r2=  3 r3=   2 pb=0 ready=1
IDLE      r1=      ? r2=  0 r3=   2 pb=0 ready=1
```

The decision involving r1>=y is part of state COMPUTE (as well as being part of state IDLE), and r1 is affected in state COMPUTE. Also there is the interdependence of r2 and r3 observed earlier. The reason why state COMPUTE works here is that all of these things occur at the same time in parallel. We have now totally left the sequential software paradigm of section 2.2.1 (one statement at a time; no dependency within a state). We are now using the dependency in the algorithm with parallelism to get the correct result much faster.

Although in this ASM r3 still serves as the place where the user can observe the quotient when the ASM returns to state IDLE, r3 accomplishes something even more important. It compensates for the fact that the loop in state COMPUTE executes one more time than the software loop would. Even though r2 becomes one more than the correct quotient, r3 is loaded with the old value of r2 each time through the loop. On the last time through the loop, r3 is scheduled to be loaded with the correct quotient.

The loop in state COMPUTE is interesting because it has a property that software loops seldom have: it either does not execute or it executes at least two times. This is because the decision r1>=y is part of both states IDLE and COMPUTE. To illustrate this, consider when x is 7 and y is 7:

```
IDLE        r1=    ? r2=  0 r3=  ? pb=0 ready=1
IDLE        r1=    7 r2=  0 r3=  ? pb=1 ready=1
COMPUTE     r1=    7 r2=  0 r3=  ? pb=0 ready=0
COMPUTE     r1=    0 r2=  1 r3=  0 pb=0 ready=0
IDLE        r1=4089 r2=  2 r3=  1 pb=0 ready=1
IDLE        r1=    ? r2=  0 r3=  1 pb=0 ready=1
```

You can see that r1 is 7 in state IDLE, and so the ASM proceeds to state COMPUTE. In state COMPUTE, r1 is scheduled to change, but it remains 7 the first time in state COMPUTE; thus the next state is state COMPUTE (it loops back to itself). Only on the second time through state COMPUTE has the scheduled change to r1 taken place; thus the next state finally becomes IDLE.

As with earlier ASMs, this ASM works for x<y only because of state ZEROR3. For example, consider when x is 5 and y is 7:

```
IDLE        r1=    ? r2=  0 r3=  ? pb=0 ready=1
IDLE        r1=    5 r2=  0 r3=  ? pb=1 ready=1
ZEROR3      r1=    5 r2=  0 r3=  ? pb=0 ready=0
IDLE        r1=    5 r2=  0 r3=  0 pb=0 ready=1
IDLE        r1=    ? r2=  0 r3=  0 pb=0 ready=1
```

The time required for this ASM is 3+quotient clock cycles.

2.2.7 Eliminate state ZEROR3

If the loop in state COMPUTE could execute one or more (rather than two or more) times, it would be possible to eliminate state ZEROR3. This would work because r2 is already 0, and the assignment of r2 to r3 would achieve the desired effect of clearing r3.

One way to describe this in ASM chart notation is to note that pb is true when making the transition from state IDLE to state COMPUTE (the first time into the loop), but pb remains false until the quotient is computed (by our original assumption about a friendly user). Let's change the decision so that it ORs the status signal pb together with the result of the r1>=y:

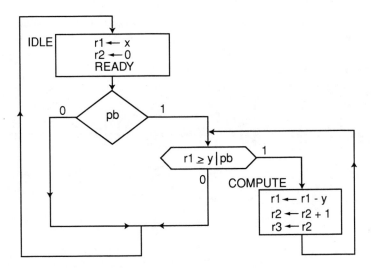

Figure 2-24. Goto-less two-state childish division ASM.

Since pb is true for only one clock cycle, this only causes the loop to go in the first time. Whether to repeat the loop subsequent times depends **only** on r1. Since for all values x>=y the loop occurs at least twice anyway, ORing pb in the decision does not change the operation of the ASM for x>=y. ORing pb only affects what happens when x<y. Rather than executing state COMPUTE zero times, ORing pb forces it to execute once. In this case, state COMPUTE will not execute more than once because x (r1) is not >= y.

For example, consider when x is 5 and y is 7:

IDLE	r1= ?	r2= 0	r3= ?	pb=0 ready=1
IDLE	r1= 5	r2= 0	r3= ?	pb=1 ready=1
COMPUTE	r1= 5	r2= 0	r3= ?	pb=0 ready=0
IDLE	r1=4094	r2= 1	r3= 0	pb=0 ready=1
IDLE	r1= ?	r2= 0	r3= 0	pb=0 ready=1

The fact that r1 and r2 are different after executing state COMPUTE than they were in the earlier ASM after executing state ZEROR3 is irrelevant since the user only looks at data output r3.

The time required for this ASM is also 3+quotient clock cycles.

The above ASM was arranged to follow the goto-less style mentioned in section 2.1.4. In essence, there is a while loop (testing r1>=y|pb) nested inside an if statement (testing pb). In order to describe this ASM in the goto-less style, pb is tested twice coming out of state IDLE. Such ASMs with redundant tests (in the same clock cycle) can be simplified into shorter equivalent ASM notation. Although this equivalent ASM is truly identical and would be implemented with the same hardware, it does not follow a style that can be thought of in terms of `if`s and `while`s:

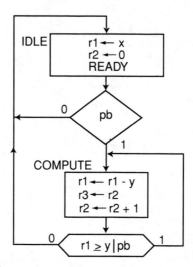

Figure 2-25. Equivalent to figure 2-24.

Also, in the above, the order of the statements within state COMPUTE were re-arranged for ease of understanding. As mentioned earlier, changing the order with a rectangle does not change the meaning. Which way you draw the ASM is both a matter of personal taste and also a matter of how you intend to use it. We will see examples where both forms of this ASM prove useful. At this stage it is important for you to be comfortable that these two ASMs mean **exactly** the same thing because under **all** possible circumstances they cause the same state transitions and computations to occur.

2.3 Mixed examples

The three stages of the top-down design process were discussed in section 2.1.5. Section 2.2 gives several alternative ways to describe the childish division algorithm in the first stage of the top-down design process (as a pure behavioral ASM). This section continues this same example into the second stage of the top-down design process. In the second stage, we partition the division machine into a controller and an architecture.

2.3.1 First example

We could use any of the ASM charts in section 2.2 as an example of translating a pure behavioral ASM into a structural architecture and a behavioral controller. For example, consider the ASM chart described in section 2.2.7, which is the simplest correct ASM chart for the division machine. We can, for the moment, ignore pb and READY because they are external ports (and will remain the same in the mixed stage). In the mixed stage, we need to eliminate RTN commands and relational decisions.

Consider the RTN commands in this ASM chart. In state IDLE, r1 is scheduled to be loaded with x, and in parallel r2 is scheduled to be cleared. In state COMPUTE, all three registers are scheduled to change. Of course, these scheduled changes take effect at the beginning of the next clock cycle.

There are many possible hardware structures that could implement these RTN commands. The designer makes an arbitrary decision (based on speed, cost, availability, ease, personal prejudice, etc.) about what hardware components to use in the architecture. The only requirement is that interconnection of the chosen components can correctly implement all RTN transfers with the precise timing indicated by the original behavioral ASM chart. The easiest (but not necessarily best) way to accomplish this is to choose register components (like those in appendix D) that internally take care of as many of the required RTN commands as possible. For example, if the designer chooses a counter register for r2, the counter can internally take care of clearing r2 (as is required in state IDLE) and also take care of incrementing r2 (as is required in state COMPUTE). We will be able to eliminate the RTN commands (such as r2 $\leftarrow$ r2+1) from the ASM chart and replace them with internal command signals (such as incr2).

If the designer were to choose a non-counter register for r2, the designer would have to provide for these actions with additional combinational devices (like those in appendix C) in the architecture. It is not wrong to choose a non-counter register for r2; it would just make the designer work harder. To keep this example as simple as possible, we will choose a counter register for r2.

On the other hand, registers r1 and r3 are loaded with values that must come from outside the register (unlike a simple counter). Therefore, it is sensible to use the simplest register component possible (the enabled register) for r1 and r3. Additional hardware will be required to make available the new values to be loaded into r1 and r3.

Having decided on the kind of registers to use in the architecture, we need to consider how those registers are interconnected. For a moment, let's concentrate only on the RTN in state COMPUTE. In this state, r1 will be loaded with a difference, r2 increments, and r3 gets loaded with the old value of r2. All three of these actions occur in parallel. This means, for example, that the difference must be computed by a dedicated combinational device (such as a subtractor). Such a combinational device is always computing the difference between r1 and y (even though that difference is loaded into r1 only when the controller is in state COMPUTE).

Loading r3 with the old value of r2 is easy. The output port of the r2 counter register is simply connected via a bus to the input port of the r3 enabled register. If the only state to mention r1, r2 or r3 were state COMPUTE, we would have the following architecture:

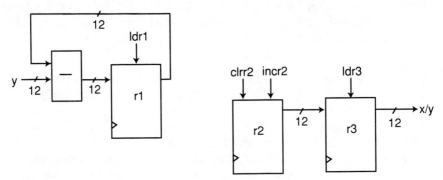

Figure 2-26. Architecture using subtractor.

but the above architecture fails to implement the RTN of state IDLE. The above architecture provides no way for r1 to be loaded with x.

One approach that often allows an architecture to deal with different kinds of RTN in different states is to use an Arithmetic Logic Unit (ALU), which is capable of many different operations, instead of a dedicated combinational device (such as a subtractor). Also, there is often a need for one or more muxes so that the proper information can be routed to the ALU. In this particular ASM, there are only two different results that might be loaded into r1: either the difference of r1 and y or passing through x unchanged. This means the ALU must be capable of at least two different operations: computing the difference of the ALU's two data inputs and passing through the ALU's second input unchanged. The ALU is commanded to do these operations by particular bit patterns on the six-bit aluctrl bus. Symbolically, we will refer to these bit patterns as 'DIFFERENCE and 'PASS. The grave accent ('), which is also known as backquote or tick, indicates a symbol that is replaced by a particular bit pattern.

On one hand, the ALU should be able to subtract y; on the other hand, the ALU should be able to pass x. To accomplish this requires a mux which can select either x or y. The output of this mux is connected as the second input of the ALU. Input 0 of this mux is connected to the external bus x. Input 1 of this mux is connected to the external bus y.

When `muxctrl` is 0, the output of the mux is the same as x. When `muxctrl` is 1, the output of the mux is the same as y. Using the mux and ALU, the architecture now appears as follows:

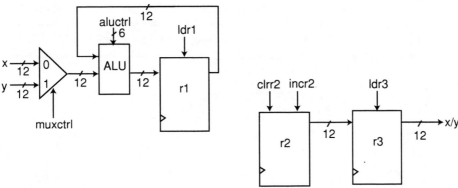

Figure 2-27. Architecture using ALU.

Although the above architecture implements all the RTN of the ASM chart in section 2.2.7, it does not consider the relational decision `r1>=y`. The simplest way to translate relational decisions into the mixed stage is to dedicate a combinational device (usually a comparator) for calculating an *internal status* signal that indicates the result of the relational comparison. In the mixed ASM, this internal status signal will be tested instead of referring to the relational decision. This is why ultimately the ASM uses only status signals. In this particular instance, we will use a comparator whose inputs are `r1` and y.

There are three outputs of a comparator: the strictly <, the exactly == and the strictly >. There is no >= output, but we can obtain that output, since it is the complement of the strictly < output. We will use the strictly < output as the input of an inverter, whose output is the internal status signal `r1gey`.

At last, we have an architecture which can correctly implement all the RTN and relational decisions of the ASM chart in section 2.2.7. Now it will be a mechanical matter to translate the pure behavioral ASM chart of section 2.2.7 into an equivalent mixed ASM chart. The purpose of this translation will be to use command signals (such as `incr2`) instead of RTN, and to use status signals (such as `r1gey`) instead of relational decisions. The ← in RTN translates to a command signal (such as `ldr1, clrr2, incr2` or `ldr3`) corresponding to the register on the left of the arrow. The computation on the right of the arrow may or may not require additional commands directed to the combinational logic units, such as the ALU and mux.

This translation from pure behavioral ASM to mixed ASM always relates to a particular architecture that the designer has in mind. Although many architectures might have been chosen for one pure behavioral ASM, each architecture will have a distinct mixed ASM. The following shows the particular architecture we have just developed and the corresponding translation of the pure behavioral ASM of section 2.2.7 into the particular mixed ASM required by this architecture. Finally, we give a system block diagram showing the interconnection of this particular controller (as described by the mixed ASM) and this particular architecture:

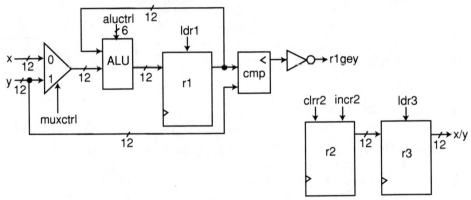

Figure 2-28. Methodical architecture.

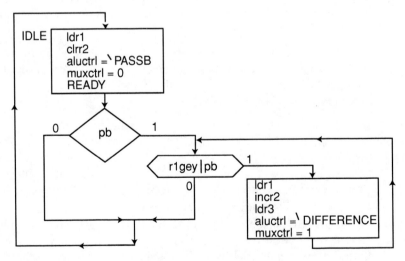

Figure 2-29. Mixed ASM corresponding to figures 2-24 and 2-28.

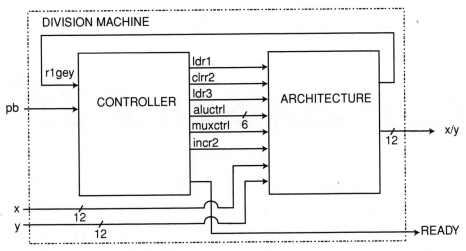

Figure 2-30. System diagram.

The external command READY and the external status pb are not affected by the translation to the mixed stage.

2.3.2 Second example

It happens in this example that the architecture just developed could have been used to implement any of the ASM charts in section 2.2. (It is rare for the same methodical architecture to work with different ASMs.) In general, this would not be the case, but it just happens that the ASM chart of section 2.2.7 requires maximal parallelism of the same register transfers as the other ASMs. All the other ASM charts of section 2.2 implement the same RTN commands with less parallelism, and so an architecture designed for maximal parallelism can implement an ASM that demands less parallelism. For example, the first ASM of section 2.2.2 could be implemented using the same architecture and system block diagram. The register r3 is not used by this ASM chart, thus the data output of the machine should be r2 instead of r3. Figure 2-31 is the mixed ASM chart that corresponds to this architecture and the ASM of section 2.2.2:

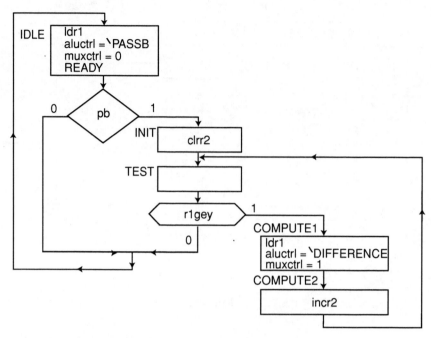

Figure 2-31. Mixed ASM corresponding to figures 2-14 and 2-28.

2.3.3 Third example

If we are content with the slower ASM chart of section 2.2.2, perhaps we can find a cheaper architecture (with less potential for parallelism) that can correctly implement the RTN given in section 2.2.2. Of course, such an architecture could not implement ASM charts, such as that in section 2.2.7, which require more parallelism. One way to reduce cost (at the expense of speed) is to use the ALU as a central unit that can do many different operations. In particular, it can output the value zero (when `aluctrl` is 'ZERO) and it can increment (when `aluctrl` is 'INCREMENT). Since the ALU is used for everything, the mux must have enough inputs to provide anything required by the ALU. In this instance, there is a three-input mux (with a two-bit `muxctrl`). This allows the mux to select `x`, `r1` or `r2` to be output to the bus which is the `a` input of the ALU.

It is no longer necessary for `r2` to be a counter register since the ALU can increment its input, and the mux can provide the value of `r2` to the ALU. The output of the ALU must be available on a central bus, from which both `r1` and `r2` can be loaded. (This ASM does not use `r3`.)

The carry out (`cout`) status signal output from the ALU is available at no cost. It can be used to determine the result of the `r1>=y` test. This is permissible because the ASM of section 2.2.2 has an empty rectangle for state TEST. In the mixed ASM, this rectangle will not be empty, although no registers will be loaded. The ALU is simply commanded to compute a difference without issuing a register load signal. As a side effect of computing the difference, `~cout` indicates `r1>=y`. The following shows the architecture and mixed ASM chart for the third example:

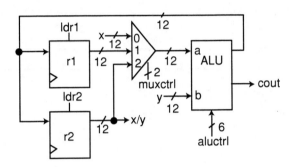

Figure 2-32. Central ALU architecture.

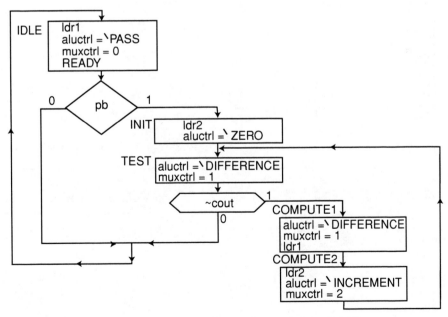

Figure 2-33. Mixed ASM corresponding to figures 2-14 and 2-32.

2.3.4 Methodical versus central ALU architectures

There is a spectrum of possible ways to choose an architecture at the mixed stage of the design process. At one extreme is the *central ALU approach*, illustrated in section 2.3.3, where one ALU does all the computation required by the entire algorithm. At the other extreme are *methodical approaches*, illustrated by sections 2.3.1 and 2.3.2, where each computation in the algorithm is done by a different hardware unit. The central ALU approach typically uses less hardware but only works with certain kinds of ASMs. For example, the ASM that works with the methodical approach in section 2.3.1 cannot work with the central ALU approach because that ASM performs more than one computation per clock cycle. The following table highlights the differences between these two approaches:

	Central ALU	Methodical
What does computation?	one ALU	registers themselves or registers tied to dedicated muxes and ALUs
What ALU output connects to?	every register	only one register
What kind of register?	enabled	all kinds[2]
Number of ← per clock cycle	one	many
Speed	slower	faster
Cost	lower	higher
Example	2.3.3	2.3.1 & 2.3.2
Figures	2-32, 2-33	2-28, 2-29, 2-31

In the methodical architecture of section 2.3.2, the output of the ALU only connects to r1, but in the central ALU architecture of section 2.3.3, the output of the ALU connects to both r1 and r2. In the ASM implemented with the methodical architecture of

---

[2] Including customized registers (other than those described in appendix D) built using muxes and combinational logic that are tailored to the specific algorithm. See section 7.2.2.1 for an example.

section 2.3.1, there are multiple ← per clock cycle. The central ALU approach is slower because it takes an ASM that uses more clock cycles to accomplish similar register transfers.

2.4 Pure structural example

In the mixed stage (described in section 2.3), the division machine was partitioned into a controller (described by a mixed ASM chart) and a structural architecture. This section continues with the same example (variations of the childish division algorithm) into the third stage (of the three stages for the top-down design process explained in section 2.1.5).

The third stage involves converting the mixed ASM chart into a hardware structure that implements the behavior described by the mixed ASM chart. At the end of the third stage, the top level of the system is completely described in terms of structure. The top level of the system is no longer described in terms of what it does (behavior) but is instead described in terms of how to build it (structure). The designer has enough information to wire together the hardware components into an operational physical machine. The algorithm has become a working piece of hardware.

2.4.1 First example

To translate the mixed ASM chart into hardware we must assign each symbolic state name a specific bit pattern. The bit patterns used are completely arbitrary. The only requirement is that the designer be consistent. One approach that is easy for the designer (when the number of states in the ASM chart is small) is to use a binary code. For an ASM with two states, a one-bit code suffices. For an ASM with three or four states, a two bit code will do. For an ASM with five to eight states, a three-bit code does the job. For an ASM with nine to sixteen states, the designer needs a four-bit code. In general, an ASM with n states requires a `ceil(log2(n))`-bit code.

The mixed ASM of section 2.3.1 is quite simple. It requires only two states. Let us say that being in state IDLE is represented by 0, and being in state COMPUTE is represented by 1.

We also need to know the bit patterns that control the ALU. As explained in appendix C, for an ALU inspired by the 74xx181, 'PASSB is 101010 and 'DIFFERENCE is 011001.

The structural controller is composed of two parts: the present state register, and the next state combinational logic. The status and the present state are the inputs of the combinational logic. The next state and the commands are the output of the combinational logic. The next state is the input to the present state register:

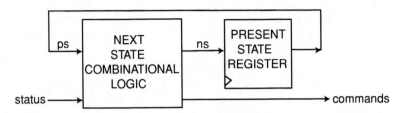

Figure 2-34. Controller.

We can describe the next state logic with a table. For the ASM chart of section 2.3.1, the corresponding table is:

| inputs | | | outputs | | | | | | | |
ps	pb	r1gey	ns	ldr1	clrr2	incr2	ldr3	muxctrl	aluctrl	ready
0	0	0	0	1	1	0	0	0	101010	1
0	0	1	0	1	1	0	0	0	101010	1
0	1	0	1	1	1	0	0	0	101010	1
0	1	1	1	1	1	0	0	0	101010	1
1	0	0	0	1	0	1	1	1	011001	0
1	0	1	1	1	0	1	1	1	011001	0
1	1	0	1	1	0	1	1	1	011001	0
1	1	1	1	1	0	1	1	1	011001	0

In the above, ps stands for the representation of the present state, and ns stands for the representation of the next state.

One possible hardware implementation of this table is a ROM. The above table can be used as is to "burn" the ROM. Since there are three bits of address input to the ROM (one bit for the present state, and two bits for the status), there are eight words (each 13-bits wide) stored in the ROM for this controller.

Another approach would be to use the above table to derive minimized logic equations for each bit of output and then use the logic equations to arrive at a structure composed of AND/OR gates. For example, the following logic equations are equivalent to the above table:

```
ns = ~ps&pb|ps&(r1gey|pb)
ldr1 = 1
clrr2 = ~ps
incr2 = ps
ldr3 = ps
muxctrl = ps
aluctrl[5] = ~ps
aluctrl[4] = ps
aluctrl[3] = 1
aluctrl[2] = 0
aluctrl[1] = ~ps
aluctrl[0] = ps
ready = ~ps
```

For a more complicated controller, using Karnaugh maps (or other logic minimization tools) could be helpful. Other approaches exist. Turning a table, such as the above, into actual hardware is a task that can be automated. Many software tools exist to aid in this job.

Several of the entries in the above table are identical to each other. For example, 000 and 001 are identical because the transition from IDLE to COMPUTE depends only on pb and not on r1gey. (The second ASM chart in section 2.2.7 makes this clear.) An abbreviated form of the table:

inputs			outputs							
ps	pb	r1gey	ns	ldr1	clrr2	incr2	ldr3	muxctrl	aluctrl	ready
0	0	-	0	1	1	0	0	0	101010	1
0	1	-	1	1	1	0	0	0	101010	1
1	0	0	0	1	0	1	1	1	011001	0
1	1	0	1	1	0	1	1	1	011001	0
1	-	1	1	1	0	1	1	1	011001	0

shows a "don't care" as a hyphen for those status inputs that do not affect a particular state transition. This table means exactly the same thing as the longer form of the table given earlier.

2.4.2   Second example

Assuming that the five states in the ASM chart of section 2.3.2 are represented as follows:

```
                    IDLE      000
                    INIT      001
                    TEST      010
                    COMPUTE1  011
                    COMPUTE2  100
```

the following table describes the controller:

inputs			outputs						
ps	pb	r1gey	ns	ldr1	clrr2	incr2	muxctrl	aluctrl	ready
000	0	–	000	1	0	0	0	101010	1
000	1	–	001	1	0	0	0	101010	1
001	–	–	010	0	1	0	0	101010	0
010	–	0	000	0	0	0	0	101010	0
010	–	1	011	0	0	0	0	101010	0
011	–	–	100	1	0	0	1	011001	0
100	–	–	010	0	0	1	0	101010	0
101	–	–	000	0	0	0	0	101010	0
11-	–	–	000	0	0	0	0	101010	0

Using the "don't care" form of the table is useful because otherwise the table would be 32-lines long. The values of muxctrl and aluctrl in states INIT, TEST and COM-PUTE2 are arbitrary. There are three extra state encodings (101, 110 and 111) that should never occur in the proper operation of the machine. On power up, the physical hardware might find itself in one of these states. To avoid problems of this kind, these states go to state IDLE but otherwise do nothing.

2.5 Hierarchical design

Even though upon completion of the third stage, the top level of the system is described with pure structure, some devices (actors) that make up part of the architecture or controller may still be described in terms of behavior. For example, if an architecture needs an adder, the block diagram for the architecture would simply show a device with two inputs (let's label them a and b) and one output (sum). There are many possible hardware structures that could implement such an adder. At the top level of the system we will view the black box of an adder as being part of a structure, even though we have not specified what internal structure (of AND/OR gates for instance) implements the adder. Whatever internal structure is used, we can describe the behavior of the adder with a one-state ASM chart. ASM charts such as this that have one state correspond to combinational logic. The following shows the black box for a two- bit adder, the ASM chart for the adder, and one of many possible internal structures (circuits) for the adder:

The above circuit diagram is *flattened*, which means it does not show any hierarchy. A flattened circuit diagram, composed entirely of logic gates, each producing a single bit of output, is equivalent to a *netlist*. A netlist is a list of gates with the corresponding

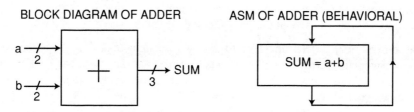

Figure 2-35. Block diagram and behavioral ASM for adder.

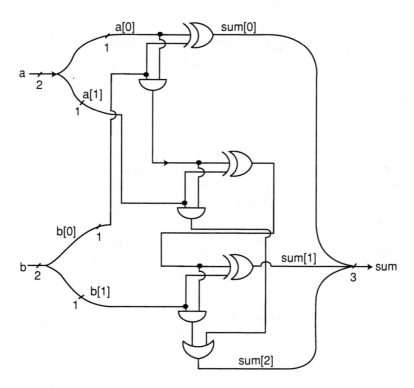

Figure 2-36. Flattened circuit diagram for adder

one-bit wires (nets) that each gate is connected to. A netlist can be submitted to a *silicon foundry* to fabricate an actual integrated circuit, or it can be used manually as directions (to a low-skilled worker) on how to hook together copper wire to form the desired machine. Although a flattened circuit diagram (or its equivalent netlist) is ultimately what is used to build a machine, it does not express much of the thought process the designer uses to arrive at the final circuit diagram.

The term *hierarchical* design refers to keeping a hierarchy of design components in the final design. Hierarchical design applies only to hardware structures, and not to behavioral ASMs. In hierarchical designs, the designer documents how to arrive at the final circuit diagram. Instead of just saying how gates are interconnected, the designer defines *modules*. For example, the two-bit adder block diagram is a module. The definition of a module occurs by *instantiating* other modules. In this example, the adder is composed of a full-adder and a half-adder, and so the designer instantiates a full-adder and also instantiates a half-adder in the definition of the adder.

The designer has to define a separate module for the full-adder. The full-adder is composed of two half-adders and an OR gate. The two half-adders that the designer instantiates in the definition of the full-adder are like tract houses in a boring suburb. Such houses are build from identical blueprints and therefore look the same. But different people live in each house. So it is for the two half-adders that compose the full-adder. They are instantiated from an identical "blueprint" (the definition of the half-adder will be discussed in the next paragraph), but different data is processed by each half-adder.

In the final circuit, the half-adder is instantiated three times. Each of the three half-adders is in turn composed of an exclusive-OR gate and an AND gate. The following diagrams show the adder module definition, full-adder module definition, half-adder module definition and a circuit diagram with dotted lines showing the instantiation of these definitions inside each other.

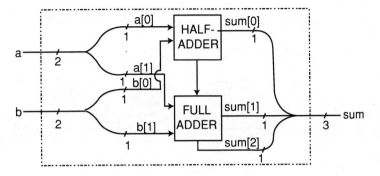

Figure 2-37. Definition of the adder module.

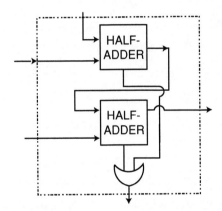

Figure 2-38. Definition of the full-adder module.

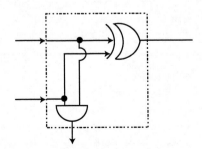

Figure 2-39. Definition of the half-adder module.

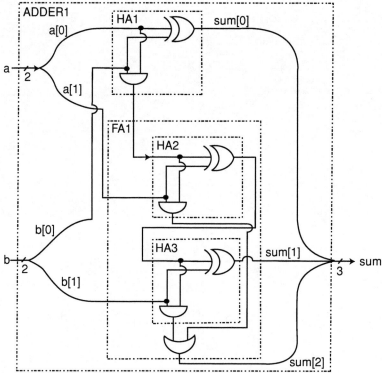

Figure 2-40. Hierarchical instantiation of modules.

Although hierarchical design can be used with either bottom-up or top-down design, it is most important with top-down design. In top-down design, upon completion of the third stage, the designer may apply the same three stages over again on any components (actors) that are not standard building blocks. Building blocks such as adders are well understood, and the designer is not normally concerned about their internal structure. Other problem-specific building blocks (such as the push button in the division machine) would be dealt with at the end of the third stage for the top-level system.

2.5.1 How pure is "pure"?

The terms "pure behavioral" and "pure structural" can be a little confusing. No matter at what level of abstraction you view a hardware system, there is some irreducible structure and some irreducible behavior. In the "pure" behavioral stage, the input and output ports are part of some larger (unspecified) structure. In the "pure" structural stage, the nature of the black boxes instantiated in the hardware diagram is known only by their behavior. Even if the designer takes the hierarchy all the way down to the gate level, the nature of each gate is known only by its behavior.

2.5.2 Hierarchy in the first example

Sections 2.2.7, 2.3.1 and 2.4.1 discuss the details for the three stages for the example two-state version of the division machine. The following three diagrams illustrate the increasing amount of structure and decreasing amount of behavior as the designer progresses through the three stages. In each of these diagrams, black boxes (and architectural devices, which are in fact black boxes themselves) represent aspects of the system whose internal nature is known only by behavior.

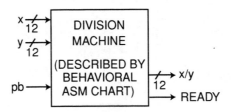

Figure 2-41. "Pure" behavioral block diagram.

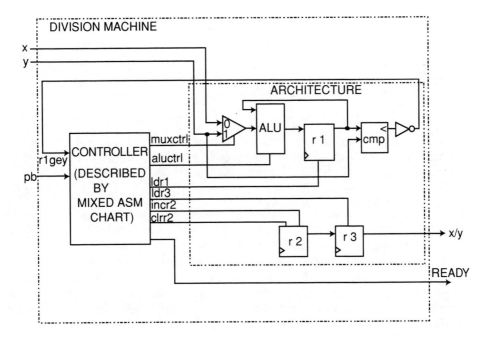

Figure 2-42. Mixed block diagram.

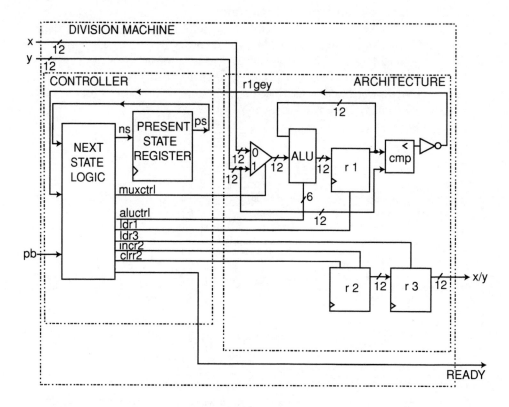

Figure 2-43. "Pure" strucural block diagram.

The first diagram (figure 2-41), which illustrates the "pure" behavioral stage, has a single black box. This box represents the complete division machine. The machine has a unified behavior which can be described by a single ASM chart (see section 2.2.7). The only structure in the first stage is the port structure that allows the machine to communicate with the outside world.

The second diagram (figure 2-42) shows the "mixed" stage. Instantiated inside the division machine are the controller and the architecture. In the mixed stage, the structure of the controller remains a mystery (i.e., a black box) which is described only in behavioral terms by the ASM chart of section 2.3.1. This ASM chart refers only to status and command signals. The architecture, on the other hand, is visible; however architectural devices (such as the ALU) remain as black boxes, known only by their behavior.

The third diagram (figure 2-43) shows the "pure" structural stage. Here the controller includes the next state logic and the present state (ps) register. The architecture remains the same as in the "mixed" stage. Even though this is the "pure" structural stage, the internal structure of the black boxes (next state logic, mux, ALU, comparator, inverter, ps, r1, r2 and r3 registers) remains hidden.

Here is where the power of hierarchical design comes to aid us in our top-down approach. We do not need to worry about the gate-level details of, say, the ALU. As the earlier full-adder example illustrates, such behavioral black boxes can be decomposed down to a gate-level netlist. For our purposes, we will be content with the third diagram. The reason we do not have to work our way down to reach the gate level is that automated tools now exist to do this dirty work.

2.6   Conclusion

This chapter illustrates two manual *graphical* notations: the ASM chart (to describe behavior) and the block diagram (to describe structure). There are three stages in the top-down design process to turn an algorithm (behavior) into hardware (structure): pure behavioral, mixed and pure structural. The mixed and pure structural stages partition the machine into a controller and an architecture.

In addition, this chapter describes instantiating structural modules (hierarchical design). This allows the pure structural stage to be described in an understandable way, without having to descend to the extreme gate-level detail of a netlist.

The next chapter introduces an automated *textual* notation that allows us to express behavioral, structural and hierarchical design in a unified notation.

2.7   Further Reading

CLAIR, C. R., *Designing Logic Systems Using State Machines,* McGraw-Hill, New York, 1973. This short but influential book was the first to explain the ASM chart notation, which T. E. Osborne had invented in the early 1960s at Hewlett Packard.

GAJSKI, DANIEL D., *Principles of Digital Design*, Prentice Hall, Upper Saddle River, NJ, 1997. Chapter 8 describes ASM charts. This book uses the term datapath to mean what is called an architecture here and uses = rather than ← for RTN.

PROSSER, FRANKLIN P. and DAVID E. WINKEL, *The Art of Digital Design: An Introduction to Top Down Design*, 2nd ed., PTR Prentice Hall, Englewood Cliffs, NJ, 2nd ed., 1987. Chapters 5-8 give several examples of ASM charts using RTN. This book uses the term architecture the way it is used here.

2.8 Exercises

2-1. Give a pure behavioral ASM for a factorial machine. The factorial of n is the product of the numbers from 1 to n. This machine has the data input, n, a push button input, pb, a data output, prod, and an external status output, READY. READY and pb are similar to those described in section 2.2.1. Until the user pushes the button, READY is asserted and prod continues to be whatever it was. When the user pushes the button, READY is no longer asserted and the machine computes the factorial by doing no more than one multiplication per clock cycle. For example, when n=5 after an appropriate number of clock cycles prod becomes $120 == 1*1*2*3*4*5 == 1*5*4*3*2*1$ and READY is asserted again.

Use a linear time algorithm in the input n, which means the exact number of clock cycles that this machine takes to compute n! for a particular value of n can be expressed as some constant times n plus another constant. (All of the childish division examples in this chapter are linear time algorithms in the quotient.) For example, a machine that takes $57*n+17$ clock cycles to compute n! would be acceptable, but you can probably do better than that.

2-2. Design an architecture block diagram and corresponding mixed ASM that is able to implement the algorithm of problem 2-1 assuming the architecture is composed of the following building blocks: up/down counter registers, multiplier, comparator and muxes. Give a system diagram that shows how the architecture and controller fit together, labeled with appropriate signal names.

2-3. Give a table that describes the structural controller for problem 2-2.

2-4. Give a pure behavioral ASM similar to problem 2-1, but use repeated addition to perform multiplication. For example, $13*14 == 0 +13 +13 +13 +13 +13 +13 +13 +13 +13 +13 +13 +13 +13 +13$. Direct multiplication in a single cycle is not allowed. The algorithm should be suitable for implementation with the central ALU approach. This will be a quadratic time algorithm in n because of nested loops.

2-5. Design an architecture block diagram and corresponding mixed ASM for problem 2-4 assuming the following building blocks: enabled registers, muxes, comparator, and the ALU described in section C.6. Give a system diagram that shows how the architecture and controller fit together, labeled with appropriate signal names. Label the a and b inputs of the ALU.

2-6. Give a table that describes the structural controller for problem 2-5.

2-7. Give a pure behavioral ASM similar to problem 2-4, but use a shift and add algorithm to perform multiplication. Direct multiplication in a single cycle is not allowed. Here is an example of multiplying 14 by 13 using the shift and add algorithm with 4-bit input representations, and an 8-bit product.

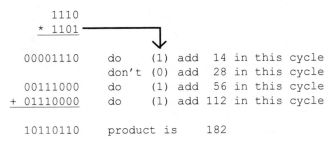

```
         1110
       * 1101

    00001110     do      (1) add  14 in this cycle
                 don't  (0) add  28 in this cycle
    00111000     do      (1) add  56 in this cycle
  + 01110000     do      (1) add 112 in this cycle

    10110110     product is    182
```

The number of cycles to perform a single multiplication by n is proportional to the number of bits used to represent n, which is roughly the logarithm of n. But you have to perform n such multiplications, and so this factorial algorithm is what is called an n log n time algorithm, which takes more clock cycles than a linear time algorithm but fewer clock cycles than a quadratic time algorithm when n is large. (Note that unlike the linear time algorithm of problem 2-1, this approach does not require an expensive multiplier.) You should use a methodical approach that exploits maximal parallelism.

2-8. Design an architecture block diagram and corresponding mixed ASM that is able to implement problem 2-7 assuming the following building blocks: enabled registers, counter registers, shift registers, muxes, adder, comparator. Give a system diagram that shows how the architecture and controller fit together, labeled with appropriate signal names.

2-9. Give a table that describes the structural controller for problem 2-8. (See section D.9 for details about controlling a shift register.)

2-10. For each of the following ASMs, draw a timing diagram. x, y and z are 8-bit registers, whose values should be shown in decimal.

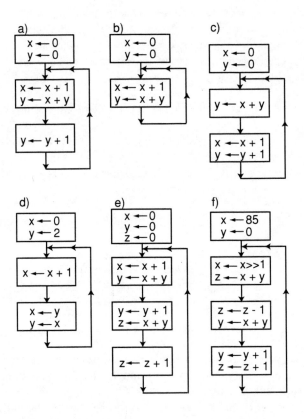

2-11. Design an architecture block diagram and corresponding mixed ASM that is able to implement problem 2-10, part a assuming the following building blocks: two 8-bit counter registers, one 8-bit adder, and any number of any kind of mux. Give a system diagram that shows how the architecture and controller fit together, labeled with appropriate signal names.

2-12. Like 2-11, except use: one 8-bit counter register, one 8-bit enabled register, one 8-bit ALU and any number of any kind of mux. Label the a and b inputs of the ALU.

2-13. Design an architecture block diagram and corresponding mixed ASM that is able to implement problem 2-10, part b assuming the following building blocks: two 8-bit enabled registers, one 8-bit adder, one 8-bit incrementor and any number of any kind of mux. Give a system diagram.

2-14. Design an architecture block diagram and corresponding mixed ASM that is able to implement problem 2-10, part c assuming the following building blocks: one 8-bit counter register, one 8-bit enabled register, one 8-bit ALU (section C.6) and any number of any kind of mux. Give a system diagram. Label the a and b inputs of the ALU.

2-15. Design an architecture block diagram and corresponding mixed ASM that is able to implement problem 2-10, part d assuming the following building blocks: two 8-bit enabled registers, one 8-bit ALU (see section C.6) and one 8-bit two-input mux. Give a system diagram. Label the a and b inputs of the ALU.

2-16. Design an architecture block diagram and corresponding mixed ASM that is able to implement problem 2-10, part e assuming the following building blocks: two 8-bit counter registers, one 8-bit enabled register, one 8-bit ALU (section C.6) and any number of any kind of mux. Give a system diagram. Label the a and b inputs of the ALU.

2-17. Like 2-16, except use: one 8-bit incrementor, three 8-bit enabled registers, one 8-bit adder, and any number of any kind of mux.

2-18. Design an architecture block diagram and corresponding mixed ASM that is able to implement problem 2-10, part f assuming the following building blocks: one 8-bit shift register (section D.9), one 8-bit counter register, one 8-bit up/down counter register (section D.8) and one 8-bit adder. You may not use any muxes. Give a system diagram.

2-19. Like 2-18, except use: one 8-bit incrementor, one 8-bit decrementor, three 8-bit enabled registers, one 8-bit adder, one 8-bit combinational shifter, and any number of any kind of mux.

3. VERILOG HARDWARE DESCRIPTION LANGUAGE

The previous chapter describes how a designer may manually use ASM charts (to describe behavior) and block diagrams (to describe structure) in top-down hardware design. The previous chapter also describes how a designer may think hierarchically, where one module's internal structure is defined in terms of the instantiation of other modules. This chapter explains how a designer can express all of these ideas in a special hardware description language known as Verilog. It also explains how Verilog can test whether the design meets certain specifications.

3.1 Simulation versus synthesis

Although the techniques given in chapter 2 work wonderfully to design small machines by hand, for larger designs it is desirable to automate much of this process. To automate hardware design requires a Hardware Description Language (HDL), a different notation than what we used in chapter 2 which is suitable for processing on a general-purpose computer. There are two major kinds of HDL processing that can occur: simulation and synthesis.

Simulation is the interpretation of the HDL statements for the purpose of producing human readable output, such as a timing diagram, that predicts approximately how the hardware will behave before it is actually fabricated. As such, HDL simulation is quite similar to running a program in a conventional high-level language, such as Java Script, LISP or BASIC, that is interpreted. Simulation is useful to a designer because it allows detection of functional errors in a design without having to fabricate the actual hardware. When a designer catches an error with simulation, the error can be corrected with a few keystrokes. If the error is not caught until the hardware is fabricated, correcting the problem is much more costly and complicated.

Synthesis is the compilation of high-level behavioral and structural HDL statements into a flattened gate-level netlist, which then can be used directly either to lay out a printed circuit board, to fabricate a custom integrated circuit or to program a programmable logic device (such as a ROM, PLA, PLD, FPGA, CPLD, etc.). As such, synthesis is quite similar to compiling a program in a conventional high-level language, such as C. The difference is that, instead of producing object code that runs on the same computer, synthesis produces a physical piece of hardware that implements the computation described by the HDL code. For the designer, producing the netlist is a simple step (typically done with only a few keystrokes), but turning the netlist into physical hardware is often costly, especially when the goal is to obtain a custom integrated circuit from a commercial silicon foundry. Typically after synthesis, but before the physical fabrication, the designer simulates the synthesized netlist to see if its behavior matches the original HDL description. Such post-synthesis simulation can prevent costly errors.

3.2 Verilog versus VHDL

HDLs are textual, rather than graphic, ways to describe the various stages in the top-down design process. In the same language, HDLs allow the designer to express both the behavioral and structural aspects of each stage in the design. The behavioral features of HDLs are quite similar to conventional high-level languages. The features that make an HDL unique are those structural constructs that allow description of the instantiation and interconnection of modules.

There are many proprietary HDLs in use today, but there are only two standardized and widely used HDLs: Verilog and VHDL. Verilog began as a proprietary HDL promoted by a company called Cadence Data Systems, Inc., but Cadence transferred control of Verilog to a consortium of companies and universities known as Open Verilog International (OVI). Many companies now produce tools that work with standard Verilog. Verilog is easy to learn. It has a syntax reminiscent of C (with some Pascal syntax thrown in for flavor). About half of commercial HDL work in the U.S. is done in Verilog. If you want to work as a digital hardware designer, it is important to know Verilog.

VHDL is a Department of Defense (DOD) mandated language that is used primarily by defense contractors. Although most of the concepts in VHDL are not different from those in Verilog, VHDL is much harder to learn. It has a rigid and unforgiving syntax strongly influenced by Ada (which is an unpopular conventional programming language that the DOD mandated defense software contractors to use for many years before VHDL was developed). Although more academic papers are published about VHDL than Verilog, less than one-half of commercial HDL work in the U.S. is done in VHDL. VHDL is more popular in Europe than it is in the U.S.

3.3 Role of test code

The original purpose of Verilog (and VHDL) was to provide designers a unified language for simulating gate-level netlists. Therefore, Verilog combines a structural notation for describing netlists with a behavioral notation for saying how to test such netlists during simulation. The behavioral notation in Verilog looks very much like normal executable statements in a procedural programming language, such as Pascal or C. The original reason for using such statements in Verilog code was to provide *stimulus* to the netlist, and to test the subsequent *response* of the netlist. The pairs of stimulus and response are known as *test vectors*. The Verilgo that creates the stimulus and observes the response is known as the *test code* or *testbench*. Snoopy's "woof" in the comic strip of section 2.2 is analougus to the role of the test codes warning us that the expected response was not observed. For example, one way to use simulation to test whether a small machine works is to do an *exhaustive test*, where the test code provides each possible combination of inputs to the netlist and then checks the response of the netlist to see if it is appropriate.

For example, consider the division machine of the last chapter. Assume we have developed a flattened netlist that implements the complete machine. It would not be at all obvious whether this netlist is correct. Since the bus width specified in this problem is small (twelve bits), we can write Verilog test code using procedural Verilog (similar to statements in C) that does an exhaustive test: A reasonable approach would be to use two nested loops, one that varies x through all its 4096 possible values, and one that varies y through all its 4095 possible values. At appropriate times inside the inner loop, the test code would check (using an `if` statement) whether the output of the netlist matches x/y. Verilog provides most of the integer and logical operations found in C, including those, such as division, that are difficult to implement in hardware. The original intent was not to synthesize such code into hardware but to document how the netlist should automatically be tested during simulation.

Verilog has all of the features you need to write conventional high-level language programs. Except for file Input/Output (I/O), any program that you could write in a conventional high-level language can also be written in Verilog. The original reason Verilog provides all this software power in a "hardware" language is because it is impossible to do an exhaustive test of a complex netlist. The 12-bit division machine can be tested exhaustively because there are only 16,773,120 combinations with the 24 bits of input to the netlist. A well-optimized version of Verilog might be able to conduct such a simulation in a few days or weeks. If the bus width were increased, say to 32-bits, the time to simulate all 2^{64} combinations would be millions of years. Rather than give up on testing, designers write more clever test code. The test code will appear longer, but will execute in much less time. Of course, if a machine has a flaw that expresses itself for only a few of the 2^{64} test patterns, the probability that our fast test code will find the flaw is usually low.

3.4 Behavioral features of Verilog

Verilog is composed of modules (which play an important role in the structural aspects of the language, as will be described in section 3.10). All the definitions and declarations in Verilog occur inside a module.

3.4.1 Variable declaration

At the start of a module, one may declare variables to be `integer` or to be `real`. Such variables act just like the software declarations `int` and `float` in C. Here is an example of the syntax:

```
integer x,y;
real Rain_fall;
```

Underbars are permitted in Verilog identifiers. Verilog is case sensitive, and so `Rain_fall` and `rain_fall` are distinct variables. The declarations `integer` and `real` are intended only for use in test code. Verilog provides other data types, such as `reg` and `wire`, used in the actual description of hardware. The difference between these two hardware-oriented declarations primarily has to do with whether the variable is given its value by behavioral (`reg`) or structural (`wire`) Verilog code. Both of these declarations are treated like `unsigned` in C. By default, `reg`s and `wire`s are only one bit wide. To specify a wider `reg` or `wire`, the left and right bit positions are defined in square brackets, separated by a colon. For example:

```
reg [3:0] nibble,four_bits;
```

declares two variables, each of which can contain numbers between 0 and 15. The most significant bit of `nibble` is declared to be `nibble[3]`, and the least significant bit is declared to be `nibble[0]`. This approach is known as *little endian notation*. Verilog also supports the opposite approach, known as *big endian notation*:

```
reg [0:3] big_end_nibble;
```

where now `big_end_nibble[3]` is the least significant bit.

If you store a signed value[1] in a `reg`, the bits are treated as though they are unsigned. For example, the following:

```
four_bits = -5;
```

is the same as:

```
four_bits = 11;
```

---

[1] In order to simplify dealing with twos complement values, many implementations allow integers with an arbitrary width. Such declarations are like `reg`s, except they are signed.

Verilog supports concatenation of bits to form a wider `wire` or `reg`, for example, `{nibble[2], nibble[1]}` is a two bit `reg` composed of the middle two bits of `nibble`. Verilog also provides a shorthand for obtaining a contiguous set of bits taken from a single `reg` or `wire`. For example, the middle two bits of `nibble` can also be specified as `nibble[2:1]`. It is legal to assign values using either of these notations.

Verilog also allows arrays to be defined. For example, an array of reals could be defined as:

```
real monthly_precip[11:0];
```

Each of the twelve elements of the array (from `monthly_precip[0]` to `monthly_precip[11]`) is a unique real number. Verilog also allows arrays of `wires` and `regs` to be defined. For example,

```
reg [3:0] reg_arr[999:0];
wire[3:0] wir_arr[999:0];
```

Here, `reg_arr[0]` is a four-bit variable that can be assigned any number between 0 and 15 by behavioral code, but `wir_arr[0]` is a four-bit value that cannot be assigned its value from behavioral code. There are one thousand elements, each four bits wide, in each of these two arrays. Although the `[]` means bit select for scalar values, such as `nibble[3]`, the `[]` means element select with arrays. It is **illegal** to combine these two uses of `[]` into one, as in `if(reg_arr[0][3])`. To accomplish this operation requires two statements:

```
nibble = reg_arr[0];
if (nibble[3]) ...
```

3.4.2   Statements legal in behavioral Verilog

The behavioral statements of Verilog include[2] the following:

```
var = expression;

if (condition)
   statement
```

---

[2] There are other, more advanced statements that are legal. Some of these are described in chapters 6 and 7.

Continued

```
if (condition)
   statement
else
   statement

while (condition)
   statement

for (var=expression; condition; var=var+expression)
   statement

forever
   statement

case (expression)
   constant: statement
   ...
   default: statement
endcase
```

where the italic *statement, var, expression, condition* and *constant* are replaced with appropriate Verilog syntax for those parts of the language. A *statement* is one of the above statements or a series of the above statements **terminated by semicolons inside** `begin` and `end`. A `var` is a variable declared as `integer`, `real`, `reg` or a concatenation of `reg`s. A *var* cannot be declared as `wire`.

3.4.3 Expressions

An *expression* involves constants and variables (including `wire`s) with arithmetic (`+`, `-`, `*`, `/`, `%`), logical (`&`, `&&`, `|`, `||`, `^`, `~`, `<<`, `>>`), relational (`<`, `==`, `===`, `<=`, `>=`, `!=`, `!==`, `>`) and conditional (`?:`) operators. A *condition* is an expression. A *condition* might be an expression involving a single bit, (as would be produced by `||`, `&&`, `!`, `<`, `==`, `===`, `<=`, `>=`, `!=`, `!==` or `>`) or an expression involving several bits that is checked by Verilog to see if it is equal to 1. Except for `===` and `!==`, these symbols have the same meaning as in C. Assuming the result is stored in a 16-bit `reg`,[3] the following table illustrates the result of these operators, for example where the left operand (if present) is ten and the right operand is three:

---

[3] Some results are different if the destination is declared differently.

symbol	name	example	16-bit unsigned result
+	addition	10+3	13
-	subtraction	10-3	7
-	negation	-10	65526
*	multiplication	10*3	30
/	division	10/3	3
%	remainder	10%3	1
<<	shift left	10<<3	80
>>	shift right	10>>3	1
&	bitwise AND	10&3	2
|	bitwise OR	10|3	11
^	bitwise exclusive OR .	10^3	9
~	bitwise NOT	~10	65525
?:	conditional operator	0?10:3	3
		1?10:3	10
!	logical NOT	!10	0
&&	logical AND	10&&3	1
||	logical OR	10||3	1
<	less than	10<3	0
==	equal to	10==30	
<=	less than or equal to	10<=3	0
>=	greater than or equal	10>=3	1
!=	not equal	10!=3	1
>	greater than	10>3	1

3.4.4 Blocks

All procedural statements occur in what are called *blocks* that are defined inside modules, after the type declarations. There are two kinds of procedural blocks: the initial block and the always block. For the moment, let us consider only the initial block. An initial block is like conventional software. It starts execution and eventually (assuming there is not an infinite loop inside the initial block) it stops execution. The simplest form for a single Verilog initial block is:

```
module top;

  declarations;

  initial
    begin
      statement;
      ...
      statement;
    end

endmodule
```

The name of the module (top in this case) is arbitrary. The syntax of the *declarations* is as described above. All variables should be declared. Each *statement* is terminated with a semicolon. Verilog uses the Pascal-like begin and end, rather than { and }. There is no semicolon after begin or end. The begin and end may be omitted in the rare case that only one procedural statement occurs in the initial block.

Here is an example that prints out all 16,773,120 combinations of values described in section 3.3:

```
module top;
  integer x,y;
  initial
    begin
      x = 0;
      while (x<=4095)
        begin
          for (y=1; y<=4095; y = y+1)
            begin
              $display("x=%d y=%d",x,y);
            end
          x = x + 1;
        end
    end
    $write("all ");
    $display("done");
endmodule
```

The loop involving x could have been written as a for loop also but was shown above as a while for illustration. Note that Verilog does not have the ++ found in C, and so it is necessary to say something like y = y + 1. This assignment statement is just like its counterpart in C: it is instantaneous. The variable changes value before the next statement executes (unlike the RTN discussed in the previous chapter). The $display is a *system task* (which begin with $) that does something similar to what printf("%d %d \n",x,y) does in C: it formats the textual output according to the string in the quotes. The system task $write does the same thing as $display, except that it does not produce a new line:

```
x=      0  y=      1
x=      0  y=      2
        . . .          . . .
x= 4095  y= 4094
x= 4095  y= 4095
all done
```

The above code would fail if the declaration had been:

```
reg [11:0] x,y;
```

because, although twelve bits are adequate for the hardware, the test code requires that x and y become 4096 in order for the loop to stop.

Since infinite loops are useful in hardware, Verilog provides the syntax forever, which means the same thing as while(1). In addition, the always block mentioned above can be described as an initial block containing only a forever loop. For simulation purposes, the following mean the same:

```
initial              initial
  begin                begin
    while(1)             forever        always
      begin                begin          begin
        . . .                . . .          . . .
      end                  end            end
  end                  end
```

For synthesis, one should use the always block form only. The statement forever is not a block and cannot stand by itself. Like other procedural statements, forever must be inside an initial or always block.

3.4.5 Constants

By default, constants in Verilog are assumed to be decimal integers. They may be specified explicitly in binary, octal, decimal, or hexadecimal by prefacing them with the syntax `'b`, `'o`, `'d`, or `'h`, respectively. For example, `'b1101`, `'o15`, `'d13`, `'hd`, and 13 all mean the same thing. If you wish to specify the number of bits in the representation, this proceeds the quote: `4'b1101`, `4'o15`, `4'd13`, `4'hd`.

3.4.6 Macros, `include` files and comments

As an aid to readability of the code, Verilog provides a way to define macros. For example, the `aluctrl` codes described in 2.3.1 can be defined with:

```
'define DIFFERENCE    6'b011001
'define PASSB         6'b101010
```

Later in the code, a reference to these macros (preceded by a backquote) is the same as substituting the associated value. The following `if`s mean the same:

```
if (aluctrl == 'DIFFERENCE)          if (aluctrl == 6'b011001)
  $display("subtracting");             $display("subtracting");
```

Note the syntax difference between variables (such as `aluctrl`), macros (such as `'DIFFERENCE`), and constants (such as `6'b011001`). Variables are not preceded by anything. Macros are preceded by backquote. Constants may include one forward single quote.

You can determine whether a macro is defined using `'ifdef` and `'endif`. This preprocessing feature should not be confused with `if`. For example, the following:

```
'ifdef DIFFERENCE
    $display("defined");
'endif
```

prints the message regardless of the value of `'DIFFERENCE`, as long as that macro is defined. The message is not printed only when there is not a `'define` for `'DIFFERENCE`.

Verilog allows you to separate your source code into more than one file (just like `#include` in C and `{$I}` in Pascal). To use code contained in another file, you say:

```
'include "filename.v"
```

There are two forms of comments in Verilog, which are the same as the two forms found in C++. A comment that extends only for the rest of the current line can occur after //. A comment that extends for several lines begins with /* and ends with */. For example:

```
          /*   a multi line comment
               that includes a declaration:
   reg a;
               which is ignored by Verilog
          */
   reg b;  //   this declaration is not ignored
```

3.5 Structural features of Verilog

Verilog provides a rich set of built-in logic gates, including and, or, xor, nand, nor, not and buf, that are used to describe a netlist. The syntax for these structural features of Verilog is quite different than for any of the behavioral features of Verilog mentioned earlier. The outputs of such gates are declared to be wire, which by itself describes a one-bit data type. (Regardless of width, an output generated by structural Verilog code must be declared as a wire.) The inputs to such gates may be either declared as wire or reg (depending on whether the inputs are themselves computed by structural or behavioral code). To instantiate such a gate, you say what kind of gate you want (xor for example) and the name of this particular instance (since there may be several instances of xor gates, let's name this example x1). Following the instance name, inside parentheses are the output and input ports of the gate (for example, say the output is a wire named c, and the inputs are a and b). The output(s) of gates are always on the left inside the parentheses:

```
          module easy_xor;
             reg a,b;
             wire c;
             xor x1(c,a,b);
             . . .
          endmodule
```

People familiar with procedural programming languages, like C, mistakenly assume this is "passing c, a and b and then calling on xor." **It is doing no such thing**. It simply says that an xor gate named x1 has its output connected to c and its inputs connected to a and b. If you are familiar with graph theory, this notation is simply a way to describe the edges (a, b, c) and vertex (x1) of a graph that represents the structure of a circuit.

3.5.1 Instantiating multiple gates

Of course, there is an equivalent structure of `and`/`or` gates that does the same thing as an `xor` gate (recall the identity `a^b == a&(~b) | (~a)&b`):

```verilog
module hard_xor;
    reg a,b;
    wire c;
    wire t1,t2,not_a,not_b;

    not i1(not_a,a);
    not i2(not_b,b);
    and a1(t1,not_a,b);
    and a2(t2,a,not_b);
    or  o1(c,t1,t2);
    ...
endmodule
```

The order in which gates are instantiated in structural Verilog code does not matter, and so the following:

```verilog
module scrambled_xor;
    reg a,b;
    wire c;
    wire t1,t2,not_a,not_b;

    or  o1(c,t1,t2);
    and a1(t1,not_a,b);
    and a2(t2,a,not_b);
    not i1(not_a,a);
    not i2(not_b,b);
    ...
endmodule
```

means the same thing, because they both represent the interconnection in the following circuit diagram:

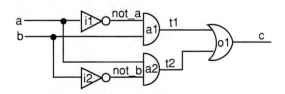

Figure 3-1. Exclusive or built with ANDs, OR and inverters.

3.5.2 Comparison with behavioral code

Structural Verilog code does not describe the order in which computations implemented by such a structure are carried out by the Verilog simulator. This is in sharp contrast to behavioral Verilog code, such as the following:

```
module behavioral_xor;

    reg a,b;
    reg c;
    reg t1,t2,not_a,not_b;

    always ...
        begin
            not_a = ~a;
            not_b = ~b;
            t1 = not_a&b;
            t2 = a&not_b;
            c = t1|t2;
        end
endmodule
```

which is a correct behavioral rendition of the same idea. (The ellipses must be replaced by a Verilog feature described later.) Also, c, t1, t2, not_a and not_b must be declared as regs because this behavioral (rather than structural) code assigns values to them.

To rearrange the order of behavioral assignment statements is incorrect:

```
module bad_xor;

    reg a,b;
    reg c;
    reg t1,t2,not_a,not_b;

    always ...
      begin
        c = t1|t2;
        t1 = not_a&b;
        t2 = a&not_b;
        not_a = ~a;
        not_b = ~b;
      end
endmodule
```

because not_a must be computed before t1 by the Verilog simulator.

3.5.3 Interconnection errors: four-valued logic

In software, a bit is either a 0 or a 1. In properly functioning hardware, this is usually the case also, but it is possible for gates to be wired together incorrectly in ways that produce electronic signals that are neither 0 nor 1. To more accurately model such physical possibilities,[4] each bit in Verilog can be one of four things: $1'b0$, $1'b1$, $1'bz$ or $1'bx$.

Obviously, $1'b0$ and $1'b1$ correspond to the logical 0 and logical 1 that we would normally expect to find in a computer. For most technologies, these two possibilities are represented by a voltage on a wire. For example, active high TTL logic would represent $1'b0$ as zero volts and $1'b1$ as five volts. Active low TTL logic would represent $1'b0$ as five volts and $1'b1$ as zero volts. Other kinds of logic families, such as CMOS, use different voltages. ECL logic uses current, rather than voltage, to represent information, but the concept is the same.

3.5.3.1 High impedance

In any technology, it is possible for gates to be miswired. One kind of problem is when a designer forgets to connect a wire or forgets to instantiate a necessary gate. This means that there is a wire in the system which is not connected to anything. We refer to this as *high impedance*, which in Verilog notation is $1'bz$. The TTL logic family will normally view high impedance as being the same as five volts. If the input of a gate to which this wire is connected is active high, $1'bz$ will be treated as $1'b1$, but if it is active low, it will be treated as $1'b0$. Other logic families treat $1'bz$ differently. Furthermore, electrical noise may cause $1'bz$ to be treated spuriously in any logic family. For these reasons, it is important for a Verilog simulator to treat $1'bz$ as distinct from $1'b0$ and $1'b1$. For example, if the designer forgets the final or gate in the example from section 3.5.1:

```
module forget_or_that_outputs_c;
   reg a,b;
   wire c;
   wire t1,t2,not_a,not_b;

   not i1(not_a,a);
   not i2(not_b,b);
   and a1(t1,not_a,b);
   and a2(t2,a,not_b);
   ...
endmodule
```

---

[4] Verilog also allows each bit to have a strength, which is an electronic concept (below gate level) beyond the scope of this book.

there is no gate that outputs the wire c, and therefore it remains 1'bz, regardless of what a and b are.

3.5.3.2 Unknown value

Another way in which gates can be miswired is when the output of two gates are wired together. This raises the possibility of *fighting outputs*, where one of the gates wants to output a 1'b0, but the other wants to output a 1'b1. For example, if we tried to eliminate the or gate by tying the output of both and gates together:

```
module tie_ands_together;
   reg a,b;
   wire c;
   wire t1,t2,not_a,not_b;

   not i1(not_a,a);
   not i2(not_b,b);
   and a1(c,not_a,b);
   and a2(c,a,not_b);
   ...
endmodule
```

the result is correct (1'b0) when a and b are the same because the two and gates both produce 1'b0 and there is no fight. The result is incorrect (1'bx) when a is 1'b0 and b is 1'b1 or vice versa, because the two and gates fight each other. Fighting gates can cause physical damage to certain families of logic (i.e., smoke comes out of the chip). Obviously, we want to be able to have the simulator catch such problems before we fabricate a chip that is doomed to blow up (literally)!

3.5.3.3 Use in behavioral code

Behavioral code may manipulate bits with the four-valued logic. Uninitialized regs in behavioral code start with a value of 'bx. (As mentioned above for structural code, disconnected wires start with a value of 'bz.) All the Boolean operators, such as &, | and ~ are defined with the four-valued logic so that the usual rules of commutativity, associativity, etc. apply.

The four-valued logic may be used with multi-bit wires and regs. When all the bits are either 1'b1 or 1'b0, such as 3'b110, the usual binary interpretation (powers of two) applies. When any of the bits is either 1'bz or 1'bx, such as 3'b1z0, the numeric value is unknown.

Arithmetic and relational operators (including == and !=) produce their usual results only when both operands are composed of 1'b0s and 1'b1s. In any other case, the result is 'bx. This relates to the fact the corresponding combinational logic required to implement such operations in hardware would not produce a reliable result under such circumstances. For example:

```
if ( a == 1'bx)
    $display("a is unknown");
```

will never display the message, even when a is 1'bx, because the result of the == operation is always 1'bx. 1'bx is not the same as 1'b1, and so the $display never executes.

There are two special comparison operators (=== and !==) that overcome this limitation. === and !== cannot be implemented in hardware, but they are useful in writing intelligent simulations. For example:

```
if ( a === 1'bx)
    $display("a is unknown");
```

will display the message if and only if a is 1'bx.

To help understand the last examples, you should realize that the following two if statements are equivalent:

```
if(expression)           if((expression)===1'b1)
    statement;               statement;
```

The following table summarizes how the four-valued logic works with common operators:

a b	a==b	a===b	a!=b	a!==b	a&b	a&&b	a|b	a||b	a^b
0 0	1	1	0	0	0	0	0	0	0
0 1	0	0	1	1	0	0	1	1	1
0 x	x	0	x	1	0	0	x	x	x
0 z	x	0	x	1	0	0	x	x	x
1 0	0	0	1	1	0	0	1	1	1
1 1	1	1	0	0	1	1	1	1	0
1 x	x	0	x	1	x	x	1	1	x
1 z	x	0	x	1	x	x	1	1	x
x 0	x	0	x	1	0	0	x	x	x
x 1	x	0	x	1	x	x	1	1	x
x x	x	1	x	0	x	x	x	x	x
x z	x	0	x	1	x	x	x	x	x
z 0	x	0	x	1	0	0	x	x	x
z 1	x	0	x	1	x	x	1	1	x
z x	x	0	x	1	x	x	x	x	x
z z	x	1	x	0	x	x	x	x	x

This table was generated by the following Verilog code:

```
module xz01;
  reg a,b,val[3:0];
  integer ia,ib;

  initial
    begin
      val[0] = 1'b0;
      val[1] = 1'b1;
      val[2] = 1'bx;
      val[3] = 1'bz;
      $display
        ("a b   a==b a===b   a!=b a!==b    a&b a&&b   a|b a||b   a^b");

      for (ia = 0; ia<=3; ia=ia+1)
        for (ib = 0; ib<=3; ib=ib+1)
          begin
            a = val[ia];
            b = val[ib];

            $display
        ("%b %b   %b  %b     %b   %b      %b   %b      %b   %b    %b ",
             a,b,a==b,a===b,a!=b,a!==b,a&b,a&&b,a|b,a||b,a^b);
          end
    end
endmodule
```

3.6 `$time`

A Verilog simulator executes as a software program on a conventional general-purpose computer. How long it takes such a computer to run a Verilog simulation, known as *real time*, depends on several factors, such as how fast the general-purpose computer is, and how efficient the simulator is. The speed with which the designer obtains the simulation results has little to do with how fast the eventual hardware will be when it is fabricated. Therefore, the real time required for simulation is not important in the following discussion.

Instead, Verilog provides a built-in variable, `$time`, which represents simulated time, that is, a simulation of the actual time required for a machine to operate when it is fabricated. Although the value of `$time` in simulation has a direct relationship to the physical time in the fabricated hardware, `$time` is not measured in seconds. Rather, `$time` is a unitless integer. Often designers map one of these units into one nanosecond, but this is arbitrary.

3.6.1 Multiple blocks

Verilog allows more than one behavioral block in a module. For example:

```
module two_blocks;
   integer x,y;

   initial
      begin
        a=1;
        $display("a is one");
      end

   initial
      begin
        b=2;
        $display("b is two");
      end
endmodule
```

The above *simulates* a system in which a and b are **simultaneously** assigned their respective values. This means, from a simulation standpoint, `$time` is the same when a is assigned one as when b is assigned two. (Since both assignments occur in `initial` blocks, `$time` is 0.) Note that this does not imply the sequence in which these assignments (or the corresponding `$display` statements) occur.

3.6.2 Sequence versus $time

In software, we often confuse the two separate concepts of time and sequence. In Verilog, it is possible for many statements to execute without $time advancing. The sequence in which statements within one block execute is determined by the usual rules found in other high-level languages. The sequence in which statements within different blocks execute is something the designer cannot predict, but that Verilog will do consistently. The advancing of $time is a different issue, discussed in section 3.7.

If you change the wires to be regs, a structural Verilog netlist is equivalent to several always blocks, where each always block computes the result output by one gate. If the design is correct, the sequence in which such always blocks execute at a particular $time is irrelevant, which helps explain why the order in which you instantiate gates in structural Verilog is also irrelevant. With Verilog, you can simulate the parallel actions of each gate or module that you instantiate, as well as the parallel actions of each behavioral block you code.

3.6.3 Scheduling processes and deadlock

Like a multiprocessing operating system, a Verilog simulator schedules several processes, one for each structural component or behavioral block. The $time variable does not advance until the simulator has given each process that so desires an opportunity to execute at that $time.

If you are familiar with operating systems concepts, such as semaphores, you will recognize that this raises a question about how Verilog operates: what are the atomic units of computation, or in other words, when does a process get interrupted by the Verilog simulator?

The behavioral statements described earlier are uninterruptible. Although it is nearly correct to model an exclusive OR with the following behavioral code:

```
module deadlock_the_simulator;
   reg a,b,c;
   always
     c = a^b;
   ... other blocks ...
endmodule
```

the Verilog simulator would never allow the other blocks to execute because the block computing c is not interruptible. Overcoming this problem requires an additional feature of Verilog, discussed in the next section.

3.7 Time control

Behavioral Verilog may include *time control* statements, whose purpose is to release control back to the Verilog scheduler so that other processes may execute and also tell the Verilog simulator at what $time the current process would like to be restarted. There are three forms of time control that have different ways of telling the simulator when to restart the current process: #, @ and wait.

3.7.1 # time control

When a statement is preceded by # followed by a number, the scheduler will not execute the statement until the specified number of $time units have passed. Any other process that desires to execute earlier than the $time specified by the # will execute before the current process resumes. If we modify the first example from section 3.6:

```
module two_blocks_time_control;
  integer x,y;


  initial
    begin
      #4
      a=1;
      $display("a is one at $time=%d",$time);
    end

  initial
    begin
      #3
      b=2;
      $display("b is two at $time=%d",$time);
    end

endmodule
```

the above will assign first to b (at $time=3) and then to a one unit of $time later. The order in which these statements execute is unambiguous because the # places them at a certain point in $time.

There can be more than one # in a block. The following nonsense module illustrates how the # works:

```
                module confusing;
                   integer a;


                initial
                   begin
                          a =  10;
                      #2  a =  20;
                      #5  a =  30;
                   end

                initial
                   begin
                      #1  a =  40;
                      #3  a =  50;
                      #4  a =  60;
                   end
             endmodule
```

In the above code, a becomes 10 at $time 0, 40 at $time 1, 20 at $time 2, 50 at $time 4, 30 at $time 7 and 60 at $time 8. The interaction of parallel blocks creates a behavior much more complex than that of each individual block.

3.7.1.1 Using # in test code

One of the most important uses of # is to generate sequences of patterns at specific $times in test code to act as inputs to a machine. The # releases control from the test code and gives the code that simulates the machine an opportunity to execute. Test code without some kind of time control would be pointless because the machine being tested would never execute.

For example, suppose we would like to test the built-in xor gate by stimulating it with all four combinations on its inputs, and printing the observed truth table:

```
module top;
   integer ia,ib;
   reg a,b;
   wire c;

   xor x1(c,a,b);

   initial
     begin
       for (ia=0; ia<=1; ia = ia+1)
         begin
           a = ia;
           for (ib=0; ib<=1; ib = ib + 1)
             begin
               b = ib;
               #10 $display("a=%d b=%d c=%d",a,b,c);
             end
         end
     end
endmodule
```

The first time through, a and b are initialized to be 0 at $time 0. When #10 executes at $time 0, the initial block relinquishes control, and x1 is given the opportunity to compute a new value (0^0=0) on the wire c. Having completed everything scheduled at $time 0, the simulator advances $time. The next thing scheduled to execute is the $display statement at $time 10. (The simulator does not waste real time computing anything for $time 2 through 9 since nothing changes during this $time.) The simulator prints out that "a=0 b=0 c=0" at $time 10 and then goes through the inner loop once again. While $time is still 10, b becomes 1. The #10 relinquishes control, x1 computes that c is now 1 and $time advances. The $display prints out that "a=0 b=1 c=1" at $time 20. The last two lines of the truth table are printed out in a similar fashion at $times 30 and 40.

3.7.1.2 Modeling combinational logic with
Physical combinational logic devices, such as the exclusive OR gate, have propagation delay. This means that a change in the input does not instantaneously get reflected in the output as shown above, but instead it takes some amount of physical time for the change to propagate through the gate. Propagation delay is a low-level detail of hardware design that ultimately determines the speed of a system. Normally, we will want to ignore propagation delay, but for a moment, let's consider how it can be modeled in behavioral Verilog with the #.

The behavioral exclusive OR example in section 3.6.3 deadlocks the simulator because it does not have any time control. If we put some time control in this `always` block (say a propagation delay of #1), the simulator will have an opportunity to schedule the test code instead of deadlocking inside the `always` block:

```
module top;
   integer ia,ib;
   reg a,b;
   reg c;

   always #1
     c = a^b;

   initial
     begin
       for (ia=0; ia<=1; ia = ia+1)
         begin
           a = ia;
           for (ib=0; ib<=1; ib = ib + 1)
             begin
               b = ib;
               #10 $display("a=%d b=%d c=%d",a,b,c);
             end
         end
       $finish;
     end
endmodule
```

As in the last example, a and b are initialized to be 0 at $time 0. When #10 executes at $time 0, the `initial` block relinquishes control, which gives the `always` loop an opportunity to execute. The first thing that the `always` block does is to execute #1, which relinquishes control until $time 1. Since no other block wants to execute at $time 1, execution of the `always` block resumes at $time 1, and it computes a new value (0^0=0) for the reg c. Because this is an `always` block, it loops back to the #1. Since no other block wants to execute at $time 2, execution of the `always` block resumes at $time 2, and it recomputes the same value for the reg c that it just computed at $time 1. The `always` block continues to waste real time by unnecessarily recomputing the same value all the way up to $time 9.

Finally, the $display statement executes at $time 10. The test code prints out "a=0 b=0 c=0" and goes through its inner loop once again. While $time is still 10, b becomes 1. The #10 relinquishes control, and the `always` block will have another ten chances to compute that c is now 1. The remaining lines of the truth table are printed out in a similar fashion.

There is an equivalent structural netlist notation for an `always` block with # time control. The following behavioral and structural code do similar things in $time:

```
reg c;                          wire c;
always #2                       xor #2 x2(c,a,b);
    c = a^b;
```

Both model an exclusive OR gate with a propagation delay of two units of $time. On many (but not all) implementations of Verilog simulators, the structural version is more efficient from a real-time standpoint. This is discussed in greater detail in chapter 6.

3.7.1.3   Generating the system clock with # for simulation

Registers and controllers are driven by some kind of a clock signal. One way to generate such a signal is to have an `initial` block give the clock signal an initial value, and an `always` block that toggles the clock back and forth:

```
reg sysclk;

initial
    sysclk = 0;

always #50
    sysclk = ~sysclk;
```

The above generates a system clock signal, `sysclk`, with a period of 100 units of $time.

3.7.1.4   Ordering processes without advancing $time

It is permissible to use a delay of #0. This causes the current process to relinquish control to other processes that need to execute at the current $time. After the other processes have relinquished control, but before $time advances, the current process will resume. This kind of time control can be used to enforce an order on processes whose execution would otherwise be unpredictable. For example, the following is algorithmically the same as the first example in 3.7.1 (b is assigned first, then a), but both assignments occur at $time 0:

```
module two_blocks_time_control;
  integer x,y;
  initial
    begin
      #0
      a=1;
      $display("a is one at $time=%d",$time);
    end
  initial
    begin
      b=2;
      $display("b is two at $time=%d",$time);
    end
endmodule
```

3.7.2   @ time control

When an @ precedes a statement, the scheduler will not execute the statement that follows until the event described by the @ occurs. There are several different kinds of events that can be specified after the @, as shown below:

```
@(expression)
@(expression or expression or ...)
@(posedge onebit)
@(negedge onebit)
@ event
```

When there is a single expression in parenthesis, the @ waits until one or more bit(s) in the result of the *expression* change. As long as the result of the *expression* stays the same, the block in which the @ occurs will remain suspended. When multiple expressions are separated by or, the @ waits until one or more bit(s) in the result of any of the *expression*s change. The word or is not the same as the operator |.

In the above, *onebit* is single-bit wire or reg (declared without the square bracket). When posedge occurs in the parenthesis, the @ waits until *onebit* changes from a 0 to a 1. When negedge occurs in the parenthesis, the @ waits until *onebit* changes from a 1 to a 0. The following mean the same thing:

```
reg a,b,c;              reg a,b,c;
@(c) a=b;               @(posedge c or negedge c) a=b;
```

An *event* is a special kind of Verilog variable, which will be discussed later.

3.7.2.1 Efficient behavioral modeling of combinational logic with @

Although you can model combinational logic behaviorally using just the #, this is not an efficient thing to do from a simulation real-time standpoint. (Using # for combinational logic is also inappropriate for synthesis.) As illustrated in section 3.7.1.2, the `always` block has to reexecute many times without computing anything new. Although physical hardware gates are continuously recomputing the same result in this fashion, it is wasteful to have a general-purpose computer spend real time simulating this. It would be better to compute the correct result once and wait until the next time the result changes.

How do we know when the output changes? Recall that perfect combinational logic (i.e., with no propagation delay) by definition changes its output whenever **any of its input(s) change**. So, we need the Verilog notation that allows us to suspend execution until any of the inputs of the logic change:

```
module top;
    integer ia,ib;
    reg a,b;
    reg c;

    always @(a or b)
        c = a^b;

    initial
        begin
            for (ia=0; ia<=1; ia = ia+1)
                begin
                    a = ia;
                    for (ib=0; ib<=1; ib = ib + 1)
                        begin
                            b = ib;
                            #10 $display("a=%d b=%d c=%d",a,b,c);
                        end
                end
            $finish;
        end
endmodule
```

At the beginning, both the `initial` and the `always` block start execution. Since neither a nor b have changed yet, the `always` block suspends. The first time through the loops in the `initial` block, a and b are initialized to be 0 at `$time` 0. When #10 executes at `$time` 0, the `initial` block relinquishes control, and the `always` block is given an opportunity to do something. Since a and b both changed at `$time` 0, the @ does not suspend, but instead allows the `always` block to compute a new value (0^0=0) for the `reg` c. The `always` block loops back to the @. Since there is no way that a or b can change anymore at `$time` 0, the simulator advances `$time`. The next thing scheduled to execute is the `$display` statement at `$time` 10. (Like the example in section 3.7.1.1, but unlike the example in section 3.7.1.2, the simulator does not waste real time computing anything for `$time` 1 through 9 since nothing changes during that `$time`.) The simulator prints out that "a=0 b=0 c=0" at `$time` 10, and then goes through the inner loop once again. While `$time` is still 10, b becomes 1. The #10 relinquishes control, and the `always` block has an opportunity to do something. Since b just changed (though a did not change), the @ does not suspend, and c is now 1. After `$time` advances, the `$display` prints out that "a=0 b=1 c=1" at `$time` 20. The last two lines of the truth table are printed out in a similar fashion at `$times` 30 and 40.

Since this is a model of combinational logic, it is very important that **every input to the logic be listed after the** @. We refer to this list of inputs to the physical gate as the *sensitivity list*.

3.7.2.2 *Modeling synchronous registers*

Most synchronous registers that we deal with use rising edge clocks. Using @ with `posedge` is the easiest way to model such devices. For example, consider an enabled register whose input (of any bus width) is `din` and whose output (of similar width as `din`) is `dout`. At the rising edge of the clock, when `ld` is 1, the value presented on `din` will be loaded. Otherwise `dout` remains the same. Assuming `din, dout, ld` and `sysclk` are taken care of properly elsewhere in the module, the behavioral code to model such an enabled register is:

```
always @(posedge sysclk)
    if (ld)
        dout = din;
```

Similar Verilog code can be written for a counter register that has `clr, ld`, and `cnt` signals:

```
always @(posedge sysclk)
  begin
    if (clr)
      dout = 0;
    else
      if (ld)
        dout = din;
      else
        begin
          if (cnt)
            dout = dout + 1;
        end
  end
```

Note that the nesting of `if` statements indicates the priority of the commands. If a controller sends this counter a command to `clr` and `cnt` at the same time, the counter will ignore the `cnt` command. At any `$time` when this `always` block executes, only one action (clearing, loading, counting or holding) occurs. Of course, improper nesting of `if` statements could yield code whose behavior would be impossible with physical hardware.

3.7.2.3 Modeling synchronous logic controllers

Most controllers are triggered by the rising edge of the system clock. It is convenient to use `posedge` to model such devices. For example, assuming that `stop`, `speed` and `sysclk` have been dealt with properly elsewhere in the module, the second ASM chart in section 2.1.1.2 could be modeled as:

```
always
  begin
    @(posedge sysclk)   //this models state GREEN
      stop = 0;
      speed = 3;
    @(posedge sysclk) //this models state YELLOW
      stop = 1;
      speed = 1;
    @(posedge sysclk)   //this models state RED
      stop = 1;
      speed = 0;
  end
```

There are several things to note about the above code. First, the indentation is used only to promote readability. Assuming the code for generating `sysclk` given in section

3.7.1.3, the `stop = 0` and `speed = 3` statements execute at `$time` 50, 350, 650, ... because there is no time control among them. The indentation simply highlights the fact that these two statements execute atomically, as a unit, without being interrupted by the simulator.

The second thing to note is that the = in Verilog is just a **software assignment statement.** (The variable is modified at the `$time` the statement executes. The variable will retain the new value until modified again.) This is different than how we use = in ASM chart notation. (The command signal is a function of the present state. The command signal does not retain the new value after the rising edge of the system clock but instead returns to its default value.) Another way of saying this is that there are no default values in standard Verilog variables as there are for ASM chart commands. Despite the distinction between Verilog and ASM chart notation, we can model an ASM chart in Verilog by fully specifying every command output in every state. For those states where a command is not mentioned in an ASM chart, one simply codes a Verilog assignment statement that stores the default value into the Verilog variable corresponding to the missing ASM chart command. The `stop=0` and `speed=0` statements above were not shown in the original ASM chart but are required for the Verilog code to model what the hardware would actually do.

The third thing is the names of the states are not yet included in the Verilog code. (The comments are of course ignored by Verilog.) Eventually, we will find a way of including meaningful state names in the actual code.

The fourth thing is that this ASM chart does not have any RTN (i.e., it is at the mixed stage). We will need an additional Verilog notation to model ASM charts that use RTN. This notation is discussed in section 3.8.

3.7.2.4 @ for debugging display

@ can also be used for causing the Verilog simulator to print debugging output that shows what happens as actions unfold in the simulation. For example,

```
always @(a or b or c)
    $display("a=%b b=%b c=%b at $time=%d",a,b,c,$time);
```

The above block would eliminate the need for the designer to worry about putting `$display` statements in the test code or in the code for the machine being tested.

With clocked systems, it is often convenient to display information shortly after each rising edge of the clock:

```
always @(posedge sysclk)
 #20 $display("stop=%b speed=%b at $time=%d",
 stop,speed,$time);
```

3.7.3 `wait`

The `wait` statement is a form of time control that is quite different than # or @. The
`wait` statement stands by itself. It does not modify the statement which follows in the
way that @ and # do (i.e., there must be a semicolon after the `wait` statement). The
`wait` statement is used primarily in test code. It is not normally used to model hardware devices in the way @ and # are used. The syntax for the `wait` statement is:

```
wait(condition);
```

The `wait` statement suspends the current process. The current process will resume
when the condition becomes true. If the condition is already true, the current process
will resume without $time advancing.

For example, suppose we want to exhaustively test one of the slow division machines
described in chapter 2. The amount of time the machine takes depends on how big the
result is. Furthermore, different ASM charts described in chapter 2 take different amounts
of $time. Therefore, the best approach is to use the `ready` signal produced by the
machine:

```
module top;
  reg pb;
  integer x,y;
  wire [11:0] quotient;
  wire sysclk;
  ...
  initial
    begin
      pb= 0;
      x = 0;
      y = 0;
      #250;
      @(posedge sysclk);
      while (x<=4095)
        begin
          for (y=1; y<=4095; y = y+1)
            begin
              @(posedge sysclk);
              pb = 1;
```

Continued

```
                @(posedge sysclk);
                pb = 0;
                @(posedge sysclk);
                wait(ready);
                @(posedge sysclk);
                if (x/y === quotient)
                  $display("ok");
                else
                  $display("error x=%d y=%d x/y=%d  quotient=%d",
                                   x,y,x/y,quotient);
            end
          x = x + 1;
        end
      $stop;
    end
endmodule
```

This test code (based on the nested loops given in section 3.4) embodies the assumptions we made in section 2.2.1. The first two @s in the loop produce the `pb` pulse that lasts exactly one clock cycle. The third @ makes sure that the machine has enough time to respond (and make `ready` 0). The `wait(ready)` keeps the test code synchronized to the division machine, so that the test code is not feeding numbers to the division machine too rapidly. The fourth @ makes sure the machine will spend the required time in state IDLE, before testing the next number.

The ellipsis shows where the code for the actual division machine was omitted in the above. The `quotient` is produced by this machine which is not shown here. The design of this code will be discussed in the next chapter.

3.8   Assignment with time control

The # and @ time control, discussed in sections 3.7.1 and 3.7.2, precede a statement. These forms of time control delay execution of the following statement until the specified `$time`. There are two special kinds of assignment statements[5] that have time control **inside the assignment statement**. These two forms are known as *blocking* and *non-blocking procedural assignment*.

_____

[5] Assignment with time control is not accepted by some commercial synthesis tools but is accepted by all Verilog simulators. Since there are problems with intra-assignment delay (section 3.8.2.1), some authors recommend against its use, but when used as recommended later in this chapter (section 3.8.2.2), it becomes a powerful tool. Chapter 7 explains a preprocessor that allows all synthesis tools to accept the use proposed in this book.

3.8.1 Blocking procedural assignment

The syntax for blocking procedural assignment has the # or @ notation (whose syntax is described in sections 3.7.1 and 3.7.2) after the = but before the expression. For example, three common forms of this are:

```
var = # delay expression;
var = @(posedge onebit) expression;
var = @(negedge onebit) expression;
```

Other variations are also legal. What distinguishes this from a normal instantaneous assignment is that the expression is evaluated at the $time the statement first executes, but the variable does not change until after the specified delay. For example, assuming temp is a reg that is not used elsewhere in the code and that temp is declared to be the same width as a and b, the following two fragments of code mean the same thing:

```
                                    initial
                                      begin
    initial                            . . .
      begin                           temp = b;
       . . .                          @(posedge sysclk) a = temp;
      a = @(posedge sysclk) b;         . . .
       . . .                          end
      end
```

Blocking procedural assignment is almost what we need to model an ASM chart with RTN. The one problem with it, as its name implies, is that it blocks the current process from continuing to execute additional statements at the same $time. We will not use blocking procedural assignment for this reason.

3.8.2 Non-blocking procedural assignment

The syntax for a non-blocking procedural assignment is identical to a blocking procedural assignment, except the assignment statement is indicated with <= instead of =. This should be easy to remember, because it reminds us of the ← notation in ASM charts. For example, the most common form of the non-blocking assignment used in later chapters is:

```
var <= @(posedge onebit) expression;
```

Typically, *onebit* is the `sysclk` signal mentioned in section 3.7.1.3. Although other forms are legal, the above `@ (posedge onebit)` form of the non-blocking assignment is the one we use in almost every case for ← in ASM charts.[6]

The expression is evaluated at the `$time` the statement first executes and further statements execute at that same `$time`, but the variable does not change until after the specified delay. For example, assuming `temp` is a `reg` that is not used elsewhere in the left-hand code and that `temp` is declared to be the same width as `a` and `b`, the following two fragments of code mean nearly the same thing:

```
                                  always @ (posedge sysclk)
                                    #0 a = temp;

    initial                       initial
      begin                        · begin
        . . .                           . . .
        a <= @ (posedge sysclk) b;        temp = b;
        . . .                           . . .
      end                          end
```

Note that, all by itself, the effect of the non-blocking assignment is like having a parallel `always` block to store into `a`. An advantage of the `<=` notation is that you do not have to code a separate `always` block for each register.

A subtle detail is that the right-hand `always` block is the last thing to execute (#0) at a given `$time`. Similarly, the `<=` causes the `reg` to change only after every other block (including the one with the `<=`) has finished execution. This subtle detail causes a problem, which is discussed in the next section, and which is solved in section 3.8.2.2.

3.8.2.1 Problem with <= for RTN for simulation

An obvious approach to translating RTN from an ASM chart into behavioral Verilog is just to put `<=` for each ← in the ASM chart. For example, assuming `stop`, `speed`, `count` and `sysclk` are taken care of properly elsewhere, one might think that the ASM chart from section 2.1.1.3 could be translated into Verilog as:

---

[6] The exceptions are when the left-hand side of the ← is a memory being changed every clock cycle, in which case `@ (negedge onebit)` is appropriate, as explained in section 6.5.2, and for post-synthesis behavorial modeling of logic equations, in which case # is appropriate, as explained in section 11.3.3.

```
always
  begin
    @(posedge sysclk)          //this models state GREEN
      stop = 0;
      speed = 3;

    @(posedge sysclk)          //this models state YELLOW
      stop = 1;
      speed = 1;
      count <= @(posedge sysclk) count + 1;

    @(posedge sysclk)          //this models state RED
      stop = 1;
      speed = 0;
      count <= @(posedge sysclk) count + 2;
  end
```

However, when one runs this code on a Verilog simulator, the following incorrect result is produced (assuming the debugging `always` block shown in section 3.7.2.4):

```
stop=0    speed=11    count=000 at    $time=      70
stop=1    speed=01    count=000 at    $time=     170
stop=1    speed=00    count=001 at    $time=     270
stop=0    speed=11    count=010 at    $time=     370
stop=1    speed=01    count=010 at    $time=     470
stop=1    speed=00    count=011 at    $time=     570
```

Recall from section 2.1.1.3 that at $time 370, count should be three instead of two. The underlying cause of this error is the subtle detail mentioned above: The <= causes the reg to change only after every other block (including the one with the <=) has finished execution.

The above Verilog starts to execute the statements for state YELLOW at $time 150. The last of these statements evaluates count+1 at $time 150 and schedules the storage of the result. Since count is still 3'b000 at $time 150, the result scheduled to be stored at the end of $time 250 is 3'b001. The @(posedge sysclk) that starts state RED causes the always block to suspend until $time 250. The problem shown above occurs at $time 250 because the assignment initiated by the <= at $time 150 will be the last thing that occurs at $time 250. Prior to the assignment, the process will resume and execute the three statements, including count <= @(posedge sysclk) count + 2. Since count is still 3'b000, this <= schedules 3'b010 to be assigned at $time 350, which is not what happens in an ASM chart. As soon as the assignment of 3'b010 has been scheduled at $time 250, 3'b001 will be stored into count (as a result of the first <=).

3.8.2.2 *Proper use of <= for RTN in simulation*

To overcome the problem described in the last section, you need to use a non-zero delay after each @(posedge sysclk) that denotes a rectangle of the ASM chart. For example, here is the complete Verilog code to model (in a primitive way) the ASM chart from section 2.1.1.3:

```verilog
module top;
  reg stop;
  reg [1:0] speed;
  reg sysclk;
  reg [2:0] count;

  initial
    sysclk = 0;
  always #50
    sysclk = ~sysclk;

  always
    begin
      @(posedge sysclk) #1      //this models state GREEN
        stop = 0;
        speed = 3;
      @(posedge sysclk) #1 //this models state YELLOW
        stop = 1;
        speed = 1;
        count <= @(posedge sysclk) count + 1;

      @(posedge sysclk) #1      //this models state RED
        stop = 1;
        speed = 0;
        count <= @(posedge sysclk) count + 2;
    end

  always @(posedge sysclk)
    #20 $display("stop=%b speed=%b count=%b at $time=%d",
         stop, speed, count, $time);

  initial
    begin
      count = 0;
      #600 $finish;
    end
endmodule
```

Let's analyze the reason why each block is required in this module. The first initial block is required to give sysclk a value other than 1'bx at $time 0. The next block toggles `sysclk` so that the clock period is 100. If `sysclk` were not initialized at `$time` 0, it would stay `1'bx` forever (`~1'bx` is `1'bx`).

The only new thing in the `always` block that models the ASM chart is the addition of #1 after each `@(posedge sysclk)`. The `always` block that follows it displays `stop`, `speed` and `count` during each state.

The test code in the final `initial` block simply initializes count to be 3'b000. (In a real machine, this would occur in a state of the ASM, but instead here it is part of the test code for the purposes of illustration only.) The test code schedules a `$finish` system task to be called at `$time` 600. This is required because the `always` blocks would otherwise tell the simulator to go on forever.

With the #1 after each @, the Verilog simulator produces the following correct output:

```
stop=0    speed=11    count=000    at    $time=       70
stop=1    speed=01    count=000    at    $time=      170
stop=1    speed=00    count=001    at    $time=      270
stop=0    speed=11    count=011    at    $time=      370
stop=1    speed=01    count=011    at    $time=      470
stop=1    speed=00    count=100    at    $time=      570
```

3.8.2.3 Translating `goto`-less ASMs to behavioral Verilog

This book concentrates on several design techniques that all begin by expressing an ASM with behavioral Verilog. Since Verilog is a `goto`-less language, only certain kinds of ASMs can be translated in this fashion. Chapters 5 and 7 explain how arbitrary ASMs can be translated into Verilog, but this section will concentrate only on ASMs that adhere to this highly desirable `goto`-less style.

3.8.2.3.1 Implicit versus explicit style

The approach of expressing a state machine with high-level statements (like `if` and `while`) is known as *implicit style* because the next state of the machine is described implicitly through the use of `@(posedge sysclk)` within the statements of an `always` block. Implicit style is the opposite of the *explicit style* table (illustrated in section 2.4.1) that requires the designer to say what state the machine goes to under all possible circumstances.

Experienced hardware designers who are new to Verilog may find the implicit style approach confusing because it requires thinking about a state machine in a different way. The implicit style is much more like software concepts, such as the distinction between `if` and `while`. On the other hand, experienced software designers may also find this approach difficult at first because the timing relationship between <= and decisions in Verilog is different than in conventional software languages. The following sections go through a series of examples that illustrate some typical kinds of ASM constructs and how they translate into implicit style Verilog.

3.8.2.3.2   Identifying the infinite loop

Unlike software, all ASMs have at least one infinite loop. Implicit style behavioral Verilog is defined by an `always` block. Many times this `always` block can also serve to implement the infinite loop of the ASM. In the following ASM, the transitions from states FIRST, SECOND, THIRD and FOURTH are implicit. The designer does not have to say anything about their next states. The transition from FIFTH to FIRST occurs because of the `always`:

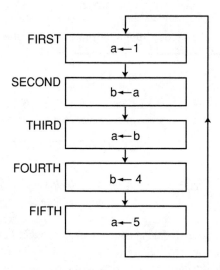

Figure 3.2 Every ASM has an infinite loop.

Inside the `always`, there is a one to one mapping of rectangles into `@(posedge sysclk)` statements. In this example, the ASM has five states, so the `always` uses five `@(posedge sysclk)`:

```
module top;
  //Following are actual hardware registers of ASM
  reg [11:0] a,b;


  //Following is NOT a hardware register
  reg sysclk;

  //The following always block models actual hardware
```

Continued

```
  always
    begin
      @(posedge sysclk) #1;                  // state FIRST
       a <= @(posedge sysclk) 1;
      @(posedge sysclk) #1;                  // state SECOND
       b <= @(posedge sysclk) a;
      @(posedge sysclk) #1;                  // state THIRD
       a <= @(posedge sysclk) b;
      @(posedge sysclk) #1;                  // state FOURTH
       b <= @(posedge sysclk) 4;
      @(posedge sysclk) #1;                  // state FIFTH
       a <= @(posedge sysclk) 5;
    end

  //Following initial and always blocks do not correspond to
  // hardware. Instead they are test code that shows what
  // happens when the above ASM executes

  always #50 sysclk = ~sysclk;
  always @(posedge sysclk) #20
    $display("%d a=%d b=%d ", $time, a, b);

  initial
    begin
      sysclk = 0;
      #1400 $stop;
    end
endmodule
```

The above is slightly more primitive than what will be used in later chapters, but the emphasis of this example is to show how an ASM translates into Verilog. In the above, there are three `always` blocks, but only the first one corresponds to hardware. The other two `always` blocks and the `initial` block are necessary for simulation (in later chapters these other blocks will be moved to other modules).

3.8.2.3.3 *Recognizing* `if else`

Most ASMs have decisions. Decisions in implicit Verilog are described either with the `if` statement (possibly followed by `else`) or with the `while` statement. For hardware designers without extensive software experience, determining whether the `if` or the `while` is appropriate for a particular decision can seem confusing at first.

The following ASM is an example where the `if else` construct is appropriate:

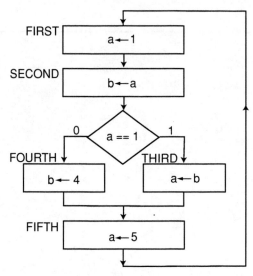

FIRST: a←1

SECOND: b←a

a == 1 (0 / 1)

FOURTH: b←4

THIRD: a←b

FIFTH: a←5

Figure 3-3. ASM corresponding to `if else`.

For brevity, only the `always` block that corresponds to the actual hardware is shown:

```
always
  begin
    @(posedge sysclk) #1;              // state FIRST
    a <= @(posedge sysclk) 1;
    @(posedge sysclk) #1;              // state SECOND
    b <= @(posedge sysclk) a;
    if (a == 1)
      begin
        @(posedge sysclk) #1;          // state THIRD
        a <= @(posedge sysclk) b;
      end
    else
      begin
        @(posedge sysclk) #1;          // state FOURTH
        b <= @(posedge sysclk) 4;
      end
    @(posedge sysclk) #1;              // state FIFTH
    a <= @(posedge sysclk) 5;
  end
```

The `if else` is appropriate here because only one of the states (THIRD or FOURTH) will execute. Because `a` is one in state SECOND, state THIRD will execute. In the following very similar Verilog, state FOURTH rather than state THIRD will execute:

```
always
  begin
    @(posedge sysclk) #1;              // state FIRST
     a <= @(posedge sysclk) 1;
    @(posedge sysclk) #1;              // state SECOND
     b <= @(posedge sysclk) a;
     if (a != 1)
       begin
         @(posedge sysclk) #1;         // state THIRD
           a <= @(posedge sysclk) b;
       end
     else
       begin
         @(posedge sysclk) #1;         // state FOURTH
           b <= @(posedge sysclk) 4;
       end
    @(posedge sysclk) #1;              // state FIFTH
     a <= @(posedge sysclk) 5;
  end
```

3.8.2.3.4   Recognizing a single alternative

Often, it is appropriate to omit the else, as in the following ASM:

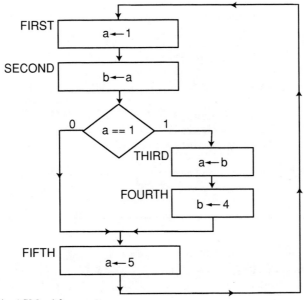

Figure 3-4. ASM without else.

which translates to the following Verilog:

```
always
  begin
    @(posedge sysclk) #1;              // state FIRST
    a <= @(posedge sysclk) 1;
    @(posedge sysclk) #1;              // state SECOND
    b <= @(posedge sysclk) a;
    if (a == 1)
      begin
        @(posedge sysclk) #1;          // state THIRD
          a <= @(posedge sysclk) b;
        @(posedge sysclk) #1;          // state FOURTH
          b <= @(posedge sysclk) 4;
      end
    @(posedge sysclk) #1;              // state FIFTH
      a <= @(posedge sysclk) 5;
  end
```

In the above, both state THIRD and state FOURTH will execute because a is one in state SECOND. The following very similar Verilog skips directly from state SECOND to state FIFTH:

```
always
  begin
    @(posedge sysclk) #1;              // state FIRST
    a <= @(posedge sysclk) 1;
    @(posedge sysclk) #1;              // state SECOND
    b <= @(posedge sysclk) a;
    if (a != 1)
      begin
        @(posedge sysclk) #1;          // state THIRD
          a <= @(posedge sysclk) b;
        @(posedge sysclk) #1;          // state FOURTH
          b <= @(posedge sysclk) 4;
      end
    @(posedge sysclk) #1;              // state FIFTH
      a <= @(posedge sysclk) 5;
  end
```

3.8.2.3.5 *Recognizing* while *loops*

The following two ASMs describe the same hardware. The first of the following two ASMs is very similar to the one in section 3.8.2.3.4, except that state FOURTH does not necessarily go to state FIFTH . Instead, state FOURTH goes to a decision which determines whether to go to state THIRD or state FIFTH. The second of the following two ASMs is a much less desirable way to describe the identical hardware. It is undesirable because the `a==1` test is duplicated; however, its meaning is exactly the same as the first of the following two ASMs:

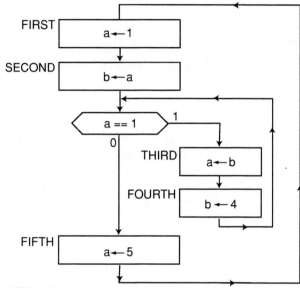

Figure 3-5. ASM with `while`.

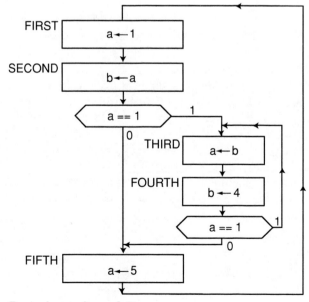

Figure 3-6. Equivalent to figure 3-5.

The reason the first of the ASMs is preferred is because it is more obvious that it translates into a `while` loop in Verilog:

```
always
  begin
    @(posedge sysclk) #1;              // state FIRST
     a <= @(posedge sysclk) 1;
    @(posedge sysclk) #1;              // state SECOND
     b <= @(posedge sysclk) a;
    while (a == 1)
      begin
        @(posedge sysclk) #1;          // state THIRD
         a <= @(posedge sysclk) b;
        @(posedge sysclk) #1;          // state FOURTH
         b <= @(posedge sysclk) 4;
      end
    @(posedge sysclk) #1;              // state FIFTH
     a <= @(posedge sysclk) 5;
  end
```

In fact, the only syntactic difference between the above Verilog and the Verilog in section 3.8.2.3.4 is that the word `if` has been changed to `while`. The advantage of looking at this particular ASM as a `while` loop is that the decision a==1 is shared by both state SECOND and state FOURTH. With the `while` loop, the designer does not have to worry that the decision is actually part of two states. Many practical algorithms that produce useful results (as illustrated in chapter 2) demand a loop of this style. The `while` in Verilog makes this easy.

3.8.2.3.6 Recognizing `forever`

Sometimes machines need initialization states that execute only once. Since synthesis tools only accept behavioral Verilog defined with `always` blocks, such ASMs still begin with the keyword `always`. However, the looping action of the `always` is not pertinent. (If the designer only wanted to simulate the machine, `initial` would work just as well as `always`, but ultimately the synthesis tool will demand `always`.)

In order to describe the infinite loop that exists beyond the initialization states, the designer must use `forever`. For example, consider the following ASM:

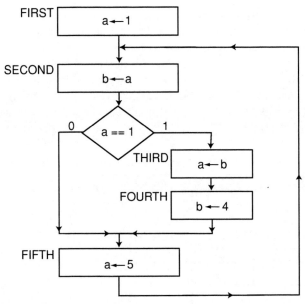

Figure 3-7. ASM needing `forever`.

It is almost identical to the one in section 3.8.2.3.4, except that state FIFTH forms an infinite loop to state SECOND instead of going to state FIRST. The corresponding Verilog implements this using `forever`:

```
always
  begin
    @(posedge sysclk) #1;              // state FIRST
    a <= @(posedge sysclk) 1;
    forever
      begin
        @(posedge sysclk) #1;          // state SECOND
          b <= @(posedge sysclk) a;
          if (a == 1)
            begin
              @(posedge sysclk) #1;    // state THIRD
                a <= @(posedge sysclk) b;
              @(posedge sysclk) #1;    // state FOURTH
                b <= @(posedge sysclk) 4;
            end
        @(posedge sysclk) #1;          // state FIFTH
          a <= @(posedge sysclk) 5;
      end
  end
```

3.8.2.3.7 *Translating into an* `if` *at the bottom of* `forever`

The following two ASMs are equivalent. Many designers would think the one on the left is more natural because it describes a loop involving only state THIRD. As long as `a==1`, the machine stays in state THIRD. The noteworthy thing about this machine is that state THIRD also forms the beginning of a separate infinite loop. (Such an infinite loop might be described with an `always` or in this case a `forever`.) Because of this, it is preferred to think of this ASM as an `if` at the bottom of a `forever`, as illustrated by the ASM on the right:

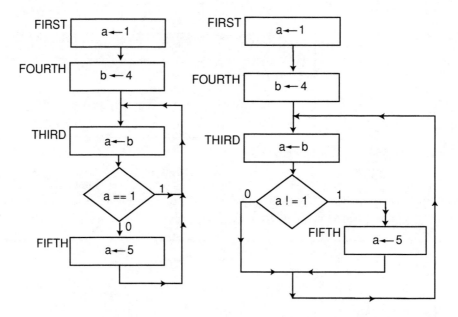

Figure 3-8. Two ways to draw `if` *at the bottom of* `forever`.

The ASM on the right tests if `a  != 1` to see whether to leave the loop involving only state THIRD and proceed to state FIFTH. The reason the ASM on the right is preferred is that its translation into Verilog is obvious:

```
always
  begin
    @(posedge sysclk) #1;                // state FIRST
      a <= @(posedge sysclk) 1;
    @(posedge sysclk) #1;                // state FOURTH
      b <= @(posedge sysclk) 4;
      forever
        begin
          @(posedge sysclk) #1;          // state THIRD
            a <= @(posedge sysclk) b;
            if (a != 1)
              begin
                @(posedge sysclk) #1;   // state FIFTH
                  a <= @(posedge sysclk) 5;
              end
        end
  end
end
```

In software, an `if` never implements a loop. This is also true in Verilog of an isolated `if`, but the combination of an `if` at the bottom of `forever` or `always` has the effect of nesting a non-infinite loop inside an infinite loop. It is the `forever` or `always` that forms the looping action, not the `if`. This example illustrates a kind of implicit behavioral Verilog that sometimes causes novice Verilog designers confusion. It is suggested that the reader should fully appreciate this example before proceeding to later chapters. Designers need to be careful not to confuse `if` with `while`.

3.9 Tasks and functions

In conventional software programming languages, it is common for a programmer to use functions and procedures (known as void functions in C) to break an algorithm apart into manageable pieces. There are two main motivations for using functions and procedures: they make the top-down design of a complex algorithm easier, and they sometimes allow reuse of the same code. Verilog provides tasks (which are like procedures) and functions, which can be called from behavioral code.

3.9.1 Tasks

The syntax for a task definition is:

```
        task name;
            input arguments;
            output arguments;
            inout arguments;
            ...
            declarations;
            begin
              statement;
              ...
            end
        endtask
```

This task definition must occur inside a module. The task is usually intended to be called only by `initial` blocks, `always` blocks and other tasks within that module. Tasks may have any behavioral statements, including time control.

Verilog lets the designer choose the order in which the `input`, `output` and `inout` definitions are given. (The order shown above is just one possibility.) The order in which `input`, `output` and `inout` definitions occur is based on the calling sequence desired by the designer. The sequence in which the formal arguments are listed in some combination of `input`, `output` and/or `inout` definitions determines how the actual arguments are bound to the formal definitions when the task is called.

The purpose of an `input` argument is to send information from the calling code into the task by value. An `input` argument may include a width (which is equivalent to a `wire` of that width) or it may be given a type of `integer` or `real` in a separate declaration. An `input` argument may not be declared as a `reg`.

The purpose of an `output` argument is to send a result from the task to the calling code by reference. An `output` argument must be declared as a `reg`, `integer` or `real` in a separate declaration.

An `inout` definition combines the roles of `input` and `output`. An `inout` argument must be declared as a `reg`, `integer` or `real` in a separate declaration.

3.9.1.1 Example task
Consider the following nonsense code:

```
          integer count,sum,prod;
          initial
            begin
              sum = 0;
              count = 1;

              sum = sum + count;
              prod = sum * count;
              count = count + 2;
              $display(sum,prod);

              sum = sum + count;
              prod = sum * count;
              count = count + 3;
              $display(sum,prod);

              sum = sum + count;
              prod = sum * count;
              count = count + 5;
              $display(sum,prod);

              sum = sum + count;
              prod = sum * count;
              count = count + 7;
              $display(sum,prod);

              $display(sum,prod,count);
            end
```

After initializing sum and count, there is a great similarity in the following four groups (each composed of four statements). Using a task allows this initial block to be shortened:

```
          integer count,sum,prod;
          initial
            begin
              sum = 0;
              count = 1;
              example(sum,prod,count,2);
              example(sum,prod,count,3);
              example(sum,prod,count,5);
              example(sum,prod,count,7);
              $display(sum,prod,count);
            end
```

The definition of the task `example` is:

```
task example;
   inout sum_arg;    //1st positional argument
   output prod_arg;  //2nd positional argument
   inout count_arg;  //3rd positional argument
   input numb_arg;   //4th positional argument

   integer count_arg, numb_arg, sum_arg, prod_arg;

   begin
     sum_arg = sum_arg + count_arg;
     prod_arg = sum_arg * count_arg;
     count_arg = count_arg + numb_arg;
     $display(sum_arg, prod_arg);
   end
endtask
```

Because the formal `inout sum_arg` is defined first, it corresponds to the actual `sum` in the `initial` block. Similarly, the formal `output prod_arg` corresponds to `prod`, and the formal `inout count_arg` corresponds to `count`. In order to pass different numbers each time to `example`, the formal `numb_arg` is defined to be `input`. The order in which the arguments are declared (in this case with the `integer` type) is irrelevant. The `$display` statements produce the following:

1	1	
4	12	
10	60	
21	231	
21	231	18

3.9.1.2 *enter_new_state task*

The translation of the ASM chart from section 2.1.1.3 into Verilog given in section 3.8.2.2 is correct but could be improved in two ways. First, this translation did not include state names as part of the Verilog code (they were only in the comments). Second, this translation did not automatically provide default values for states where command signals were not mentioned, as occurs in ASM chart notation.

To overcome both of these limitations, we will define a task, which is arbitrarily given the name `enter_new_state`. The purpose of this task is to do things that occur whenever the machine enters any state. This includes storing into `present_state` a representation of a state (which is passed as an input argument, `this_state`), doing the #1 (which is legal in a task) to allow the `<=` to work properly and giving default values to the command outputs. In order to use this task, the designer needs to define several arbitrary bit patterns for the state names, define the `present_state` as a `reg` and indicate the number of bits in the `present_state`:

```
           `define NUM_STATE_BITS     2
           `define GREEN              2'b00
           `define YELLOW             2'b01
           `define RED                2'b10


           . . .


           reg ['NUM_STATE_BITS-1:0] present_state;


           . . .
```

The `always` block that implements the ASM chart is similar to the one given in section 3.8.2.2:

```
always
  begin
    @(posedge sysclk) enter_new_state('GREEN);
      speed = 3;

    @(posedge sysclk) enter_new_state('YELLOW);
      stop = 1;
      speed = 1;
      count <= @(posedge sysclk) count + 1;

    @(posedge sysclk) enter_new_state('RED);
      stop = 1;
      count <= @(posedge sysclk) count + 2;
  end
```

The only differences are that the state names are passed as arguments to `enter_new_state`, and default values do not have to be mentioned. For example, state GREEN uses the default value 0 for `stop`, and state RED uses the default value 0 for `speed`.

The task that accomplishes these things for this particular ASM is:

```
task enter_new_state;
    input ['NUM_STATE_BITS-1:0] this_state;
    begin
      present_state = this_state;
      #1 stop = 0;
          speed = 0;
    end
  endtask
```

Even though default values are assigned for every state, since no time control occurs in this task after the assignment of default values, those states where non-default values are assigned work correctly. For example, assume the machine enters state GREEN at $time 50. At that $time, present_state will be assigned 2'b00. At $time 51, stop and speed will assigned their defaults of 0, but since there is no more time control, the always block which called on the task is not interruptable. At the same $time 51 speed changes to 3. Any other module concerned about speed at $time 51 would only observe a change to a value of 3. To understand this, we need to distinguish between sequence and $time. Because the task was called, two changes occurred to speed in sequence, but since they happened at the same $time, the outside world can only observe the last change. This creates exactly the effect we want. We are now ready to model ASM charts that do practical things with behavioral Verilog. Examples of translating ASM charts into Verilog using tasks like this are given in chapter 4.

3.9.2   Functions

The syntax for a function is similar to a task:

```
function type name;
    input arguments;
    ...
    declarations;
    begin
      statement;
      ...
      name = expression;
    end
  endfunction
```

except only input arguments are allowed. In the function definition, *type* is either integer, real or a bit width defined in square brackets. The statement(s) in a **function never include any time control**. The name of the function must be assigned the result to be returned (like the syntax of Pascal). These restrictions on functions exist so that every use of a function could, in theory, be synthesized as combinational logic.

3.9.2.1   Real function example

Verilog does not provide built-in trigonometric functions, but it is possible to define a function that approximates such a function using a polynomial:

```
function real sine;
    input x;
    real x;
    real y,y2,y3,y5,y7;
    begin
      y = x*2/3.14159;
      y2 = y*y;
      y3 = y*y2;
      y5 = y3*y2;
      y7 = y5*y2;
      sine = 1.570794*y - 0.261799*y3 +
             0.0130899*y5 - 0.000311665*y7;
    end
endfunction
```

Such a function might be useful if a designer needs to test the Verilog model of a machine, such as a math coprocessor, that implements an ASM to approximate transcendental functions.

3.9.2.2   Using a function to model combinational logic

A more common use of a function in Verilog is as a behavioral way to describe combinational logic. For example, rather than being described by the logic gates given in section 2.5, a half-adder can also be described by a truth table:

inputs		output	
a	b	c	s
0	0	0	0
0	1	0	1
1	0	0	1
1	1	1	0

Such a table can be written in Verilog as a function defined with a `case` statement. Since the result of the function is composed of more than one bit, the function is better documented by using local variables (`c` and `s` in this example), which are concatenated to form the result:

```
function [1:0] half_add;
    input a,b;
    reg c,s;  //local for documentation

    begin
      case ({a,b})
        2'b00:  begin
                    c = 0;
                    s = .0;
                end
        2'b01:  begin
                    c = 0;
                    s = 1;
                end
        2'b10:  begin
                    c = 0;
                    s = 1;
                end
        2'b11:  begin
                    c = 1;
                    s = 0;
                end
        default:begin
                    c = 1'bx;
                    s = 1'bx;
                end
      endcase
      half_add = {c,s};
    end
endfunction
```

So `half_add(0,0)` returns 2'b00 and `half_add(1,1)` returns 2'b10. Both `half_add(1,0)` and `half_add(0,1)` return 2'b01. All other possibilities, such as `half_add(1'bx,0)` return 2'bx. In order to use this function to model the combinational logic of a half-adder, the designer would define an `always` block with @ time control as explained in section 3.7.2.1:

```
reg C,S;
. . .
always @(A or B)
    {C,S} = half_add(A,B);
```

The actual argument A in the `always` block is bound to the formal a in `half_add`, and the actual argument B is bound to the formal b. The locals c and s are concatenated to form a two-bit result (hence the [1:0] declaration for the function.) This two bit result is stored in the two-bit concatenation {C,S}.

3.10 Structural Verilog, modules and ports

The preceding sections have covered many behavioral and a few structural (built-in gate), features of Verilog. This section discusses the most central aspect of Verilog: how the designer can define and instantiate Verilog modules to achieve hierarchical design.

Verilog code is composed of one or more modules. Each module is either a *top-level module* or an *instantiated module*. A top-level module is one (like all the earlier examples in this chapter) which is not instantiated elsewhere in the source code. There is only one copy of a top-level module. The definition of a top-level module is the same as the code that executes. The `reg`s and `wire`s in a top-level module are unique.

An instantiated module, on the other hand, is a unique executable copy of the definition. There may be many such copies. The definition is a "blueprint" for each of these instances. For example, section 2.5 illustrates an adder that needs three instances of a half-adder. It is only necessary to define the half-adder once. It can be instantiated as many times as required. Each instance of an instantiated module has its own copy of the `reg`s and `wire`s specified by the designer. For example, the value stored in a particular `reg` in one instance of a module need not be the same as the value stored in the `reg` of the same name in another instance of that module.

Instantiated modules should have ports that allow outside connections with each instance. It is this interconnection (i.e., structure) with the system external to the instance that gives each instance its unique role in the total system. Normally, each instance is internally identical to other instances derived from the same module definition, and how an instance is connected within the system gives that instance its characteristics.

The syntax for a module definition with ports is:

```
      module name (port1,port2, ... );
        input ...  ;
        output ... ;
        inout ... ;
        declarations;
        structural instance;
        ...
        behavioral instance;
        ...
        tasks
        ...
        functions
        ...
      endmodule
```

An example of a *structural instance* is given using built-in gates in section 3.5.1. Examples of designer supplied (rather than built-in) *structural instances* will be given later in section 3.10.6. A *behavioral instance* is either an always or initial block, as explained in section 3.4. (Tasks and functions are local to a module, and may be called by a *behavioral instance*, but are not by themselves *behavioral instances*.) The *declarations* include specifying either wire or reg of an appropriate width for each port listed in parentheses, as well as any local variables used internally within the module.

The order in which ports appear in the parentheses on the first line of the module definition is the order which matters elsewhere when this module is instantiated. Every one of the ports listed in the parentheses must be defined as one of the following: input, output or inout. Unlike tasks, the order in which the ports of a module appears in the input, output or inout definitions themselves is irrelevant. Although there is some vague similarity, the meaning of the words input, output and inout for a module is quite different than for a task. The designer makes the choice among these three alternatives based on the direction of information flow relative to the module in question. When making this decision, the designer looks at the system from the viewpoint of this one module.

3.10.1 input ports

An input port is one through which information comes into the module in question from the outside world. An input port must be declared within the module to have a size, or else Verilog will treat the input port as a one-bit wire, which is often incorrect. There are two ways to declare the size: either as a `wire` of some size (regardless of whether the module uses a *behavioral instance* or a *structural instance*) or with the `input` definition.[7]

Failure to declare an `input` port as a `wire` will cause it to be treated as a single-bit `wire`.

3.10.2 output ports

An `output` port is one through which information goes out of the module in question to the outside world. When the module in question uses a *behavioral instance* to produce the `output` port, the `output` port **must** be declared as a `reg` of some size. When the module in question uses a *structural instance*, the `output` port should be declared as a `wire` of some size. In other words, whether to declare an `output` port to be a `wire` or `reg` depends on whether it is generated by structural or behavioral code within the module in question.

3.10.3 inout ports

An `inout` port is one that is used to send information both directions. The advantage of an `inout` port is that the same port can do two separate things (at different times). The Verilog code for using an `inout` port is more complex than for simple `input` and `output` ports. An `inout` port corresponds to a hardware device known as a *tristate buffer*. The details of `inout` ports and tristate buffers are discussed in appendix E.

3.10.4 Historical analogy: pins versus ports

Consider the analogy that "ports are like the doors of a building." For buildings like a store in a shopping center, some doors are labeled "IN," meaning that customers who wish to enter the store in question should go through that door. Those who are finished shopping leave through a different door labeled "OUT." It would be possible to look at the world from the viewpoint of the parking lot, but it is more convenient to look at things relative to the store in question (since there may be many stores in the shopping center to choose from).

There is another analogy for ports: ports are like the pins on an integrated circuit. Some pins are inputs and some pins are outputs. This is a very good analogy, but it is a little dangerous because when a large design is fabricated by a modern silicon foundry, most of the ports in the design do not correspond to a physical pin on the final integrated circuit.

---

[7] Some synthesis tools require that the `input` definition have the size.

To understand this pin analogy, let's digress for a moment and look at the history of hierarchical design and integrated circuit technology. Before the mid-1960s, all digital computers were built using *discrete electronic devices* (such as relays, vacuum tubes or transistors). It takes several such devices, wired together by hand in a certain structure, to make a gate, and of course, as we have seen in section 2.5, it takes many such gates to make anything remotely useful. In the early 1960's, photographic technologies became practical to mass-produce entire circuits composed of several devices on a wafer of semiconductor material (typically silicon). The wafer is sliced into "chips," which are mounted in epoxy (or similar material) with metal pins connecting the circuitry on the chip to the outside. There are several standard sizes for the number and placement of pins. For example, one of the oldest and smallest configurations is the 16-Pin Dual Inline Package (DIP). It is a rectangle with seven data pins on each side, and no pins on the top or bottom. (Two pins are reserved for power and ground.) A notch or dot at the top of the chip indicates where pin one is.

Designers in the 1960s and 1970s were limited by the number of devices that fit onto the chip and also by the number of pins allowed in these standard sizes. Realizing the power of hierarchical design, these designers built chips that contain standard building blocks that fit within the number of pins available. An example is a four-bit counter in one chip, TTL part number 74xx163, which is still widely used. Whenever designers needed a four-bit counter, they could simply specify a 74xx163, without worrying about its internal details. This, of course, is hierarchical design and provides the same mental simplification as instantiating a module. Physically, the pins of the 74xx163 chip would be soldered into the final circuit.

The relationship between these early integrated circuits and hierarchical design is not perfect, hence the danger of saying ports are like pins. If a design needs one 13-bit counter, a designer in the 1970s would have to specify that four 74xx163s be soldered into the final circuit to act as a single counter. There is an interconnection between these four chips so that they collectively count properly. From a hierarchical standpoint, we want to see only one black box, with a 13-bit bus, but this counter is fabricated as four 74xx163s wired together. Some of the physical pins (connected to another one of the 74xx163s) have nothing to do with the ports of a 13-bit counter.

With modern silicon fabrication technologies, the limitations on the number of devices on a chip have been eased, but the limitations on physical pins have become even more severe. Although chips can contain millions of gates, the number of pins allowed is seldom more than a few hundred. Hierarchical design should be driven by the problem being solved (which is the fundamental principle of all top-down design) and not by the limitations (such as pins) of the technology used. Every physical pin on a chip is (part of) a Verilog port, but not every Verilog port necessarily gets fabricated as a physical pin(s). Even so, the **analogy** is a good one: ports are **like** pins.

3.10.5 Example of a module defined with a behavioral instance

Section 2.5 defines an adder several ways. The simplest way to explain what an adder does is to describe it behaviorally. Since an adder is combinational logic, we can use the @ time control technique discussed in section 3.7.2.1 to model its behavior. However, since an adder is used in a larger structure, we should make the `always` block that models the adder's behavior part of a module definition. Those ports (a and b) that are physical inputs to the fabricated adder will be `input` ports to this module, and are exactly the variables listed in the sensitivity list. The port that is a physical output (`sum`) is, of course, defined to be an `output` port. Since this module computes `sum` with behavioral code, `sum` is declared to be a `reg`. (There are no "registers" in combinational logic, but a Verilog `reg` is used in a behavioral model of combinational logic. A `reg` is not a "register" as long as the sensitivity list has all the inputs listed.) As in the example of section 2.5, the widths of a and b are two bits each, and the width of `sum` is three bits:

```
module adder(sum,a,b);
   input [1:0] a,b;
   output [2:0] sum;
   wire [1:0] a,b;
   reg [2:0] sum;

   always @(a or b)
      sum = a + b;
endmodule
```

The widths shown on input and output definitions are optional for simulation purposes.[8]

To exhaustively test this small adder, test code similar to section 3.7.2.1 enumerates all possible combinations of a and b:

---

[8] The width will not be shown on later examples in this chapter, although describing the width on `input` and `output` definitions would be legal in simulation. The width might be required to overcome the limitations of some commercial simulation tools.

```
module top;
  integer ia,ib;
  reg [1:0] a,b;
  wire [2:0] sum;

  adder adder1(sum,a,b);

  initial
    begin
      for (ia=0; ia<=3; ia = ia+1)
        begin
          a = ia;
          for (ib=0; ib<=3; ib = ib + 1)
            begin
              b = ib;
              #1 $display("a=%d b=%d sum=%d",a,b,sum);
            end
        end
    end
endmodule
```

The important thing in this top-level test module is that adder (the name of the module definition) is instantiated in top with the name adder1. In the top-level module, a and b are regs because, within this module (top), a and b are supplied by behavioral code. On the other hand, sum is supplied by adder1, and so top declares sum to be a wire. The syntax for instantiating a user defined module is similar to instantiating a built-in gate. In this example, the local sum of top corresponds to the output port (coincidentally named sum) of an instance of module adder. If the names (such as sum) in module adder were changed to other names (such as total), the module would work the same:

```
module adder(total,alpha,beta);
  input alpha,beta;
  output total;
  wire [1:0] alpha,beta;
  reg [2:0] total;

  always @(alpha or beta)
    total = alpha + beta;
endmodule
```

It is the position within the parentheses, and not the names, that matter[9] when the module is instantiated in the test code.

3.10.6 Example of a module defined with a structural instance

Of course, in hierarchical design, we need a structural definition of the module. As described in section 2.5, the module `adder` can be defined in terms of instantiation of an instance of a `half_adder` (which we will call `ha1`) and an instance of a `full_adder` (which we will call `fa1`):

```
module adder(sum,a,b);
   input a,b;
   output sum;
   wire [1:0] a,b;
   wire [2:0] sum;

   wire c;

   half_adder ha1(c,sum[0],a[0],b[0]);
   full_adder fa1(sum[2],sum[1],a[1],b[1],c);
endmodule
```

Since the adder is defined with two *structural instances* (named `ha1` and `fa1`), all of the ports, including the `output` port, `sum`, are `wires`. The local wire `c` sends the carry from the half-adder to the full-adder. Of course, we need identical test code as in the last example, and we also need module definitions for `full_adder` and `half_adder`.

3.10.7 More examples of behavioral and structural instances

Even though `half_adder` and `full_adder` are instantiated structurally in section 3.10.6, they can be defined either behaviorally or structurally. For example, a behavioral definition of these modules is:

---

[9] Verilog provides an alternative syntax, described in chapter 11, that allows the name, rather than the position, to determine how the module is instantiated.

```
module half_adder(c,s,a,b);
   input a,b;
   wire a,b;
   output c,s;
   reg c,s;

   always @(a or b)
     {c,s} = a+b;
endmodule

module full_adder(cout,s,a,b,cin);
   input a,b,cin;
   wire a,b,cin;
   output cout,s;
   reg cout,s;

   always @(a or b or cin)
     {cout,s} = a+b+cin;
endmodule
```

Once again, notice that the outputs are regs. Concatenation is used on the left of the =
to make the definition of the module simple. {cout,s} is a two-bit reg capable of
dealing with the largest possible number (2'b11) produced by a+b+cin.

An alternative would be to define the half_adder and full_adder modules with
structural instances, which means all outputs are wires:

```
module half_adder(c,s,a,b);
   input a,b;
   wire a,b;
   output c,s;
   wire c,s;

   xor x1(s,a,b);
   and a1(c,a,b);
endmodule

module full_adder(cout,s,a,b,cin);
   input a,b,cin;
   wire a,b,cin;
   output cout,s;
   wire cout,s;
   wire cout1,cout2,stemp;

   half_adder ha2(cout1,stemp,a,b);
   half_adder ha3(cout2,s,cin,stemp);
   or         o1(cout,cout1,cout2);
endmodule
```

There are two instances of `half_adder` (ha2 and ha3). The only difference between these two instances is how they are connected within `full_adder`. There are three local wires (`cout1`, `cout2` and `stemp`) that allow internal interconnection within the module.

At this point, we have reduced the problem down to Verilog primitive gates (`and`, `or`, `xor`) whose behavior is built into Verilog.

3.10.8   Hierarchical names

Although ports are intended to be the way in which modules communicate with each other in a properly functioning system, Verilog provides a way for one module to access the internal parts of another module. Conventional high-level languages, like C and Pascal, have *scope rules* that absolutely prohibit certain kinds of access to local information. Verilog is completely different in this regard. The philosophy of Verilog for accessing variables is very similar the philosophy of the NT or UNIX operating systems for accessing files: if you know the path to a file (within subdirectories), you can access the file. Analogously in Verilog: if you know the path to a variable (within modules), you can access the variable.

For example, using the definition of `adder` given in section 3.10.6, and the instance `adder1` shown in the test code of section 3.10.5, `adder1` has a local wire c that is not accessible to the outside world. The following statement **in the test code** would allow the designer to observe this wire, even though there is no port that outputs c:

```
$display(adder1.c);
```

A name, such as `adder1.c` is known as a *hierarchical name*, or *path*.

The following statement allows the designer to observe `cout2` from the test code:

```
$display(adder1.fa1.cout2);
```

which happens to be the same as:

```
$display(adder1.fa1.ha3.c);
```

The parts of a hierarchical name are separated by periods. Every part of a hierarchical name, except the last, is the name of an instance of a module. The names of the corresponding module definitions (`adder`, `full_adder` and `half_adder` in the above example) **never** appear in a hierarchical name.

3.10.9 Data structures

The term "structure" has three distinct meanings in computer technology. Elsewhere in this book, "structure" takes on its hardware meaning: the interconnection of modules using wires. But you have probably heard of the other two uses of this word: "structured programming," and "data structures." The concept of "structured programming" is a purely behavioral software concept which is closely related to what we call goto-less programming (see section 2.1.4). "Data structures" are software objects that allow programmers to solve complex problems in a more natural way.

The period notation used in Verilog for hierarchical names is reminiscent of the notation used in conventional high-level languages for accessing components of a "data structure" (record in Pascal, struct in C, and class in C++). In fact, you can create such software "data structures" in Verilog by defining a portless module that has only data, but that is intended to be instantiated. Such a portless but instantiated module is worthless for hardware description, but is identical to a conventional software "data structure." Such a module has no behavioral instances or structural instances. For example, a data structure could be defined to contain payroll information about an employee:

```
module payroll;
    reg [7:0] id;
    reg [5:0] hours;
    reg [3:0] rate;
endmodule
```

Suppose we have two employees, joe and jane. Each employee has a unique instance of this module:

```
payroll joe();
payroll jane();

    initial
      begin
        joe.id=254;
        joe.hours=40;
        joe.rate=14;
        jane.id=255;
        jane.hours=63;
        jane.rate=15;
      end
```

The empty parentheses are a syntactic requirement of Verilog. In this example, the fields of jane contain the largest possible values.

Data structures usually have a limited set of operations that manipulate the fields of the data. For example, the `hours` and `rate` fields can be combined to display the corresponding total pay. This operation is defined as a local task of the module. However, since there are no behavioral instances in this module, this task sits idle until it is called from the outside (using a hierarchical name):

```verilog
module payroll;
  reg [7:0] id;
  reg [5:0] hours;
  reg [3:0] rate;

  task display_pay;
    integer pay;   //local
    begin
      if (hours>40)
        pay = 40*rate + (hours-40)*rate*3/2;
      else
        pay = hours*rate;
      $display("employee %d earns %d",id,pay);
    end
  endtask
endmodule

module top;
  payroll joe();
  payroll jane();
  initial
    begin
      joe.id=254;
      joe.hours=40;
      joe.rate=14;
      joe.display_pay;
      jane.id=255;
      jane.hours=63;
      jane.rate=15;
      jane.display_pay;
    end
endmodule
```

This is very close to the software concept of *object-oriented programming* in languages like C++, except the current version of Verilog lacks the inheritance feature found in C++.

Data structures are a powerful use of hierarchical names, but they are somewhat afield from the central focus of this book: hardware structures. Application of hierarchical names are useful in test code, and so it is important to understand them. Also, the above example helps illustrate what instantiation really means in Verilog.

3.10.10   Parameters

Verilog modules allow the definition of what are known as parameters. These are constants that can be different for each instance. For example, suppose you would like to define a module behaviorally that models an enabled register of arbitrary width:

```
module enabled_register(dout, din, ld, sysclk);
   parameter WIDTH = 1;
   input din,ld,sysclk;
   output dout;
   wire [WIDTH-1:0] din;
   reg [WIDTH-1:0] dout;
   wire ld,sysclk;

   always @(posedge sysclk)
     if (ld)
        dout = din;
endmodule
```

By convention, we use capital letters for parameters, but this is not a requirement. Note that parameters do not have a backquote preceding them.

If you instantiate this module without specifying a constant, the default given in the parameter statement (in this example, 1) will be used as the WIDTH, and so the instance R1 will be one bit wide:

```
wire ldR1,sysclk;
wire R1dout,R1din;
enabled_register       R1(R1dout,R1din,ldR1,sysclk);
```

To specify a non-default constant, the syntax is a # followed by a list of constants in parentheses. Since there is only one parameter in this example, there can be only one constant in the parentheses. For example, to instantiate a 12-bit register for R12:

```
wire ldR2,sysclk;
wire [11:0] R12dout,R12din;
enabled_register #(12) R12(R12dout,R12din,ldR12,sysclk);
```

Verilog requires that the width of a `wire` that attaches to an `output` port match the `reg` declaration within the module. In this example, R12dout is a wire twelve bits wide, the parameter `WIDTH` in the instance R12 is twelve, and the corresponding output port, dout, is declared as `reg[WIDTH-1:0]`, which is the same as `reg [11:0]`.

Since there is only one constant in the parentheses above, it is legal to omit the parentheses:

```
enabled_register #12    R12(R12dout,R12din,ldR12,sysclk);
```

Sometimes, you need more than one constant in the definition of a module. For example, a combinational multiplier has two input buses, whose widths need not be the same:

```
module multiplier(prod,a,b);
   parameter WIDTHA=1,WIDTHB=1;
   output prod;
   input a,b;
   reg [WIDTHA+WIDTHB-1:0] prod;
   wire [WIDTHA-1:0] a;
   wire [WIDTHB-1:0] b;

   always @(a or b)
      prod = a*b;
endmodule
```

Here is an example of instantiating this:

```
wire [5:0] hours;
wire [3:0] rate;
wire [9:0] pay;

multiplier #(6,4) m1(pay,hours,rate);
```

3.11   Conclusion

Modules are the basic feature of the Verilog hardware description language. Modules are either top-level or instantiated. Top-level modules are typically used for test code. Instantiated modules have ports, which can be defined to be either `input`, `output` or `inout`. Constants in modules may be defined with the `parameter` statement. A module is either defined with a *behavioral instance* (always or initial block(s) or with a *structural instance* (built-in gates or instantiation of other designer-provided modules). Behavioral and structural instances may be mixed in the same module.

Variables produced by behavioral code, including outputs from the module, are declared to be `regs`. Behavioral modules have the usual high-level statements, such as `if` and `while`, as well as time control (#, @ and `wait`) that indicate when the process can be suspended and resumed. The `$time` variable simulates the passage of time in the fabricated hardware. Verilog makes a distinction between algorithmic sequence and the passage of `$time`. The most important forms of time control are # followed by a constant, which is used for generating the clock and test vectors; @ (`posedge sysclk`), which is used to model controllers and registers; and @ followed by a sensitivity list, which is used for combinational logic. Verilog provides the non-blocking assignment statement, which is ideal for translating ASM charts that use RTN into behavioral Verilog. Verilog also provides tasks and functions, which like similar features in conventional high-level languages, simplify coding.

Structural modules have a simple syntax. They may instantiate other designer-provided modules to achieve hierarchical design. They may also instantiate built-in gates. The syntax for both kinds of instantiation is identical. All variables in a structural module, including outputs, are `wires`.

Hierarchical names allow access to tasks and variables from other modules. Use of hierarchical names is usually limited to test code.

The next chapter uses the features of Verilog described in this chapter to express the three stages (pure behavioral, mixed and pure structural) of the design process for the childish division machine designed manually in chapter 2. The advantage of using Verilog at each of these stages is that the designer can simulate each stage to be sure it is correct before going on to the next stage. Also, the final Verilog code can be synthesized into a working piece of hardware, without the designer having to toil manually to produce a flattened circuit diagram and netlist.

3.12   Further reading

LEE, JAMES M., *Verilog Quickstart*, Kluwer, Norwell, MA, 1997. Gives several examples of implicit style.

PALNITKAR, S., *Verilog HDL: A Guide to Digital Design and Synthesis,* Prentice Hall PTR, Upper Saddle River, NJ, 1996. An excellent reference for all aspects of Verilog.

SMITH, DOUGLAS J., *HDL Chip Design: A Practical Guide for Designing, Synthesizing, and Simulating ASICs and FPGAs Using VHDL or Verilog*, Doone Publications, Madison, AL, 1997. A Rosetta stone between Verilog and VHDL.

STERNHEIM, ELIEZER, RAJVIR SINGH and YATIN TRIVEDI, *Digital Design with Verilog HDL*, Automata Publishing, San Jose, CA, 1990. Has several case studies of using Verilog.

THOMAS, DONALD E. and PHILIP R. MOORBY, *The Verilog Hardware Description Language*, Third edition, Kluwer, Norwell, MA., 1996. Explains how a simulator works internally.

3.13   Exercises

3-1. Design behavioral Verilog for a two-input 3-bit wide mux using the technique described in section 3.7.2.1. The port list for this module should be:

```
module mux2(i0, i1, sel, out);
```

3-2. Design a structural Verilog module (`mux2`) equivalent to problem 3-1 using only instances of `and`, `or`, `not` and `buf`.

3-3. Modify the solution to problem 3-1 to use a parameter named `SIZE` that allows instantiation of an arbitrary width for `i0`, `i1` and `out` as explained in section 3.10.10. For example, the following instance of this device would be useful in the architecture drawn in section 2.3.1:

```
wire muxctrl;
wire [11:0] x,y,muxbus;
mux2 #12 mx(x,y,muxctrl,muxbus);
```

3-4. Given the instance (`mx`) of the module (`mux2`) shown in problem 3-3, what hierarchical names are equivalent to `x`, `y`, `muxctrl` and `muxbus`?

3-5. Design behavioral Verilog for combinational incrementor and decrementor modules using the technique described in section 3.7.2.1. Use a parameter named `SIZE` that allows instantiation of an arbitrary width for the ports as explained in section 3.10.10.

3-6. Design behavioral Verilog for an up/down counter (section D.8) using the technique described in section 3.7.2.2. The port list for this module should be:

```
module updown_register(din,dout,ld,up,count,clk);
```

3-7. Modify the solutions to problem 3-6 to use a parameter named SIZE that allows instantiation of an arbitrary width for the ports as explained in section 3.10.10.

3-8. Design behavioral Verilog for a simple D-type register (section D.5) using the technique described in section 3.7.2.2. Use a parameter named SIZE that allows instantiation of an arbitrary width for the ports as explained in section 3.10.10. The port list for this module should be:

```
module simpled_register(din,dout,clk);
```

3-9. Design a structural Verilog module (`updown_register`) equivalent to problem 3-7 using only instances of the modules defined in problems 3-3, 3-5 and 3-8.

3-10. For each of the ASM charts given in problem 2-10, translate to implicit style Verilog using non-blocking assignment for ← and @(posedge sysclk) #1 for each rectangle, as explained in section 3.8.2.3.1. As in that example, there should be one `always` that models the hardware, one `always` for the `$display` and an `always` and `initial` for `sysclk`. Compare the result of simulation with the manually produced timing diagram of problem 2-10.

3-11. Without using a Verilog simulator, give a timing diagram for the machine described by the ASM chart of section 3.8.2.3.3. Show the values of a and b in the first twelve clock cycles, and label each clock cycle to indicate which state the machine is in. Next, run the **original** implicit style Verilog code equivalent to the ASM and make a printout of the .log file. On this printout, write the name of the state that the machine is in during each clock cycle. The manually created timing diagram should agree with the Verilog .log file. Finally, modify the following:

```
@(posedge sysclk) #1;              // state FIRST
    a <= @(posedge sysclk) 1;
```

to become:

```
@(posedge sysclk) #1;              // state FIRST
    a = 1;
```

Run the modified Verilog code and make a printout of its .log file. On this printout, circle the differences, if any, that exist between the correct timing diagram and the .log file for the modified Verilog. In no more than three sentences, explain why there are or are not any differences between = and <=.

3-12. Without using a Verilog simulator, give a timing diagram for the machine described by the ASM of section 3.8.2.3.4. Show the values of a and b in the first twelve clock cycles, and label each clock cycle to indicate which state the machine is in. Next, run the **original** implicit style Verilog code equivalent to the ASM and make a printout of the .log file. On this printout write the name of the state that the machine is in during each clock cycle. The manually created timing diagram should agree with the Verilog .log file. Finally, modify the code to change the if to a while. Run the modified Verilog code and make a printout of its .log file. On this printout, circle the differences, if any, that exist between the correct timing diagram and the .log file for the modified Verilog. In no more than three sentences, explain why there are or are not any differences between if and while.

3-13. Without using a Verilog simulator, give a timing diagram for the machine described by the ASM of section 3.8.2.3.5. Show the values of a and b in the first twelve clock cycles, and label each clock cycle to indicate which state the machine is in. Next, run the **original** implicit style Verilog code equivalent to the ASM and make a printout of the .log file. On this printout write the name of the state that the machine is in during each clock cycle. The manually created timing diagram should agree with the Verilog .log file. Finally, modify the code to eliminate all #1s. Run the modified Verilog code and make a printout of its .log file. On this printout, circle the differences, if any, that exist between the correct timing diagram and the .log file for the modified Verilog. In no more than three sentences, explain why there are or are not any differences between using and omitting #1s.

4. THREE STAGES FOR VERILOG DESIGN

The design of the childish division machine described in chapter 2 can be expressed in Verilog using the language features discussed in chapter 3. This chapter uses variations on the childish division machine as a simple but practical example of how to design hardware with Verilog.

As described in sections 2.1.5.1 through 2.1.5.3, the three-stages of the top-down design process are: pure behavioral, mixed behavior/structure and pure structural. Because Verilog allows both behavioral and structural constructs, one can transform a design through these three stages by minor editing of the Verilog source code. Section 4.1 gives several examples of how the pure behavioral ASMs of chapter 2 can be written in pure behavioral Verilog using the implicit style of section 3.8.2.3. Section 4.2 uses one of these behavioral examples to illustrate translation into the mixed stage. Section 4.3 translates the mixed example from section 4.2 into the pure structural stage (the "explicit style" often used by Verilog designers). Section 4.4 shows that, having completed the three phases of our design process, Verilog allows additional structure (in this case, dealing with the controller) to be instantiated in place of behavior using the hierarchical design process described in section 2.5.

Chapter 7 and appendix F describe an alternate way (which can be automated) for translating pure behavioral (implicit style) Verilog directly to a form that can be synthesized into physical hardware. The manual technique in this chapter is more intricate and involved than the one in chapter 7, but an understanding of the three-stage technique in this chapter will give the reader a better appreciation for the behavioral and structural aspects of Verilog simulation.

4.1 Pure behavioral examples

This section gives examples of modeling various ASM charts for the childish division algorithm with pure behavioral Verilog.

4.1.1 Four-state division machine

Let's consider translating the second ASM chart of section 2.2.3 (the one that has four-states, with state COMPUTE1 at the top of the loop) into Verilog, using the `enter_new_state` approach described in section 3.9.1.2.

4.1.1.1 Overview of the source code

First, we have to define the arbitrary bit patterns that represent the states and indicate the number of bits required:

```
'define NUM_STATE_BITS 2
'define IDLE      2'b00
'define INIT      2'b01
'define COMPUTE1  2'b10
'define COMPUTE2  2'b11
```

These definitions occur outside any modules. Next, we need to include the definition of a module that generates the clock in a fashion similar to 3.7.1.3, except the clock is output as a port of the module:

```
module cl(clk);
    parameter TIME_LIMIT = 110000;
    output clk;
    reg clk;

    initial
      clk = 0;

    always
      #50 clk = ~clk;

    always @(posedge clk)
      if ($time > TIME_LIMIT) #70 $stop;
endmodule
```

The purpose of the parameter TIME_LIMIT is to stop the simulation after a certain amount of $time has elapsed. Without scheduling such a $stop (or $finish) statement, the simulation could continue forever.

Third, the module that implements the behavioral simulation of the actual hardware needs to be defined with an appropriate portlist:

```
module slow_div_system(pb,ready,x,y,r2,sysclk);
   input pb,x,y,sysclk;
   output ready,r2;
   wire ...
   reg...
   ...

endmodule
```

This module corresponds to the first block diagram in section 2.5.2. At this stage, we are leaving the details of this module ambiguous. The discussion of the details inside this module will continue in section 4.1.1.2.

Finally, the last module that needs to be defined is the top-level test code. Using a technique involving #, @ and wait that is similar to the one described in section 3.7.3, the test code checks if the simulated hardware can divide values of x that vary from 0 to 14 properly when y is held fixed at 7:

```verilog
module top;
   reg pb;
   reg [11:0] x,y;
   wire [11:0] quotient;
   wire ready;
   integer s;
   wire sysclk;

   cl #20000 clock(sysclk);
   slow_div_system  slow_div_machine(pb,ready,x,y,
                                      quotient,sysclk);
   initial
     begin
       pb= 0;
       x = 0;
       y = 7;
       #250;
       @(posedge sysclk);
       for (x=0; x<=14; x = x+1)
         begin
           @(posedge sysclk);
           pb = 1;
           @(posedge sysclk);
           pb = 0;
           @(posedge sysclk);
           wait(ready);
           @(posedge sysclk);
           if (x/y === quotient)
            $display("ok");
           else
            $display("error x=%d y=%d x/y=%d quotient=%d",
                     x,y,x/y,quotient);
         end
       $stop;
     end
 endmodule
```

Even for a small machine like this, an exhaustive test (such as was given in section 3.7.3) would be too time consuming to use as an illustration.

In the above code, note that `slow_division_system` is instantiated with the instance name `slow_division_machine`. The first port in the module, the input `pb`, corresponds to a `reg` in the test code of the same name. The second port in the module, the output `ready`, corresponds to a `wire` of the same name in the test code. The third and fourth ports of the module, the inputs `x` and `y`, correspond to 12-bit `reg`s of the same names in the test code. The fifth port of the module, an output named `r2`, corresponds to a 12-bit `wire` whose name is `quotient` in the test code. The final port of the module, an input named `sysclk`, corresponds to a `wire` of the same name in the test code.

The reason `sysclk` is a `wire` is because it happens to be an output port of the instance of the `cl` module named `clock`. This means `clock`, rather than the test code, supplies `sysclk`. The test code therefore has a little bit of structure connecting `clock` to `slow_div_machine`. The variables in the test code that are `reg`s are so declared because the test code must supply them to `slow_division_machine` using the behavioral =. The remaining `wire`s are so declared because the behavioral test code does not supply them.

4.1.1.2   Details on `slow_division_system`

Let's return to the definition of the `slow_division_system` module. We need to declare the types of the inputs and outputs. Also, we need to declare local variables, such as those that model the physical registers of the hardware (`r1` and `r2`), and also the `present_state`:

```
module slow_div_system(pb,ready,x,y,r2,sysclk);
   input pb,x,y,sysclk;
   output ready,r2;
   wire pb,sysclk;
   wire [11:0] x,y;
   reg ready;
   reg [11:0] r1,r2;
   reg ['NUM_STATE_BITS-1:0] present_state;
   ...
endmodule
```

These declarations were constrained by the portlist, how the module was instantiated in the test code and by the description of the problem given in chapter 2.

As described in section 3.9.1.2, we should define a task that simplifies the code for the sequence of statements that must occur when the machine enters each state:

```verilog
task enter_new_state;
  input ['NUM_STATE_BITS-1:0] this_state;
  begin
    present_state = this_state;
    #1 ready=0;
  end
endtask
```

The definition of this task will be nearly identical for every pure behavioral ASM. The only distinction from one problem to another is the list of external command outputs (see section 2.1.3.2.1) specific to the particular machine. In this case, the only external command output is `ready`. It has a default value of 0; thus this task must initialize it at the beginning of every clock cycle.

Having defined the above task within the `slow_division_system` module, it is possible to translate the ASM from section 2.2.3 into Verilog:

```verilog
always
  begin
    @(posedge sysclk) enter_new_state('IDLE);
    r1 <= @(posedge sysclk) x;
    ready = 1;
    if (pb)
      begin
        @(posedge sysclk) enter_new_state('INIT);
        r2 <= @(posedge sysclk) 0;
        while (r1 >= y)
          begin
            @(posedge sysclk) enter_new_state('COMPUTE1);
            r1 <= @(posedge sysclk) r1 - y;
            @(posedge sysclk) enter_new_state('COMPUTE2);
            r2 <= @(posedge sysclk) r2 + 1;
          end
      end
  end
```

The only other thing that would be desirable to put in this module is a debugging display, as described in section 3.7.2.4:

```
always @(posedge sysclk) #20
    $display("%d r1=%d r2=%d pb=%b ready=%b",
                $time, r1,r2, pb, ready);
```

The net effect of the other modules defined in section 4.1.1.1 and all the details inside the module given above is to produce the following simulation output from Verilog:

```
        70 r1=    x r2=    x pb=0 ready=1
       170 r1=    0 r2=    x pb=0 ready=1
       270 r1=    0 r2=    x pb=0 ready=1
       370 r1=    0 r2=    x pb=1 ready=1
       470 r1=    0 r2=    x pb=0 ready=0
       570 r1=    0 r2=    0 pb=0 ready=1
ok
       670 r1=    0 r2=    0 pb=0 ready=1
       770 r1=    1 r2=    0 pb=1 ready=1
       870 r1=    1 r2=    0 pb=0 ready=0
       970 r1=    1 r2=    0 pb=0 ready=1
ok
  . . .
      6670 r1=   12 r2=    1 pb=0 ready=1
      6770 r1=   13 r2=    1 pb=1 ready=1
      6870 r1=   13 r2=    1 pb=0 ready=0
      6970 r1=   13 r2=    0 pb=0 ready=0
      7070 r1=    6 r2=    0 pb=0 ready=0
      7170 r1=    6 r2=    1 pb=0 ready=1
ok
      7270 r1=   13 r2=    1 pb=0 ready=1
      7370 r1=   14 r2=    1 pb=1 ready=1
      7470 r1=   14 r2=    1 pb=0 ready=0
      7570 r1=   14 r2=    0 pb=0 ready=0
      7670 r1=    7 r2=    0 pb=0 ready=0
      7770 r1=    7 r2=    1 pb=0 ready=0
      7870 r1=    0 r2=    1 pb=0 ready=0
      7970 r1=    0 r2=    2 pb=0 ready=1
ok
```

The regs r1 and r2 are not initialized at $time 0, and so the value 12'bx is printed by the $display simply as x (not to be confused with the variable x). Each time the machine returns to state IDLE (ready=1), the outputs of the machine from r2 (which is the same as quotient) are highlighted above. These are the values that the test code uses to determine that everything is "ok" each time.

4.1.2   Verilog catches the error

The first ASM chart in section 2.2.3 (with state COMPUTE2 at the top of the loop) has an error. An advantage of using Verilog to simulate such a machine while it is still in the behavioral stage is that Verilog can usually catch such errors before they become costly. In this case, the Verilog code would be identical to section 4.1.1.2, except the ASM chart would be translated as:

```
always
  begin
      . . .
        while (r1 >= y)
          begin
            @(posedge sysclk) enter_new_state('COMPUTE2);
              r2 <= @(posedge sysclk) r2 + 1;
            @(posedge sysclk) enter_new_state('COMPUTE1);
              r1 <= @(posedge sysclk) r1 - y;
          end
      end
  end
```

The output from the Verilog simulator makes the problem obvious:

```
        . . .
              2670 r1=    5 r2=    0 pb=0 ready=1
              2770 r1=    6 r2=    0 pb=1 ready=1
              2870 r1=    6 r2=    0 pb=0 ready=0
              2970 r1=    6 r2=    0 pb=0 ready=1
        ok
              3070 r1=    6 r2=    0 pb=0 ready=1
              3170 r1=    7 r2=    0 pb=1 ready=1
              3270 r1=    7 r2=    0 pb=0 ready=0
              3370 r1=    7 r2=    0 pb=0 ready=0
              3470 r1=    7 r2=    1 pb=0 ready=0
              3570 r1=    0 r2=    1 pb=0 ready=0
              3670 r1=    0 r2=    2 pb=0 ready=0
              3770 r1=4089 r2=    2 pb=0 ready=1
        error x=    7 y=    7 x/y=    1 quotient=    2
        . . .
              8670 r1=   13 r2=    2 pb=0 ready=1
              8770 r1=   14 r2=    2 pb=1 ready=1
              8870 r1=   14 r2=    2 pb=0 ready=0
              8970 r1=   14 r2=    0 pb=0 ready=0
              9070 r1=   14 r2=    1 pb=0 ready=0
              9170 r1=    7 r2=    1 pb=0 ready=0
```

Continued.

```
              9270 r1=    7 r2=    2 pb=0 ready=0
              9370 r1=    0 r2=    2 pb=0 ready=0
              9470 r1=    0 r2=    3 pb=0 ready=0
              9570 r1=4089 r2=    3 pb=0 ready=1
        error x=   14 y=    7 x/y=    2 quotient=    3
```

Because of the error (which causes the loop to execute an extra time), the time to complete the test is longer. The `wait` statement in the test code compensates for this; thus the test code is checking `r2` via `quotient` at the proper time, but when `y>=7`, `quotient` is just plain wrong.

Rather than spending thousands of dollars actually fabricating a faulty computer, the designer can observe the problem simply from the behavioral Verilog code.

4.1.3   Importance of test code

Do not be **deluded** that because Verilog **can sometimes** catch errors such as described in section 4.1.2 that it **will always** catch such errors. Just because a designer uses an automated tool like a Verilog is not the same as saying that the designer does not have to think. In fact, using such tools requires a higher level of thought process. The designer is responsible not just for trying to design a machine that works, but also for designing test code that aggressively checks to see if the design fails.

When we began with this example in section 2.2.1, we stated in informal English some assumptions about the environment in which this machine operates. We assumed two things: there is a friendly user (who provides inputs in a specific sequence), and that this user needs the output of the machine to remain constant when the machine is in state IDLE (because the user is a person who cannot perceive events that happen within a single clock cycle, which are typically less than a millionth of a second).

The test code given in section 4.1.1.1 satisfies the assumption that the inputs are provided in the order demanded in section 2.2.1 (in particular the test code lets the machine stay in state IDLE two clock cycles before the button is pressed). There is, however, a problem with the above test code with regard to the second assumption, as is illustrated below.

Consider the ASM chart of section 2.2.4. The Verilog code to simulate this machine is identical to the previous examples, except for the following:

```
always
  begin
  @(posedge sysclk) enter_new_state('IDLE);
  r1 <= @(posedge sysclk) x;
  r2 <= @(posedge sysclk) 0;
  ready = 1;
  if (pb)
    begin
      while (r1 >= y)
        begin
          @(posedge sysclk) enter_new_state('COMPUTE1);
          r1 <= @(posedge sysclk) r1 - y;
          @(posedge sysclk) enter_new_state('COMPUTE2);
          r2 <= @(posedge sysclk) r2 + 1;
        end
    end
  end
```

When Verilog simulates this with the test code given earlier, it detects no errors:

```
          70 r1=   x r2=   x pb=0 ready=1
         170 r1=   0 r2=   0 pb=0 ready=1
         270 r1=   0 r2=   0 pb=0 ready=1
         370 r1=   0 r2=   0 pb=1 ready=1
         470 r1=   0 r2=   0 pb=0 ready=1
         570 r1=   0 r2=   0 pb=0 ready=1
  ok
  ...
        6070 r1=  12 r2=   0 pb=0 ready=1
        6170 r1=  13 r2=   0 pb=1 ready=1
        6270 r1=  13 r2=   0 pb=0 ready=0
        6370 r1=   6 r2=   0 pb=0 ready=0
        6470 r1=   6 r2=   1 pb=0 ready=1
  ok

        6570 r1=  13 r2=   0 pb=0 ready=1
        6670 r1=  14 r2=   0 pb=1 ready=1
        6770 r1=  14 r2=   0 pb=0 ready=0
        6870 r1=   7 r2=   0 pb=0 ready=0
        6970 r1=   7 r2=   1 pb=0 ready=0
        7070 r1=   0 r2=   1 pb=0 ready=0
        7170 r1=   0 r2=   2 pb=0 ready=1
  ok
```

But as was discussed in section 2.2.4, it is unacceptable from a user interface stand-point because it throws away the correct answer after only one clock cycle. For example, at $time 6470, the correct answer 1 is present in r2, and the test code notes this. But at $time 6570, the machine has thrown away the correct answer even though ready still indicates that the correct answer should be displayed.

The proper approach is to take all details in the informal English specification and make these details part of the test code. The test code should be a formal specification of what the machine is supposed to do under all circumstances. This raises the interesting, and somewhat unsolvable dilemma: how does the designer test the test code?[1] Nevertheless, it is important to put reasonable effort into creating robust test code.

4.1.4   Additional pure behavioral examples

The first ASM chart of 2.2.5 can be translated into Verilog as:

```
always
  begin
    @(posedge sysclk) enter_new_state('IDLE);
    r1 <= @(posedge sysclk) x;
    r2 <= @(posedge sysclk) 0;
    ready = 1;
    if (pb)
      begin
        while (r1 >= y)
          begin
            @(posedge sysclk) enter_new_state('COMPUTE1);
            r1 <= @(posedge sysclk) r1 - y;
            @(posedge sysclk) enter_new_state('COMPUTE2);
            r2 <= @(posedge sysclk) r2 + 1;
            @(posedge sysclk) enter_new_state('COMPUTE3);
            r3 <= @(posedge sysclk) r2;
          end
      end
  end
```

---

[1] See J. Cooley, *Integrated System Design*, July 1995, pp. 56-60 for a description of a Verilog contest where the test code provided to the contestants was erroneous. The "winning design" would not actually work correctly because the test code could not detect a flaw in the design.

where everything else is the same as earlier examples, except the states are represented as:

```
'define NUM_STATE_BITS 2
'define IDLE        2'b00
'define COMPUTE1    2'b01
'define COMPUTE2    2'b10
'define COMPUTE3    2'b11
```

and the portlist now has r3 rather than r2 as the output:

```
module slow_div_system(pb,ready,x,y,r3,sysclk);
   input pb,x,y,sysclk;
   output ready,r3;
   wire pb;
   wire [11:0] x,y;
   reg ready;
   reg [11:0] r1,r2,r3;
   reg ['NUM_STATE_BITS-1:0] present_state;
```

Also, r3 must be mentioned in the debugging $display statement. As was described in chapter 2, this machine fails for x<y (the quotient is unknown):

```
         70 r1=    x r2=    x r3=x pb=0 ready=1
        170 r1=    0 r2=    0 r3=x pb=0 ready=1
        270 r1=    0 r2=    0 r3=x pb=0 ready=1
        370 r1=    0 r2=    0 r3=x pb=1 ready=1
        470 r1=    0 r2=    0 r3=x pb=0 ready=1
        570 r1=    0 r2=    0 r3=x pb=0 ready=1
error x=    0 y=    7 x/y=    0 quotient=x
...
```

although it does work for larger values:

```
. . .
        3070 r1=   6 r2=   0 r3=   x pb=0 ready=1
        3170 r1=   7 r2=   0 r3=   x pb=1 ready=1
        3270 r1=   7 r2=   0 r3=   x pb=0 ready=0
        3370 r1=   0 r2=   0 r3=   x pb=0 ready=0
        3470 r1=   0 r2=   1 r3=   x pb=0 ready=0
        3570 r1=   0 r2=   1 r3=   1 pb=0 ready=1
ok
. . .
```

The second ASM chart of section 2.2.5 can be translated as:

```
always
  begin
   @(posedge sysclk) enter_new_state('IDLE);
    r1 <= @(posedge sysclk) x;
    r2 <= @(posedge sysclk) 0;
    ready = 1;
    if (pb)
      begin
       if (r1 >= y)
         while (r1 >= y)
           begin
            @(posedge sysclk) enter_new_state('COMPUTE1);
             r1 <= @(posedge sysclk) r1 - y;
            @(posedge sysclk) enter_new_state('COMPUTE2);
             r2 <= @(posedge sysclk) r2 + 1;
            @(posedge sysclk) enter_new_state('COMPUTE3);
             r3 <= @(posedge sysclk) r2;
           end
       else
         begin
          @(posedge sysclk) enter_new_state('ZEROR3);
           r3 <= @(posedge sysclk) 0;
         end
      end
  end
```

where the states are represented as:

```
'define NUM_STATE_BITS 3
'define IDLE       3'b000
'define COMPUTE1   3'b001
'define COMPUTE2   3'b010
'define COMPUTE3   3'b011
'define ZEROR3     3'b100
```

which does work correctly for all values, including x<y:

```
        70  r1=   x r2=   x r3=   x pb=0 ready=1
       170  r1=   0 r2=   0 r3=   x pb=0 ready=1
       270  r1=   0 r2=   0 r3=   x pb=0 ready=1
       370  r1=   0 r2=   0 r3=   x pb=1 ready=1
       470  r1=   0 r2=   0 r3=   x pb=0 ready=0
       570  r1=   0 r2=   0 r3=   0 pb=0 ready=1
ok
...
```

The second ASM of section 2.2.6 can be translated as:

```
always
  begin
   @(posedge sysclk) enter_new_state('IDLE);
   r1 <= @(posedge sysclk) x;
   r2 <= @(posedge sysclk) 0;
   ready = 1;
   if (pb)
     begin
      if (r1 >= y)
        while (r1 >= y)
          begin
           @(posedge sysclk) enter_new_state('COMPUTE);
            r1 <= @(posedge sysclk) r1 - y;
           @(posedge sysclk) enter_new_state('COMPUTE23);
            r2 <= @(posedge sysclk) r2 + 1;
            r3 <= @(posedge sysclk) r2;
          end
        else
          begin
           @(posedge sysclk) enter_new_state('ZEROR3);
```

```
              r3 <= @(posedge sysclk) 0;
          end
      end
  end
```

The above correctly models the design error due to inappropriate use of parallelism in state COMPUTE23 that causes $r3$ to be assigned a value too early:

```
...
      3070 r1=    6 r2=    0 r3=    0 pb=0 ready=1
      3170 r1=    7 r2=    0 r3=    0 pb=1 ready=1
      3270 r1=    7 r2=    0 r3=    0 pb=0 ready=0
      3370 r1=    0 r2=    0 r3=    0 pb=0 ready=0
      3470 r1=    0 r2=    1 r3=    0 pb=0 ready=1
error x=    7 y=    7 x/y=    1 quotient=    0
...
```

The corrected ASM chart of 2.2.6 that has all three computations happening in parallel in state COMPUTE123 can be translated into Verilog as:

```
always
  begin
    @(posedge sysclk) enter_new_state('IDLE);
    r1 <= @(posedge sysclk) x;
    r2 <= @(posedge sysclk) 0;
    ready = 1;
    if (pb)
      begin
        if (r1 >= y)
          while (r1 >= y)
            begin
              @(posedge sysclk) enter_new_state('COMPUTE123);
              r1 <= @(posedge sysclk) r1 - y;
              r2 <= @(posedge sysclk) r2 + 1;
              r3 <= @(posedge sysclk) r2;
            end
        else
          begin
            @(posedge sysclk) enter_new_state('ZEROR3);
            r3 <= @(posedge sysclk) 0;
          end
      end
  end
```

where the states are defined as:

```
`define NUM_STATE_BITS 2
`define IDLE        2'b00
`define COMPUTE123 2'b01
`define ZEROR3     2'b11
```

The simulator shows that this version works correctly:

```
...
     3070 r1=    6 r2=   0 r3=   0 pb=0 ready=1
     3170 r1=    7 r2=   0 r3=   0 pb=1 ready=1
     3270 r1=    7 r2=   0 r3=   0 pb=0 ready=0
     3370 r1=    0 r2=   1 r3=   0 pb=0 ready=0
     3470 r1=4089 r2=   2 r3=   1 pb=0 ready=1
ok
...
```

4.1.5   Pure behavioral stage of the two-state division machine

The best correct design proposed in chapter 2 for the division machine is described by
the ASM chart in section 2.2.7. It has the advantage that it takes only one clock cycle
each time it goes through the loop, and it only needs an ASM with two states. Here is
how this ASM chart can be translated into a pure behavioral module, similar to the
earlier examples:

```
`define NUM_STATE_BITS 1
`define IDLE         1'b0
`define COMPUTE      1'b1

`include "clock.v"

module slow_div_system(pb,ready,x,y,r3,sysclk);
  input pb,x,y,sysclk;
  output ready,r3;
  wire pb;
  wire [11:0] x,y;
  reg ready;
  reg [11:0] r1,r2,r3;
  reg [`NUM_STATE_BITS-1:0] present_state;

  always
    begin
```

Continued

```
      @(posedge sysclk) enter_new_state('IDLE);
      r1 <= @(posedge sysclk) x;
      r2 <= @(posedge sysclk) 0;
      ready = 1;
      if (pb)
        begin
          while (r1 >= y | pb)
            begin
             @(posedge sysclk) enter_new_state('COMPUTE);
              r1 <= @(posedge sysclk) r1 - y;
              r2 <= @(posedge sysclk) r2 + 1;
              r3 <= @(posedge sysclk) r2;
            end
        end
    end

  task enter_new_state;
    input ['NUM_STATE_BITS-1:0] this_state;
    begin
      present_state = this_state;
      #1 ready=0;
    end
  endtask

  always @(posedge sysclk) #20
    $display("%d r1=%d r2=%d r3=%d pb=%b ready=%b",
             $time, r1,r2,r3, pb, ready);
endmodule
```

For brevity, the `cl` module has been placed in the "`clock.v`" file. Here is the Verilog simulation that shows it working:

```
. . .
      2670 r1=    5 r2=    0 r3=    0 pb=0 ready=1
      2770 r1=    6 r2=    0 r3=    0 pb=1 ready=1
      2870 r1=    6 r2=    0 r3=    0 pb=0 ready=0
      2970 r1=4095 r2=    1 r3=    0 pb=0 ready=1
ok
      3070 r1=    6 r2=    0 r3=    0 pb=0 ready=1
      3170 r1=    7 r2=    0 r3=    0 pb=1 ready=1
      3270 r1=    7 r2=    0 r3=    0 pb=0 ready=0
      3370 r1=    0 r2=    1 r3=    0 pb=0 ready=0
      3470 r1=4089 r2=    2 r3=    1 pb=0 ready=1
```

Three Stages for Verilog Design   *149*

Continued

```
ok
...
        6570  r1=  13  r2=   0  r3=   1  pb=0  ready=1
        6670  r1=  14  r2=   0  r3=   1  pb=1  ready=1
        6770  r1=  14  r2=   0  r3=   1  pb=0  ready=0
        6870  r1=   7  r2=   1  r3=   0  pb=0  ready=0
        6970  r1=   0  r2=   2  r3=   1  pb=0  ready=0
        7070  r1=4089  r2=   3  r3=   2  pb=0  ready=1
ok
```

This two-state machine will be the basis for the examples that show how to translate from pure behavioral Verilog into mixed Verilog (section 4.2) and into pure structural Verilog (section 4.3). This example is also used to illustrate the hierarchical refinement of the controller to become a netlist (section 4.4).

4.2 Mixed stage of the two-state division machine

As was explained in section 2.3.1, to translate an algorithm into hardware eventually requires that the designer decide upon a particular architecture. The mixed stage is the point in the design process when the designer decides how the registers and combinational logic devices of the architecture are to be interconnected. The only constraint is that this interconnection be able to implement all of the RTN commands used in the pure behavioral stage at the times required.

4.2.1 Building block devices

Verilog only provides built-in primitives for elementary gates, such as `and`. In order to model the mixed stage of the design, we need to define modules that simulate the bus width devices that are instantiated in the architecture. The devices outlined here are a few of the common ones from appendixes C and D.

The details of just exactly how these modules work internally need not concern us now. We just need to know the (arbitrary) order and definition of ports in the portlist for devices we want to use in the architecture. We will assume these devices are fully defined in a file, "`archdev.v`," with behavioral code.

All of these modules have a parameter, SIZE, that indicates the bus width of the data inputs and outputs of the module. (The widths of command and status ports, if any, on the device are determined by the nature of the device.)

4.2.1.1 `enabled_register` *portlist*

This module has a data input bus, `di`, of a chosen SIZE (the default is 1), and a similar sized data output bus, `do`. (See section D.6 for a description of the hardware being modeled by this module.) It also has the `enable` input, which, when it is 1, causes the current value of `di` to be loaded into the register (thereby changing `do`) at the next rising edge of `sysclk`:

```
module enabled_register(di,do,enable,clk);
     parameter SIZE = 1;
     input di,enable,clk;
     output do;
     reg [SIZE-1:0] do;
     wire [SIZE-1:0] di;
     wire enable;
     wire clk;
     ...
endmodule
```

The ellipsis indicates where the behavioral definition of the register (see sections 3.7.2.2 and 3.10.10) goes. This module was inspired by the 74xx377 (eight-bit enabled register) and 74xx378 (six-bit enabled register) TTL chips, which have an active low enable signal. This means, in a clock cycle where the physical 74xx377 is scheduled to change value at the next rising edge, the 74x377 will have zero volts representing the $1'b1$ used in the simulation for `enable`. Other than this minor detail, any architecture instantiating the above module can be constructed from these chips.

4.2.1.2 `counter_register` *portlist*

This module has a data input bus, `di`, and a similar sized data output bus, `do`. (See section D.7 for a description of the hardware being modeled by this module.) It has an output, `tc`, that is 1 when the current output is at its maximal value. It also has `load`, `count` and `clr` inputs that determine what the value of the counter will be after the next rising edge of `sysclk`:

```
module counter_register(di,do,tc,load,count,clr,clk);
     parameter SIZE = 1;
     input di,load,count,clr,clk;
     output do,tc;
     reg [SIZE-1:0] do;
     wire [SIZE-1:0] di;
     reg tc;
```

Continued

```
        wire load,count,clr;
        wire clk;
        . . .
    endmodule
```

This module was inspired by the 74xx163 (4-bit up counter), which has active low `clr` and `load` signals (see the discussion in section 4.2.1.1.) Also, this chip has two inputs that must both simultaneously be one to cause counting. The reason for having two inputs, rather than just the one `count` shown above, is to simplify the connections required to cascade the four-bit chip to form larger counters. Since at this stage of the design we are not at all concerned with such physical details, the above module was simplified to have a single `count` signal.

4.2.1.3   `alu181` *portlist*

This module models a combinational ALU inspired by the 74xx181. (See section C.6 for a description of the hardware being modeled by this module.) It has two data inputs, a and b, and a similar sized data output bus, f. It also has status outputs, `cout` (1 when addition and similar operations produce a carry) and `zero` (1 when f is zero). It is controlled by the commands: s, m and cin:

```
    module alu181(a,b,s,m,cin,cout,f,zero);
        parameter SIZE = 1;
        input a,b,s,m,cin;
        output cout,f,zero;
        wire [SIZE-1:0] a,b;
        wire  m,cin;
        wire [3:0] s;
        reg [SIZE-1:0] f;
        reg cout,zero;
        . . .
    endmodule
```

In chapter 2, the ALU was considered to have a six-bit command input, `aluctrl`. When this module is instantiated, this input should be subdivided in the following fashion:

```
    alu181 #size instancename(a,b,aluctrl[5:2],
                    aluctrl[1],aluctrl[0],cout,f,zero);
```

4.2.1.4 *comparator portlist*

This module models a comparator inspired by the 74xx85. (See section C.7 for a description of the hardware being modeled by this module.) It has two data inputs, a and b, and three status outputs, a_lt_b, a_eq_b, a_gt_b. At any time, only one of the outputs will be 1, depending on a and b:

```
module comparator(a_lt_b, a_eq_b, a_gt_b, a, b);
    parameter SIZE = 1;
    output a_lt_b, a_eq_b, a_gt_b;
    input a, b;
    wire [SIZE-1:0] a,b;
    reg a_lt_b, a_eq_b, a_gt_b;
    ...
endmodule
```

4.2.1.5 *mux2 portlist*

This module models a multiplexor inspired by the 74xx157. (See section C.4 for a description of the hardware being modeled by this module.) It has two data inputs, i0 and i1, and a similarly sized data output, out. When the command input, sel, is 0, the output is i0. When sel is 1, the output is i1:

```
module mux2(i0, i1, sel, out);
    parameter SIZE = 1;
    input i0, i1, sel;
    output out;
    wire [SIZE-1:0] i0, i1;
    wire sel;
    reg [SIZE-1:0] out;
    ...
endmodule
```

4.2.2 Mixed stage

As discussed in chapter 2, the system is no longer described simply in terms of its behavior. Instead, in the mixed stage, there is a specific structure that interconnects the controller and the architecture:

```
module slow_div_system(pb,ready,x,y,r3,sysclk);
   input pb,x,y,sysclk;
   output ready,r3;
   wire pb;
   wire [11:0] x,y;
   wire ready;
   wire [11:0] r3;
   wire sysclk;

   wire [5:0] aluctrl;
   wire muxctrl,ldr1,clrr2,incr2,ldr3,r1gey;

   slow_div_arch a(aluctrl,muxctrl,ldr1,clrr2,
               incr2,ldr3,r1gey,x,y,r3,sysclk);
   slow_div_ctrl c(pb,ready,aluctrl,muxctrl,ldr1,
               clrr2,incr2,ldr3,r1gey,sysclk);
endmodule
```

This version of `slow_div_system` replaces the behavioral version of this module discussed in section 4.1. The test code that instantiates `slow_div_system` should not notice any difference between this mixed stage module and the earlier pure behavioral stage. Note that all ports and locals in this module are now declared to be `wire` since this module is composed simply of two structural instances, and there are no behavioral assignment statements.

4.2.3 Architecture for the division machine

At the end of section 2.3.1, a particular architecture for the division machine was chosen that handles all the RTN and decisions required by the division machine. To aid in understanding the Verilog that is equivalent to this architecture, figure 4-1 shows the same architecture from section 2.3.1 redrawn with the names to be used in the Verilog code:

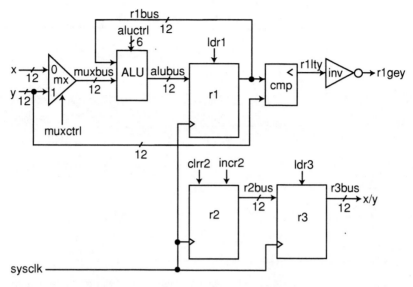

Figure 4-1. Architecture with names used for Verilog coding.

Here is how this architecture can be described as a structural Verilog module:

```
module slow_div_arch(aluctrl,muxctrl,ldr1,
              clrr2,incr2,ldr3,r1gey,x,y,r3bus,sysclk);
  input aluctrl,muxctrl,ldr1,clrr2,incr2,ldr3,x,y,sysclk;
  output r1gey,r3bus;
  wire [5:0] aluctrl;
  wire muxctrl,ldr1,clrr2,incr2,ldr3,r1gey,sysclk;
  wire [11:0] x,y,r3bus;
  wire [11:0] muxbus,alubus,r1bus,r2bus;

  enabled_register #12 r1(alubus,r1bus,ldr1,sysclk);
  mux2             #12 mx(x,y,muxctrl,muxbus);
  alu181           #12 alu(r1bus,muxbus,aluctrl[5:2],
                      aluctrl[1],aluctrl[0],,alubus,);

  comparator       #12 cmp(r1lty,,,r1bus,y);
  not                  inv(r1gey,r1lty);

  counter_register #12 r2(,r2bus,,1'b0,incr2,clrr2,sysclk);
  enabled_register #12 r3(r2bus,  r3bus,ldr3,sysclk);
```

Continued

```
always @(posedge sysclk) #20
  begin
    $display("%d r1=%d r2=%d r3=%d pb=%b ready=%b", $time,
             r1bus,r2bus,r3bus,
             slow_div_machine.pb,slow_div_machine.ready);
    $write("               %b        %b         %b",
             ldr1,{clrr2,incr2},ldr3);
    $display("     muxbus=%d alubus=%d",muxbus,alubus);
    $write("                                  ");
    $display("     muxctrl=%b    aluctrl=%b",
             muxctrl,aluctrl);
    $write("                                  ");
    $display("     x=%d r1gey=%b",x,r1gey);
  end
endmodule
```

The portlist for this module includes the commands that are input to this architecture (that were output from the controller). These commands include the six-bit aluctrl as well as muxctrl, ldr1, clrr2, incr2 and ldr3. Also, the portlist has the status output r1gey. The portlist has the twelve bit data inputs x and y and the 12-bit data output r3bus. Of course, since there are clocked registers in the architecture, they must be supplied with sysclk. The order in this portlist matches the order where this is instantiated in section 4.2.2.

The first three structural instances (r1, mx and alu) define the portion of the block diagram that relates to register r1. This name is no longer a reg as it was in the pure behavioral stage, but is instead the instance name for an enabled_register, whose portlist is defined in section 4.2.1.1. This instance is for a twelve bit wide register (because the parameter is instantiated with 12). The input to this enabled_register comes from alubus, which is described below. The output from this enabled_register is known as r1bus. Of course both alubus and r1bus are wires since this module is defined only with structure. The load signal for r1 is ldr1, and as is necessary in synchronous design, r1 is connected to sysclk.

There is an instance (named mx) of mux2 (see section 4.2.1.5) instantiated to be 12 bits wide. It selects the data input x when muxctrl is 0, and y when muxctrl is 1. Its output is muxbus. All of these buses are of course 12-bits wide. The instance of alu181 (see section 4.2.1.3) named alu takes its inputs from r1bus and muxbus. The aluctrl is provided to the appropriate ports. The cout and zero ports of alu181 are left disconnected. The f output connects to the alubus (mentioned in the last paragraph) that provides the input to r1.

The next two structural instances (cmp and inv) produce r1gey. The instance of comparator (see section 4.2.1.4) produces the output r1lty when r1bus is less than y. The a_eq_b and a_gt_b outputs of comparator are left disconnected. The built-in not produces r1gey from r1lty.

There is a 12-bit instance of counter_register (see section 4.2.1.2) named r2. Its data input is left disconnected. Normally this is a bad idea to leave an input (as opposed to an output) disconnected. However, in this case, the counter is only cleared and incremented, so the data input is not needed. The output of this counter is r2bus. The tc output of this register is not used. The load port of this instance is not utilized, but it **must** be specified as being tied to 1'b0. (In the physical TTL active low logic of the 74xx163, this would be equivalent to tying this to the five-volt supply. In active high logic, this would be equivalent to tying it to ground.) The load port **cannot be left disconnected**. The commands incr2 and clrr2 are provided to r2, as is sysclk.

Finally, there is another instance (r3) of enabled_register whose input comes from r2bus and whose output is r3bus, which is the data output of this entire module. The command ldr3 is provided to r3, as is sysclk.

At the bottom of this module are debugging $display statements. Some of the names have changed from those used in the pure behavioral version of this machine. We now refer to r1bus, r2bus and r3bus. Also, this module does not have pb or ready in its portlist, and so to produce the same style output as earlier simulations, the $display statement must use hierarchical names. (Since the $display statement does not correspond to synthesized hardware, and is only there for the convenience of the designer, it would be inappropriate to use ports to access such information. Hierarchical names are exactly what we need for a situation like this.) Two extra $display statements show signals that were not present in the pure behavioral stage. The command signals (ldr1, {clrr2, incr2} and ldr3) are displayed directly below the values of the registers they affect. Also, the values of internal buses are shown. Directly below them are the command signals that affect them. Also, the status signal output to the test code (r1gey) and the data input from the test code (x) are displayed. The other data input (y) is not displayed because in this test code it remains unchanged at 7.

4.2.4   Controller for the division machine

Although instantiating the components for the architecture as described in section 4.2.3 is a creative task that could have been done in a variety of ways, translating the pure behavioral stage into the mixed stage controller is a straightforward process that does not involve any creativity. For a particular pure behavioral machine, and for a chosen architecture, there is only one correct controller. To arrive at this controller,

1. Define a module (`slow_div_ctrl` in this example) for the controller with appropriate declarations that match how it was instantiated in the mixed system module (in this example, it is instantiated inside `slow_div_system` as shown in section 4.2.2).
2. Initialize all the internal command signals (`aluctrl`, `muxctrl`, `ldr1`, `clrr2`, `incr2` and `ldr3` in this example) used by the architecture to their default values in the `enter_new_state` task.
3. Move the `always` block from the pure behavioral system (in this example, `slow_div_system` of section 4.1.5) to inside the controller module, and comment out all non-blocking assignment statements (using `//`).
4. Following each non-blocking assignment that is commented out, put assignment statement(s) that assert the equivalent command signals. For example, `//r2 <= @(posedge sysclk) 0;` is followed by `clrr2 = 1`.
5. Replace relational conditions, such as `r1 >= y`, with status signals, such as `r1gey`.

Here is the module that is instantiated in section 4.2.2 and corresponds to pure behavioral Verilog of section 4.1.5, and that is equivalent to the mixed ASM chart of section 2.3.1:

```
module slow_div_ctrl(pb, ready, aluctrl, muxctrl, ldr1,
                     clrr2, incr2, ldr3, r1gey, sysclk);
   input pb, r1gey, sysclk;
   output ready, aluctrl, muxctrl, ldr1, clrr2, incr2, ldr3;

   reg ['NUM_STATE_BITS-1:0] present_state;
   wire pb;
   reg ready;
   reg [5:0] aluctrl;
   reg muxctrl, ldr1, clrr2, incr2, ldr3;
   wire r1gey, sysclk;

   always
     begin
       @(posedge sysclk) enter_new_state('IDLE);
         //r1 <= @(posedge sysclk) x;
         //r2 <= @(posedge sysclk) 0;
         ready = 1;
         aluctrl = 'PASSB;
         muxctrl = 0;
         ldr1 = 1;
         clrr2 = 1;
         if (pb)
```

Continued

```
          begin
           while (r1gey | pb)
             begin
              @(posedge sysclk) enter_new_state('COMPUTE);
              ready = 0;
              //r1 <= @(posedge sysclk) r1 - y;
              //r2 <= @(posedge sysclk) r2 + 1;
              //r3 <= @(posedge sysclk) r2;
              aluctrl = 'DIFFERENCE;
              muxctrl = 1;
              ldr1 = 1;
              incr2 = 1;
              ldr3 = 1;
             end
           end
      end

    task enter_new_state;
     input ['NUM_STATE_BITS-1:0] this_state;
     begin
      present_state = this_state;
      #1{ready,aluctrl,muxctrl,ldr1,clrr2,incr2,ldr3}=0;
     end
    endtask
  endmodule
```

The boldface above shows the editing done to transform the pure behavioral Verilog
into this mixed stage. Of some interest is the fact that the pure behavioral while,
which has the condition ((r1>=y)|pb), is translated above into (r1gey | pb).
Use of single bit & and | (or perhaps more clearly && and ||) is permitted inside a
mixed controller. This notation is not a data computation that must occur in the architecture (although the designer could have chosen to put a single or gate in the architec-
ture to accomplish this). It is important to distinguish this decision-making use of |
from a data manipulation use of |, such as r1|y, which should be performed by com-
binational logic (such as the ALU) in the architecture. In the case of ((r1>=y)|pb),
there are two reasonable ways to translate this into the mixed stage: the way that was
shown above, and the way that requires introducing an extra signal in the architecture
to represent the or of r1gey and pb. (pb would then be classified as an external data
input to the architecture, in addition to being an external status input to the controller.)
Since we would like to minimize the number of wires that interconnect the controller to
the architecture, we chose the former approach where pb is simply an external status
signal.

Of course, the bit patterns for controlling the ALU must be defined outside this module:

```
'define DIFFERENCE   6'b011001
'define PASSB        6'b101010
```

The test code is the same as the pure behavioral system. Here is the output from the completed mixed stage:

```
   70 r1=   x r2=   x r3=   x pb=0 ready=1
         1      10         0     muxbus=   0 alubus=   0
                                 muxctrl=0    aluctrl=101010
                                 x=   0 r1gey=x
  170 r1=   0 r2=   0 r3=   x pb=0 ready=1
         1      10         0     muxbus=   0 alubus=   0
                                 muxctrl=0    aluctrl=101010
                                 x=   0 r1gey=0
  270 r1=   0 r2=   0 r3=   x pb=0 ready=1
         1      10         0     muxbus=   0 alubus=   0
                                 muxctrl=0    aluctrl=101010
                                 x=   0 r1gey=0
  370 r1=   0 r2=   0 r3=   x pb=1 ready=1
         1      10         0     muxbus=   0 alubus=   0
                                 muxctrl=0    aluctrl=101010
                                 x=   0 r1gey=0
  470 r1=   0 r2=   0 r3=   x pb=0 ready=0
         1      01         1     muxbus=   7 alubus=4089
                                 muxctrl=1    aluctrl=011001
                                 x=   0 r1gey=0
  570 r1=4089 r2=   1 r3=   0 pb=0 ready=1
         1      10         0     muxbus=   0 alubus=   0
                                 muxctrl=0    aluctrl=101010
                                 x=   0 r1gey=1
ok
...
 6570 r1=  13 r2=   0 r3=   1 pb=0 ready=1
         1      10         0     muxbus=  14 alubus=  14
                                 muxctrl=0    aluctrl=101010
                                 x=  14 r1gey=1
 6670 r1=  14 r2=   0 r3=   1 pb=1 ready=1
         1      10         0     muxbus=  14 alubus=  14
                                 muxctrl=0    aluctrl=101010
                                 x=  14 r1gey=1
```

Continued

```
6770 r1=   14 r2=    0 r3=    1 pb=0 ready=0
         1         01         1    muxbus=    7 alubus=    7
                                   muxctrl=1    aluctrl=011001
                                   x=   14 r1gey=1
6870 r1=    7 r2=    1 r3=    0 pb=0 ready=0
         1         01         1    muxbus=    7 alubus=    0
                                   muxctrl=1    aluctrl=011001
                                   x=   14 r1gey=1
6970 r1=    0 r2=    2 r3=    1 pb=0 ready=0
         1         01         1    muxbus=    7 alubus=4089
                                   muxctrl=1    aluctrl=011001
                                   x=   14 r1gey=0
7070 r1=4089 r2=    3 r3=    2 pb=0 ready=1
         1         10         0    muxbus=   14 alubus=   14
                                   muxctrl=0    aluctrl=101010
                                   x=   14 r1gey=1
ok
```

4.3 Pure structural stage of the two state division machine

Translating from the mixed stage to the "pure" structural stage is an easy and mechanical (although somewhat tedious) process. All modules except the controller remain the same. As explained in section 2.4.1, the controller module becomes a structure composed of a present state register (which is an instance of an actual register module, and not a `reg`) and the next state logic.

In the "pure" structural stage, the definition of the next state logic may remain as behavioral code (a function) that is a transformation of the code inside the `always` block of the mixed stage. In section 4.4, we will see how the next state logic could also be defined in terms of built-in gates, using hierarchical design. Fortunately, it is not normally necessary to worry about the details given later in section 4.4, because synthesis tools exist that can automatically transform the behavioral next state function described in this section into a netlist. For this reason, we consider this section to be the final step that the designer has to be involved with. Section 4.4 is presented later only to motivate the kind of transformations that synthesis tools are capable of.

4.3.1 The pure structural controller

The structure of the controller is quite simple. The instance name of the `next_state_logic` is `nsl`. As shown in the diagram in section 2.4.1, `r1gey` and `pb` are the inputs to `nsl`. Also, `next_state`, `ldr1`, `incr2`, `clrr2`, `ldr3`, `muxctrl`, `aluctrl`, `ready` and `next_state` are outputs of `nsl`. The input to `ps_reg` is `next_state` and its output is `present_state`. The portlist of the controller module is identical to the mixed stage, except the outputs are declared to be `wires`:

```
module slow_div_ctrl(pb,ready,aluctrl,muxctrl,ldr1,
                     clrr2,incr2,ldr3,r1gey,sysclk);
   input pb,r1gey,sysclk;
   output ready,aluctrl,muxctrl,ldr1,clrr2,incr2,ldr3;

   wire ['NUM_STATE_BITS-1:0] present_state;
   wire pb;
   wire ready;
   wire [5:0] aluctrl;
   wire muxctrl,ldr1,clrr2,incr2,ldr3;
   wire r1gey,sysclk;

   next_state_logic  nsl(next_state,
                         ldr1,incr2,clrr2,ldr3,
                         muxctrl,aluctrl,ready,
                         present_state, r1gey, pb);
   enabled_register #('NUM_STATE_BITS) ps_reg(next_state,
                         present_state,1'b1,sysclk);
endmodule
```

For simplicity, we are using an `enabled_register` for `ps_reg` with its enable tied to `1'b1`.

4.3.2 `next_state_logic` module

The combinational logic that computes the next state needs to be defined. Following the technique outlined in section 3.7.2.1, there is an `always` block with an @ sensitivity list which has all the inputs to this module. Since the calculation of the next state and corresponding outputs is quite lengthy, this calculation is isolated in a function, `state_gen`, that is defined in the file "`divbookf.v`":

```
module next_state_logic(next_state,
                        ldr1,incr2,clrr2,ldr3,
                        muxctrl,aluctrl, ready,
                        present_state, r1gey, pb);
    output next_state,ldr1,incr2,clrr2,ldr3,muxctrl,
           aluctrl,ready;
    input present_state, r1gey, pb;
    reg ['NUM_STATE_BITS-1:0] next_state;
    reg ldr1,incr2,clrr2,ldr3,muxctrl,ready;
    reg [5:0] aluctrl;
    wire ['NUM_STATE_BITS-1:0] present_state;
    wire r1gey,pb;

    'include "divbookf.v"

    always @ (present_state or r1gey or pb)
      {next_state,ldr1,clrr2,incr2,ldr3,muxctrl,aluctrl,
        ready} = state_gen(present_state, pb, r1gey);
endmodule
```

4.3.3 `state_gen` function

To create the file that contains the `state_gen` function is a simple matter of editing a portion of the mixed code:

1. Create a function header that returns the proper number of bits and has the appropriate input arguments. The number of bits to be returned is `'NUM_STATE_BITS-1` plus the number of command output bits. The input arguments of the function are the same as the input ports of `next_state_logic`. (In this example, the inputs are `ps`, `pb` and `r1gey`.)

2. The output `regs` of the mixed controller become local `regs` in the function. (In this example, `ldr1`, `incr2`, `clrr2`, `ldr3`, `muxctrl`, `aluctrl` and `ready` are local `regs` in the function.) A local `ns` `reg` is also defined to hold the next state within the function. Remember that local `regs` of a function would never be synthesized as physical registers. They are `regs` simply because the function uses behavioral assignment.

3. The assignment of default values that occurs in the `enter_new_state` task of the mixed stage becomes the first executable statement of the function. Also, `ns` is also given a default value (the starting point of the algorithm).

4. The next executable statement of the function is a `case` statement based on the present state, `ps`. This `case` statement is equivalent to a truth table, as explained

in section 3.9.2.2. The advantage of the `case` statement for this purpose is that it is more compact than a truth table, and it documents some of the thought process of the earlier stages through meaningful identifiers (such as `aluctrl`).

5. The statements that follow each @ (posedge sysclk) inside the `always` block of the mixed controller are moved to a place within the `case` statement that corresponds to that state. The @ and the call to `enter_new_state` are eliminated. The commented-out non-blocking assignment statements are retained for documentation.

6. In each block of code that corresponds to a state, `ns` is computed. This computation, in effect, acts like a `goto`. It says which state in the `case` statement will execute when this function is called during the next clock cycle (after `ns` becomes `ps`). Unlike the mixed stage, the order in which the designer types the states into the `case` statement has no effect on the order in which the states execute. The `ns` computation determines the order in which they execute. By putting the ASM into a function, we have lost the perfect correspondence to the `goto`-less style that we had in the mixed stage.

7. The next state and the outputs are concatenated to be returned from this function in the order needed (see section 4.3.2).

Here is the `state_gen` function for the two-state division machine:

```
function ['NUM_STATE_BITS-1+12:0] state_gen;
  input ['NUM_STATE_BITS-1:0] ps;
  input pb,r1gey;
  reg ready;
  reg [5:0] aluctrl;
  reg muxctrl,ldr1,clrr2,incr2,ldr3;
  reg ['NUM_STATE_BITS-1:0] ns;

begin
  {ns,ready,aluctrl,muxctrl,ldr1,clrr2,incr2,ldr3}=0;
  case (ps)
    'IDLE:      begin
                  //r1 <= @(posedge sysclk) x;
                  //r2 <= @(posedge sysclk) 0;
                  ready = 1;
                  aluctrl = 'PASSB;
                  muxctrl = 0;
                  ldr1 = 1;
                  clrr2 = 1;
                  if (pb)
                    if (r1gey|pb)
```

Continued

```
                           ns = `COMPUTE;
                   else
                       ns = `IDLE;
                   else
                       ns = `IDLE;
               end
        `COMPUTE: begin
                   ready = 0;
                   //r1 <= @(posedge sysclk) r1 - y;
                   //r2 <= @(posedge sysclk) r2 + 1;
                   //r3 <= @(posedge sysclk) r2;
                   aluctrl = `DIFFERENCE;
                   muxctrl = 1;
                   ldr1 = 1;
                   incr2 = 1;
                   ldr3 = 1;
                   if (r1gey|pb)
                      ns = `COMPUTE;
                   else
                       ns = `IDLE;
               end
        endcase
        state_gen = {ns,ldr1,clrr2,incr2,ldr3,
                     muxctrl,aluctrl,ready};
    end
 endfunction
```

Here boldface shows some changes that were made to make this work as a function.

4.3.4 Testing `state_gen`

Since `state_gen` is isolated in a file by itself, we can test `state_gen` by writing some trivial Verilog, unrelated to any of the earlier code:

```
`define DIFFERENCE   6'b011001
`define PASSB        6'b101010

`define NUM_STATE_BITS 1
`define IDLE           1'b0
`define COMPUTE        1'b1

module test;
```

Continued.

```
`include "divbookf.v"


integer i;
reg ps,pb,r1gey;


initial
  begin
    for (i=0;i<=7;i=i+1)
      begin
        {ps,pb,r1gey} = i;
        $display("%b %b %b %b",ps,pb,r1gey,
                    state_gen(ps,pb,r1gey));
      end
  end
endmodule
```

This produces an output which agrees with the manual derivation given in section 2.4.1. The bit patterns used in the function must be defined, as shown above, because they are not defined in "divbookf.v".

4.3.5 It *seems* to work

Having tested the function to see that it behaves as expected, we can now put the controller code from sections 4.3.1 through 4.3.3 together with the architecture code from section 4.2.3 to obtain the pure structural version of the two-state division machine. The simulation of this produces the same output as that of the mixed stage, shown in section 4.2.4.

There is one additional detail, described in the next section, that we will want to consider in all future designs. Since the code (including this additional detail) that the designer develops in the pure structural stage can be run through a synthesis tool, it is not necessary for the designer to manually transform Verilog code after reaching the pure structural stage. In the next section, we will see how the synthesis tool continues to transform the Verilog code for the designer automatically.

4.4 Hierarchical refinement of the controller

Except for one small physical problem to be explained in this section, the Verilog code in section 4.3 can be submitted to a synthesis tool, which produces the netlist that can be used to fabricate the chip. Assuming we recognize and fix this little problem, we do not need to write any more Verilog. The synthesis tool can do the rest of the job of creating the netlist.

On the other hand, since the two-state division machine, with its childish algorithm, is so simple, it makes a good example to illustrate what a synthesis tool does. Also, by looking at the netlist, we will discover this small physical problem alluded to earlier. This problem, which occurs with most controllers, has a simple solution; however, it was previously hidden from view because behavioral Verilog usually does not model all the physical details of a circuit. The power of top-down design is to hide those details until the last moment. We have reached this moment of truth when we get down to gates!

This section illustrates the need to do post-synthesis simulation prior to fabrication. All three-stages of design (including the pure structural stage) have some behavioral aspects that are well above the gate level. To predict more accurately what the fabricated circuit will do, we need to simulate the synthesized netlist. With this simple machine, the netlist for the controller is simple enough that we can generate it manually, but in most machines, an automatically produced netlist would be incomprehensible to the designer.

4.4.1 A logic equation approach

The `state_gen` function defined in section 4.3.3 can be replaced by a series of logic equations representing the low-level behavior of `nsl`, as explained in section 2.5:

```
module next_state_logic(next_state,
                        ldr1,incr2,clrr2,ldr3,
                        muxctrl,aluctrl, ready,
                        present_state, r1gey, pb);
   output next_state,ldr1,incr2,clrr2,ldr3,muxctrl,
          aluctrl,ready;
   input present_state, r1gey, pb;
   reg ['NUM_STATE_BITS-1:0] next_state;
   reg ldr1,incr2,clrr2,ldr3,muxctrl,ready;
   reg [5:0] aluctrl;
   wire ['NUM_STATE_BITS-1:0] present_state;
   wire r1gey,pb;

   always @(present_state or r1gey or pb)
```

Continued

```
     begin
       next_state =
        ~present_state&pb|present_state&(r1gey|pb);
       ldr1 = 1;
       clrr2 = ~present_state;
       incr2 = present_state;
       ldr3 = present_state;
       muxctrl = present_state;
       aluctrl[5] = ~present_state;
       aluctrl[4] = present_state;
       aluctrl[3] = 1;
       aluctrl[2] = 0;
       aluctrl[1] = ~present_state;
       aluctrl[0] = present_state;
       ready = ~present_state;
     end
 endmodule
```

Even though this is a rather tedious way to describe the controller, it is still a behavioral description as explained in section 3.7.2.1. This module has an `always` block with @ followed by a sensitivity list naming all the inputs to the combinational logic, and the outputs are defined as `regs`. Although using logic equations in this fashion is perfectly legal, this is not the preferred style of Verilog coding for three reasons: it is not a netlist, a designer can easily make a mistake when deriving the logic equations manually, and even if the logic equations are correct, they are usually meaningless to the designer. One reason why we practice top-down and hierarchical design is to minimize our exposure to details, especially details as tedious as a page full of logic equations.

Algorithms to manipulate Boolean equations are some of the most studied aspects of computer design. Even prior to the introduction of Verilog, software existed to automatically produce logic equations from truth tables. It would be a giant leap backward to use Verilog with manually produced logic equations. We will not consider writing modules of this style again.

4.4.2   At last: a netlist
Even though logic equations should not normally be your first choice when designing behavioral code, it is important for you to be aware of the properties of Boolean algebra. Ultimately, at the lowest levels, all computation carried out on digital computers are the result of iterative application of Boolean equations.

It was popular to use logic equations manually before the introduction of HDLs because there is a one-to-one mapping between logic equations and a netlist. For the same reason, a synthesis tool internally manipulates logic equations as it explores the vast space of possible hardware structures that correctly implement a combinational function. When the synthesis tool decides what the optimum logic equation is, it is trivial for it to produce the netlist. Using the logic equations from section 4.4.1, we have at last gotten all the way down to the netlist:

```
module next_state_logic(next_state,
                        ldr1,incr2,clrr2,ldr3,
                        muxctrl,aluctrl, ready,
                        present_state, r1gey, pb);
   output next_state,ldr1,incr2,clrr2,ldr3,muxctrl,
          aluctrl,ready;
   input present_state, r1gey, pb;
   wire ['NUM_STATE_BITS-1:0] next_state;
   wire ldr1,incr2,clrr2,ldr3,muxctrl,ready;
   wire [5:0] aluctrl;
   wire ['NUM_STATE_BITS-1:0] present_state;
   wire r1gey,pb;

   buf b0(ldr1,aluctrl[3],1'b1);
   buf b1(aluctrl[2],1'b0);
   buf b2(incr2,ldr3,muxctrl,aluctrl[4],aluctrl[0],
          present_state[0]);
   not i1(not_ps,clrr2,aluctrl[5],aluctrl[1],ready,
          present_state[0]);
   and a1(not_ps_and_pb,not_ps,pb);
   and a2(ps_and_or,present_state[0],r1gey_or_pb);
   or  o1(r1gey_or_pb,r1gey,pb);
   or  o2(next_state[0],not_ps_and_pb,ps_and_or);
endmodule
```

Figure 4-2 shows the corresponding circuit diagram.

Of course, many other possible solutions exist that produce the same truth table. The built-in Verilog gate buf (non-inverting buffer) passes through its last port unchanged to all the other ports, which are outputs. The only difference between buf and not is that the latter inverts its outputs.

We will ignore discussing the netlist for the architectural devices, such as mux2, since this is a trivial but tedious task. The synthesis tool would do this identically to the way the controller was synthesized.

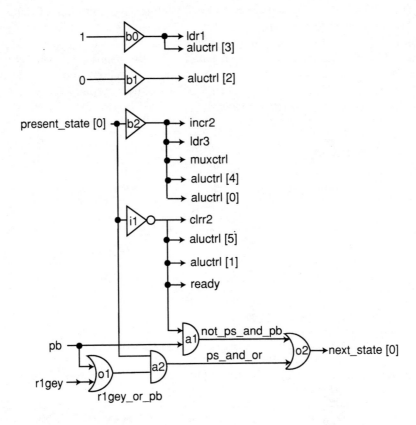

Figure 4-2. Netlist for the childish division controller.

4.4.3 Post-synthesis simulation

Even though up to this point all the simulation results indicate that our synthesized division machine should work, it is wise to conduct post-synthesis simulation prior to fabricating the hardware. This is why Verilog was developed in the first place: to simulate gate level netlists. Synthesis from behavioral code is now the predominate use of Verilog, but synthesis tools appeared later in the history of Verilog than simulators for netlists.

So let's use Verilog in the way it was originally intended to be used and simulate the netlist given in section 4.4.2. When this simulation runs, we get a very interesting but discouraging output:

```
 70 r1=x r2=x r3=x pb=0 ready=x
     1       xz       z     muxbus=x alubus=x
                            muxctrl=z    aluctrl=xz10xz
                            x=   0 r1gey=x
170 r1=x r2=x r3=x pb=0 ready=x
     1       xz       z     muxbus=x alubus=x
                            muxctrl=z    aluctrl=xz10xz
                            x=   0 r1gey=x
270 r1=x r2=x r3=x pb=0 ready=x
     1       xz       z     muxbus=x alubus=x
                            muxctrl=z    aluctrl=xz10xz
                            x=   0 r1gey=x
. . .
```

that continues like this for as long as you are willing to let the simulator run. This is the physical problem alluded to earlier: as the controller is currently interconnected, the gate level netlist does not seem to work. The simulator's output is splattered with 1'bxs and 1'bzs.

Although you might think something is wrong with the logic equations (given in section 4.4.1) or the equivalent netlist (given in section 4.4.2), there is not. The logic equations and equivalent netlist are correct. What's the problem?

To understand the problem, you need to remember the intent behind having the four-valued logic system. When 1'bxs or 1'bzs appear where you were expecting a 1 or a 0, this is an indication of some flaw in the design. Although major interconnection errors can cause this (see section 3.5.3), more subtle problems can cause this as well. Since everything is interconnected properly in this netlist, we need to understand what the 1'bxs and 1'bzs are trying to tell us here.

At $time 0, all regs start as 'bxs and all wires start as 'bzs. If the simulation does not change these values, that is how they will stay. The ps_reg of the controller has an internal reg that holds the present state. At $time 0, it is 1'bx. The next state that the machine computes from ps_reg is also unknown. A Boolean function of 'bx is usually 'bx. Therefore, the ps_reg is reloaded with 1'bx, rather than the proper sequence of states. The four-valued logic of the simulation has detected a potential flaw in the design: we do not know what state the controller starts out in, so we cannot predict what happens next.

Why didn't the pure structural version (see section 4.3.3) detect this problem? The reason is found in the definition of the state_gen function. The first statement of this function initializes ns (which is what becomes next_state) to be 1'b0:

```
{ns,ready,aluctrl,muxctrl,ldr1,clrr2,incr2,ldr3}=0;
case (ps)
  ...
endcase
```

Unfortunately, this also means that even though `ps` is `'bx` at `$time` 0, `ns` will become 0 after the first rising edge. This is a disguised way of saying:

```
{ready,aluctrl,muxctrl,ldr1,clrr2,incr2,ldr3}=0;
if (ps === 1'bx | ps === 1'bz)
  ns = 1'b0;
else
case (ps)
  ...
endcase
```

The reason this is cheating is because hardware cannot implement the `===` operation. Each physical wire can only carry one bit of information. Each simulated wire tested with `===` carries two bits of information (to represent 0, 1, `1'bx` and `1'bz`). At `$time` 0, all `reg`s are intialiazed to `'bx`. As soon as the above function detects `'bx`, it turns it into 0. In this indirect way, the function is informed when `$time` is 0. The fabricated hardware has no way to know when `$time` is 0, because a wire is simply either 0 or 1.

We can make the pure structural stage act more like the netlist by omitting `ns` from the default initialization:

```
{ready,aluctrl,muxctrl,ldr1,clrr2,incr2,ldr3}=0;
case (ps)
  ...
```

By omitting `ns` here, the simulation using `state_gen` fails similarly to the netlist.

4.4.4 Resetting the present state

What post-synthesis simulation discovered is a problem that all state machines exhibit, and that happily has a simple solution. When the power is first turned on, we do not know what state the machine will be in. The pure behavioral and mixed stages use position within behavioral code (rather than the `ps_reg`) to indicate the current state, so they start at the top of the code at $time 0. As explained above, the way we did the pure structural stage (with the `state_gen` function) tricked Verilog into disguising the problem. At the netlist level, the problem cannot be hidden any longer.

If you are at all familiar with state machines, you probably spotted this problem a long time ago. We need an asynchronous reset for the `ps_reg` that is activated soon after $time 0, and left inactive thereafter. Here is a behavioral model of such a simple D-type resettable register:

```
module resetable_register(di,do,reset,sysclk);
   parameter SIZE=1;
   input di,reset,sysclk;
   output do;
   wire [SIZE-1:0] di;
   reg [SIZE-1:0] do;
   wire reset,sysclk;

   always @(posedge sysclk or posedge reset)
     begin
       if (reset)
         do = 0;
       else
         do = di;
     end
endmodule
```

The above is patterned after the 74xx175 (six-bit resettable D-type register), except as is typical with TTL logic, the `reset` signal on the 74xx175 is active low.

This is the first and only time that we will admit an asynchronous signal into our design. *Asynchronous* means that a change happens in a register at a $time other than the rising edge of `sysclk`. Notice the difference between the `clr` signal used in the synchronous `counter_register` (described in sections 3.7.2.2 and 4.2.1.2) and the `reset` signal described here. Although both signals cause the register to become zero at some point in $time, the `clr` signal simply schedules the change to happen at the next rising edge, but the `reset` signal causes the clearing to happen instantly. The register is continually rezeroed for as long as `reset` is asserted because of the `if`, even should a rising edge of the clock occur. Without the `posedge reset`, the register would be a synchronous, clearable D-type register.

The `ps_reg` needs to have an asynchronous reset so that it is zero prior to the arrival of the first rising edge of `sysclk`. The `reset` signal must be provided by our friendly user, which means for the Verilog simulation that `reset` becomes a port of several modules and must be provided by the test code. It is a input port of the controller:

```
module slow_div_ctrl(pb,ready,aluctrl,muxctrl,ldr1,
                clrr2,incr2,ldr3,r1gey,reset,sysclk);
   input pb,r1gey,sysclk,reset;
   output ready,aluctrl,muxctrl,ldr1,clrr2,incr2,ldr3;

   wire ['NUM_STATE_BITS-1:0] present_state;
   wire pb;
   wire ready;
   wire [5:0] aluctrl;
   wire muxctrl,ldr1,clrr2,incr2,ldr3;
   wire r1gey,sysclk,reset;

   next_state_logic  nsl(next_state,
                            ldr1,incr2,clrr2,ldr3,
                            muxctrl,aluctrl,ready,
                            present_state, r1gey, pb);
   resetable_register #('NUM_STATE_BITS) ps_reg(next_state,
                            present_state,reset,sysclk);
endmodule
```

and of the system that instantiates the controller:

```
module slow_div_system(pb,ready,x,y,r3,reset,sysclk);
   input pb,x,y,sysclk,reset;
   output ready,r3;
   wire pb;
   wire [11:0] x,y;
   wire ready;
   wire [11:0] r3;
   wire sysclk,reset;

   wire [5:0] aluctrl;
   wire muxctrl,ldr1,clrr2,incr2,ldr3,r1gey;

   slow_div_arch a(aluctrl,muxctrl,ldr1,clrr2,
                incr2,ldr3,r1gey,x,y,r3,sysclk);
```

Continued

```
    slow_div_ctrl c(pb,ready,aluctrl,muxctrl,ldr1,
                clrr2,incr2,ldr3,rlgey,reset,sysclk);
endmodule
```

It must also appear in the test code:

```
module top;
  reg pb;
  reg [11:0] x,y;

  wire [11:0] quotient;
  wire ready;

  integer s;
  wire sysclk;
  reg reset;

  cl #20000 clock(sysclk);
  slow_div_system
slow_div_machine(pb,ready,x,y,quotient,reset,sysclk);

  initial
    begin
      pb= 0;
      x = 0;
      y = 7;
      reset = 0;
      #30 reset = 1;
      #10 reset = 0;
      #210;
      ...
endmodule
```

The test code issues a reset pulse that lasts for 30 units of $time, which causes the present state to become zero. When the netlist for the controller (section 4.4.2) is re-simulated with the above, it produces the same correct answers we obtained for the mixed stage.

4.5 Conclusion

The three stages of top-down design can be expressed in Verilog. The pure behavioral stage requires writing a single system module to model the machine. The only structure at this stage is the portlist of the module. During this stage, the designer also develops test code that instantiates the pure behavioral module. The test code is important because it is a specification of what the machine is supposed to do. By translating the ASM that describes the machine with the `enter_new_state` approach, the pure behavioral Verilog is organized so that it will be easy to translate to the mixed stage.

In the mixed stage, the designer develops structural Verilog code for the architecture. The modules instantiated inside the architecture may themselves be defined behaviorally. The architecture and a controller are instantiated as the system module. The controller is behavioral code derived from the pure behavioral stage. The <= statements are commented out and replaced with appropriate command signals for the chosen architecture, and relational conditions are replaced with appropriate status signals. The default values for the command signals are indicated in the `enter_new_state` task.

In the pure structural stage, the controller is edited to become a structural module that instantiates a next state generator and a resettable register. The test code and portlists must be edited to include a `reset` signal. The next state generator has an `always` block modeling combinational logic which calls a function defined with a `case` statement. The cases in this statement are copied from the mixed controller code. Additional statements must be provided in each state to describe the calculation of the next state. Default values are indicated prior to the `case` statement.

The next state logic and all of the architectural building blocks of the pure structural stage could be refined down to the gate level by using hierarchical design. However, synthesis tools can take a pure structural description (with all of its instantiated modules still defined behaviorally) and produce a gate-level netlist. In most cases, the pure structural stage is equivalent to working hardware. It is important to do post-synthesis simulation before fabrication to insure the netlist solves the problem correctly because it is cheaper to find flaws before fabrication.

4.6 Exercises

4-1. Use a simulator to take problems 2-1, 2-2 and 2-3 through the three-stages in Verilog.

4-2. Use a simulator to take problems 2-4, 2-5 and 2-6 through the three stages in Verilog.

4-3. Use a simulator to take problems 2-7, 2-8 and 2-9 through the three-stages in Verilog.

5.   ADVANCED ASM
TECHNIQUES

Although the ASM techniques illustrated in chapter 2 and the corresponding Verilog notation given in chapter 4 are adequate to solve any problem, they may yield a hardware solution that is not optimal in terms of speed (number of clock cycles) and cost (number of gates.) Despite all the marketing hype one hears, neither speed nor cost should be the primary concern of the designer. The primary responsibility of the designer is producing a **correct** design. (Intel Corporation illustrated the wisdom that before one produces a fast chip one ought to design a correct chip when they sold a version of the Pentium in 1994 whose division algorithm was incorrect.)

Despite the fact speed should not be our first concern, in many problems a correct solution demands that an algorithm find its answer by a certain time. Consider, for example, the onboard computers of the space shuttle. They need to compute the correct result in a timely enough fashion that the shuttle may correct its course. In such a context, a machine that computes a correct answer too late is not a solution at all.

Our search for a faster solution should always begin in the abstract world of algorithms, and not in the gruesome world of gates. The best way to speed up a machine is not to do some trickery with gates; instead the best way is to describe a better algorithm that solves the same problem. Chapter 2 illustrates this point with several different variations on the division machine with the final solution being three times faster than the slowest solution. One difficulty is that certain faster algorithms cannot be expressed with the notations discussed in chapters 2 and 4. This chapter discusses an additional ASM feature that helps us describe more efficient algorithms. Also, this chapter explains how any ASM can be written in Verilog, including those that use such notations, as well as those that have complex branches that do not follow the goto-less style we adhered to in earlier examples.

5.1   Moore versus Mealy

Recall that an ASM chart is composed of diamonds, rectangles and ovals connected by arrows. Chapter 2 ignored the use of ovals. Such ASM charts that do not have ovals are referred to as Moore machines. All the ASM charts discussed previously were for Moore machines. ASM charts that also include ovals are known as Mealy machines. Mealy ASM charts provide a way to express algorithms that are faster (and in some instances less costly) than Moore ASM charts.

Commands (both RTN and signal names) that occur inside a rectangle are known as *unconditional* commands, because the commands are issued regardless of anything when the machine is in the state corresponding to the rectangle. Ovals are used to describe *conditional* commands, that are sometimes (but not always) issued when a machine is in a particular state. Ovals are not by themselves a state, but rather they are the children of some parent state that corresponds to a rectangle in the ASM chart. Because ovals represent a conditional command, they must occur after one or more diamond(s). If you follow the arrows of a Mealy ASM, you would first come to a rectangle, then you would come to one or more diamond(s) and finally you would come to the oval. After the oval, the arrow might go to another diamond or a rectangle.

The actions in one rectangle and all ovals and diamonds that are connected (without any intervening rectangles) to that one rectangle occur in the same clock cycle. In essence, a combination of diamonds and ovals allows the designer to implement an arbitrarily nested if else construct that executes in a single clock cycle. Large numbers of decisions can be carried out in parallel by such ASM charts, allowing some algorithms to be sped up considerably.

5.1.1 Silly example of behavioral Mealy machine

Suppose we take the silly ASM of section 2.1.2.1 and include two conditional command signals: STAY and LEAVE. STAY is supposed to be 1 while the machine stays in state YELLOW, and LEAVE is 1 during the last cycle that the machine is in state YELLOW. STAY and LEAVE are never asserted at the same time. Here is the ASM chart:

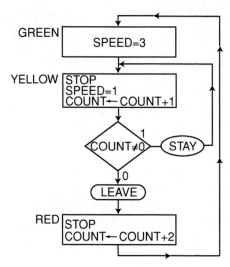

Figure 5-1. Behavioral Mealy ASM.

Assuming, as was the case for the example in section 2.1.2.1, that the period of the clock is 0.5 seconds, the following illustrates what would be observed physically when the hardware corresponding to this ASM operates:

```
present
time    state
 0.0    GREEN    stop=0   speed=11   count=000   stay=0   leave=0
 0.5    YELLOW   stop=1   speed=01   count=000   stay=0   leave=1
 1.0    RED      stop=1   speed=00   count=001   stay=0   leave=0
 1.5    GREEN    stop=0   speed=11   count=011   stay=0   leave=0
 2.0    YELLOW   stop=1   speed=01   count=011   stay=1   leave=0
 2.5    YELLOW   stop=1   speed=01   count=100   stay=1   leave=0
 3.0    YELLOW   stop=1   speed=01   count=101   stay=1   leave=0
 3.5    YELLOW   stop=1   speed=01   count=110   stay=1   leave=0
 4.0    YELLOW   stop=1   speed=01   count=111   stay=1   leave=0
 4.5    YELLOW   stop=1   speed=01   count=000   stay=0   leave=1
 5.0    RED      stop=1   speed=00   count=001   stay=0   leave=0
 5.5    GREEN    stop=0   speed=11   count=011   stay=0   leave=0
 6.0    YELLOW   stop=1   speed=01   count=011   stay=1   leave=0
 ...    ...       ...       ...         ...       ...       ...
```

Between 0.5 and 1.0, the machine is in state YELLOW, but because COUNT is zero, the decision does not go on the path that loops back to state YELLOW but instead goes on the path where the next state is state RED. On this path is an oval that asserts the LEAVE signal. This conditional signal is asserted during the entire clock cycle, just as the unconditional signal STOP is asserted during the same time. The signal STAY, which is on a different path, is not asserted during this clock cycle.

Between 2.0 and 2.5, the machine again is in state YELLOW, but because COUNT is non-zero, the decision goes on the path that loops back to the same state. On this path is the oval that asserts the STAY signal. The signal LEAVE is not asserted in this clock cycle.

Because COUNT is three bits, COUNT is zero again between 4.5 and 5.0, and so this is the last clock cycle that the machine loops in state YELLOW. This means that STAY is not asserted but that LEAVE is asserted.

5.1.2 Silly example of mixed Mealy machine

An architecture that implements the silly example of section 5.1.1 is a counter register attached to an adder. One input of the adder is tied to the constant two. There is a comparator to see when COUNT is equal to zero. Here are the architecture and corresponding mixed ASM:

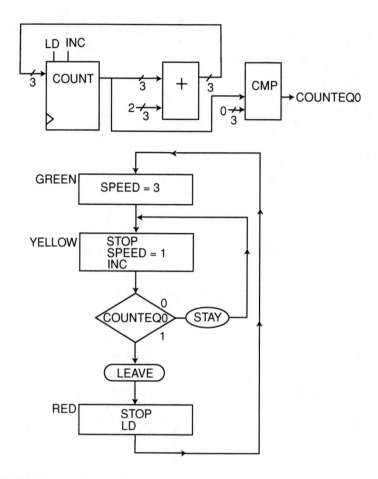

Fig 5-2. Mixed Mealy ASM.

5.1.3 Silly example of structural Mealy machine

The generic diagram of the pure structural controller given in section 2.4.1 applies to any machine, whether it is a Mealy or Moore machine. The next state combinational logic will be a little different when the machine is a Mealy machine than when it is a Moore machine. With a Moore machine, only the next state bits (and not the command bits) are a function of both the present state and the status inputs. With a Moore machine, the commands are a function of the present state only. In other words, for a Moore machine, every line of the truth table where ps is the same has the same command outputs.

A Mealy machine is completely general. The commands as well as the next state are a function of both the present state and the status. To illustrate this, consider the truth table of the next state combinational logic for the machine described by the ASM chart and architecture of section 5.1.2:

ps	COUNTEQ0	ns	STOP	SPEED	**LEAVE**	**STAY**	INC	LD
00	–	01	0	11	0	0	0	0
01	0	10	1	01	**1**	**0**	1	0
01	1	01	1	01	**0**	**1**	1	0
10	–	00	1	00	0	0	0	1

SPEED, STOP, INC and LD are unconditional commands that are a function of ps only. LEAVE and STAY are a function of both ps and COUNTEQ0. The conditional signals LEAVE and STAY are the only things here that make this a Mealy machine.

5.2 Mealy version of the division machine

Section 2.2 gives many variations of Moore ASMs that implement the childish division algorithm. This section describes how this algorithm can be improved by including ovals in the ASM.

5.2.1 Eliminating state INIT again

Section 2.2.3 describes a correct four-state version of the division machine that uses only two registers (r1 and r2) in the architecture. Section 2.2.4 describes an unsuccessful attempt to remove state INIT from this ASM. Register r3 was introduced in section 2.2.5 to compensate for the user interface bug that exists in the ASM of section 2.2.4.

The problem in section 2.2.4 is that the assignment to r2 was written as an unconditional command in state IDLE. By using a Mealy ASM, it is possible to eliminate state INIT without destroying the contents of register r2 when the machine is waiting in state IDLE for pb to be pressed. Only when the machine is leaving state IDLE to begin computing the quotient does r2 get cleared. In other words, when the machine makes a transition from state IDLE to state COMPUTE1 is the time when r2 becomes zero. Here is the ASM:

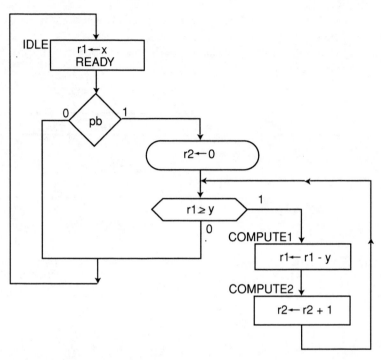

Figure 5-3. Mealy division machine with two states in loop.

Here is an example that shows the machine works when x=14 and y=7:

IDLE	r1=	?	r2=	?	pb=0	ready=1
IDLE	r1=	?	r2=	?	pb=0	ready=1
IDLE	**r1=**	**14**	**r2=**	**?**	**pb=1**	**ready=1**
COMPUTE1	r1=	14	**r2=**	**0**	pb=0	ready=0
COMPUTE2	r1=	7	r2=	0	pb=0	ready=0
COMPUTE1	r1=	7	r2=	1	pb=0	ready=0
COMPUTE2	r1=	0	r2=	1	pb=0	ready=0
IDLE	r1=	0	r2=	2	pb=0	ready=1
IDLE	r1=	?	r2=	2	pb=0	ready=1

The highlighted line shows where the conditional command to clear r2 occurs. This takes effect at the next rising edge of the clock, which is when the machine enters state COMPUTE1 (r2=0 on the next line is also highlighted to illustrate this).

Based on the assumptions used throughout all of the chapter 2 examples, the above ASM executes in 2+2*quotient clock cycles, which is one clock cycle faster than the correct ASM of section 2.2.3.

5.2.2 Merging states COMPUTE1 and COMPUTE2

The above ASM requires about twice as long as the best solution discussed in chapter 2. To achieve the same kind of speed up with the Mealy ASM, we need to do the same thing we did in chapter 2: the operations in the loop need to occur in parallel. Consider the following **incorrect** ASM:

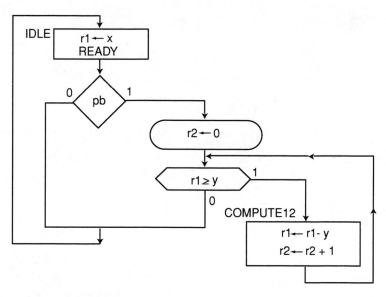

Figure 5-4. Incorrect Mealy division ASM.

To illustrate how this ASM fails, consider when x=14, and y=7:

IDLE	r1=	?	r2=	?	pb=0	ready=1
IDLE	r1=	?	r2=	?	pb=0	ready=1
IDLE	r1=	14	r2=	?	pb=1	ready=1
COMPUTE12	r1=	14	r2=	0	pb=0	ready=0
COMPUTE12	r1=	7	r2=	1	pb=0	ready=0
COMPUTE12	r1=	0	r2=	2	pb=0	ready=0
IDLE	r1=	4089	**r2=**	**3**	pb=0	ready=1
IDLE	r1=	?	r2=	3	pb=0	ready=1

By the point when the machine returns to state IDLE, r2 has been incremented one time too many. In section 2.2.5, this problem was solved by using the r3 register to save the correct quotient. However, since we are striving for a faster and cheaper solution here, it would be better to avoid introducing the r3 register in this design.

5.2.3 Conditionally loading r2

To solve the bug illustrated in section 5.2.2, we need to load r2 only when the machine stays in the loop, and to keep the old value of r2 when the machine leaves the loop to return to state IDLE. This of course requires another oval in the ASM:

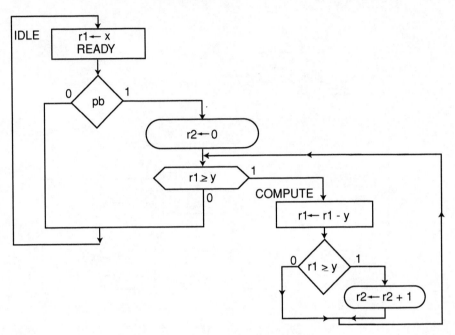

Figure 5-5. Mealy division ASM with conditional load.

To illustrate that this ASM works correctly, consider the case we looked at in the last section:

IDLE	r1=	?	r2=	?	pb=0	ready=1
IDLE	r1=	13	r2=	?	pb=0	ready=1
IDLE	r1=	14	r2=	?	pb=1	ready=1
COMPUTE	r1=	14	r2=	0	pb=0	ready=0
COMPUTE	r1=	7	r2=	1	pb=0	ready=0
COMPUTE	r1=	0	r2=	2	pb=0	ready=0
IDLE	r1=	4089	**r2=**	**2**	pb=0	ready=1
IDLE	r1=	?	r2=	2	pb=0	ready=1

This machine can achieve the correct result in 3+quotient clock cycles using only two (instead of three) registers. Therefore, it is as fast as the fastest Moore machine in chapter 2 using fewer registers.

5.2.4 Asserting READY early

The reason that the Mealy machine in section 5.2.3 is no faster than the the Moore machine in 2.2.7 is because of the assumption that the user waits at least two clock cycles while the machine asserts READY. In the Moore machines of chapter 2, asserting READY was the same as being in state IDLE, but with a Mealy machine, it would be possible to assert READY one clock cycle earlier. There are two reasons why asserting READY one clock cycle early works. First, during the last clock cycle of state COMPUTE in the ASM of section 5.2.3, r2 already contains the correct quotient, and so r2 is not scheduled to be incremented again. Second, the user is unaware of what state the machine is in and instead relies on READY to indicate the proper time when pb can be pressed again.

The following machine asserts READY in state IDLE and in the last clock cycle of state COMPUTE. When the machine is in state COMPUTE, the r1 >= y test at the bottom of the loop is evaluated at the same time as the r1 >= y test at the top of the loop. When the machine stays in the loop another time, r2 is incremented. When the machine will leave the loop, READY is asserted instead.

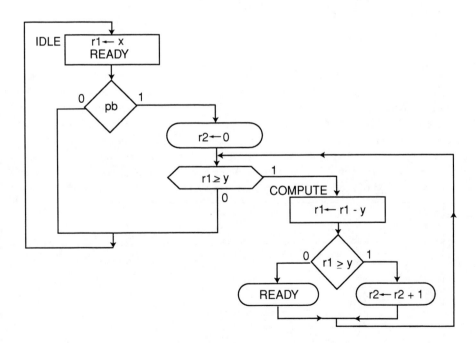

Figure 5-6. Mealy division ASM with conditional READY.

It would not be legal to attempt to assert READY after leaving the loop because READY is already asserted in state IDLE. The conditional assertion of READY as shown above is legal because it can only happen in state COMPUTE. The following illustrates how READY is asserted for the example when x=14 and y=7:

```
IDLE       r1=       ?    r2=     ?    pb=0   ready=1
IDLE       r1=      14    r2=     ?    pb=1   ready=1
COMPUTE    r1=      14    r2=     0    pb=0   ready=0
COMPUTE    r1=       7    r2=     1    pb=0   ready=0
COMPUTE    r1=       0    r2=     2    pb=0   ready=1
IDLE       r1=    4089    r2=     2    pb=0   ready=1
```

The last time in state COMPUTE, r2 is already the correct quotient, and the conditional signal READY is asserted. The user can begin counting clock cycles from this moment, rather than having to wait an extra clock cycle.

This machine can achieve the correct result in 2+quotient clock cycles using only two (instead of three) registers. Therefore, this Mealy machine is cheaper and faster than any of the Moore machines given in chapter 2.

5.3   Translating Mealy ASMs into behavioral Verilog

Pure behavioral Mealy ASMs that use only RTN can usually be translated into Verilog quite easily. Often it is as simple as using an if statement, where the first statement inside the if is not time control. (A Moore machine that has an if decision must have time control as the first statement inside the body of the if. Of course nested ifs are allowed in both Moore and Mealy machines.) The statements inside an if without intervening time control correspond to ovals in a Mealy ASM. For example, consider how the ASM of section 5.2.1 translates into Verilog:

```
always
  begin
     @(posedge sysclk) enter_new_state('IDLE);
       r1 <= @(posedge sysclk) x;
       ready = 1;
       if (pb)
         begin
           r2 <= @(posedge sysclk) 0;
           while (r1 >= y)
             begin
           @(posedge sysclk) enter_new_state('COMPUTE1);
         r1 <= @(posedge sysclk) r1 - y;
```

Continued

```
@(posedge sysclk) enter_new_state('COMPUTE2);
   r2 <= @(posedge sysclk) r2 + 1;
     end
        end
   end
```

Although there is time control later inside the if, the statement r2 <= @(posedge sysclk) 0 occurs directly after the if with no intervening time control, and so is conditional during state IDLE.

As another example, consider translating the Mealy ASM of section 5.2.3 into Verilog:

```
always
  begin
      @(posedge sysclk) enter_new_state('IDLE);
        r1 <= @(posedge sysclk) x;
        ready = 1;
        if (pb)
          begin
            r2 <= @(posedge sysclk) 0;
            while (r1 >= y)
              begin
                @(posedge sysclk) enter_new_state('COMPUTE);
                  r1 <= @(posedge sysclk) r1 - y;
                  if (r1 >= y)
                  r2 <= @(posedge sysclk) r2 + 1;
              end
          end
  end
```

The condition (r1 >= y) always produces identical results in the if and in the while because no $time passes from when it is evaluated by the if and when it is later reevaluated by the while.

As a final example, consider translating the Mealy ASM of section 5.2.4:

```
always
   begin
       @(posedge sysclk) enter_new_state('IDLE);
         r1 <= @(posedge sysclk) x;
         ready = 1;
         if (pb)
           begin
```

Continued.

```
                        r2 <= @(posedge sysclk) 0;
                  while (r1 >= y)
                     begin
                        @(posedge sysclk) enter_new_state('COMPUTE);
                        r1 <= @(posedge sysclk) r1 - y;
                        if (r1 >= y)
                           r2 <= @(posedge sysclk) r2 + 1;
                        else
                           ready = 1;
                     end
            end
  end
```

The conditional command simply translates into an `else`.

5.4   Translating complex (`goto`) ASMs into behavioral Verilog

Section 2.1.4 discusses the `goto`-less style for ASM charts, where every decision is described in terms of high-level `while`, `if` and `case` constructs. Since Verilog has statements that correspond to these constructs, it is usually straightforward to translate such an ASM chart into behavioral Verilog, regardless of whether it is a Moore or Mealy ASM.

On the other hand, because Verilog does not provide a `goto` statement, there are three situations when translating an ASM chart into Verilog is more difficult. First, translation is difficult when an ASM chart uses a bottom testing loop construct, similar to the `repeat ... until` of Pascal or `do ... while()` of C. Second, translation is difficult when an ASM chart has intervening time control before the loop exit decision (as in the ASM of section 2.2.2). Third, translation is difficult when the decision can only be described with `goto`s.

The general solution to these difficulties involves using the `present_state` variable inside `if`s and `while`s. In the behavioral Verilog model of an ASM, the `present_state` variable indicates which algorithmic state the ASM is currently performing. By testing the `present_state` inside `if`s and `while`s with the `!==` operator, it is possible to implement arbitrary (`goto`-like) decisions without needing a `goto` statement. Such tests are not part of what the hardware does. Mentioning `present_state` in an ASM chart is unnecessary since an ASM chart allows arbitrary `goto`s to any state. Such decisions are required only to overcome a limitation of Verilog, and so using `!==` (rather than `!=`) is appropriate. The need for using `!==`

comes from the fact that `present_state` may be `'bx`. The following three sections illustrate the three kinds of difficulties, and how testing `present_state` with `!==` can solve the problem.

5.4.1 Bottom testing loop

A bottom testing loop is, technically speaking, "goto-less," but such a loop is still difficult to translate because Verilog does not provide a bottom testing loop construct in the language. In essence, since such a construct does not exist, the decision at the bottom of the loop has to be thought of as a conditional `goto` that branches to the top of the loop. (In the pure structural stage, this is how the loop would be implemented by the computation of the next state in the `state_gen` function.) In the pure behavioral stage, since Verilog lacks a `goto` statement, the only choice is to describe such a loop using a `while`.

As an illustration, consider the nonsense ASM chart from section 2.1.2.1. Suppose the states are assigned the following representations:

```
'define     NUM_STATE_BITS   2
'define     GREEN     2'b00
'define     YELLOW    2'b01
'define     RED       2'b10
```

It is **incorrect** to translate the loop involving state YELLOW using just a while:

```
always
  begin
    @ (posedge sysclk) enter_new_state('GREEN);
      stop = 0;
      speed = 3;
    while(count != 0)
      begin
        @ (posedge sysclk) enter_new_state('YELLOW);
          stop = 1;
          speed = 1;
          count <= @ (posedge sysclk) count + 1;
      end
    @ (posedge sysclk) enter_new_state('RED);
      stop = 1;
      speed = 0;
      count <= @ (posedge sysclk) count + 2;
  end
```

because the `count != 0` test in the while occurs as part of both states GREEN as well YELLOW, but in the original ASM the test is part of state YELLOW only. To illustrate this problem, consider the simulation of the above while loop:

```
$time=      70   ps=00   stop=0   speed=11   count=000
$time=     170   ps=10   stop=1   speed=00   count=000
$time=     270   ps=00   stop=0   speed=11   count=010
$time=     370   ps=00   stop=1   speed=01   count=010
$time=     470   ps=01   stop=1   speed=01   count=011
$time=     570   ps=01   stop=1   speed=01   count=100
$time=     670   ps=01   stop=1   speed=01   count=101
$time=     770   ps=01   stop=1   speed=01   count=110
$time=     870   ps=01   stop=1   speed=01   count=111
$time=     970   ps=01   stop=1   speed=01   count=000
$time=    1070   ps=10   stop=1   speed=00   count=001
. . .
```

The condition in the `while` is evaluated in state GREEN at `$time` 51, which is the problem. At `$time` 170, the machine has gone directly to state RED, rather than to state YELLOW where it is supposed to be. This error occurs because the `while` loop has inserted an extra (incorrect) test whether `count != 0` in state RED. Because `count` is zero, the machine avoids state YELLOW altogether, which is not the desired behavior.

It is an unavoidable feature of Verilog that the behavioral construct that we have to implement this loop is the `while` statement. A `while` loop always tests at the top of the loop, but in hardware, as is the case here, we often want to test at the bottom of the loop. To implement the ASM of section 2.1.2.1 correctly for Verilog simulation requires nullifying the fact that this `while` loop executes both in state GREEN and in state YELLOW. To overcome this problem, the correct code ORs the original ASM condition with a `!==` test that mentions the bottom state of the loop. In this example, because there is only the state YELLOW in the loop, the `while` condition ORs `count != 0` with `present_state !== 'YELLOW`:

```
always
  begin
    @(posedge sysclk) enter_new_state('GREEN);
    speed = 3;
    while(count != 0 | present_state !== 'YELLOW)
      begin
        @(posedge sysclk) enter_new_state('YELLOW);
          stop = 1;
          speed = 1;
          count <= @(posedge sysclk) count + 1;
      end
```

Continued.

```
    @(posedge sysclk) enter_new_state('RED);
        stop = 1;
        count <= @(posedge sysclk) count + 2;
end
```

When the machine is in the bottom state of the loop (YELLOW in this example), the !== will be false, and the original ASM condition will be the only thing that decides whether the `while` loop continues. When the machine is in any other state, such as state GREEN, the `!==` will be true (even if present_state is `'bx`), and so the ORed condition is true. The loop will begin execution, regardless of the original ASM condition. It does not matter whether `count != 0`; the loop will execute at least once. The simulation of this code shows that it correctly models the behavior of the original ASM:

```
$time=        70    ps=00    stop=0    speed=11    count=000
$time=       170    ps=01    stop=1    speed=01    count=000
$time=       270    ps=10    stop=1    speed=00    count=001
$time=       370    ps=00    stop=0    speed=11    count=011
$time=       470    ps=01    stop=1    speed=01    count=011
$time=       570    ps=01    stop=1    speed=01    count=100
$time=       670    ps=01    stop=1    speed=01    count=101
$time=       770    ps=01    stop=1    speed=01    count=110
$time=       870    ps=01    stop=1    speed=01    count=111
$time=       970    ps=01    stop=1    speed=01    count=000
$time=      1070    ps=10    stop=1    speed=00    count=001
. . .
```

At $time 170, the machine has entered state YELLOW as it should. Since during that clock cycle `count` is zero, the machine proceeds to state RED.

5.4.2 Time control within a decision

An ASM chart that is translated directly from a software paradigm `while` loop (see section 2.2.2 for an example) would appear to be `goto`-less, but in fact it is not. The problem with such an ASM chart is that it must have @ time control between the algorithmic top of the loop and the place where the decision occurs. Despite the fact that Verilog has a `while` loop, the testing of the condition required by a software "`while`" loop occurs in the middle of the loop (after the @), rather than at the algorithmic top of the loop (as required by Verilog's `while` construct). The reason for this @ time control is to give the algorithm $time to make the decision before any computations occur that could effect the outcome of the decision.

Again, it is possible to overcome this kind of problem using a present_state !== test. For example, consider translating the Moore ASM chart of section 2.2.2 into behavioral Verilog:

```
module slow_div_system(pb,ready,x,y,r2,sysclk);
  input pb,x,y,sysclk;
  output ready,r2;
  wire pb;
  wire [11:0] x,y;
  reg ready;
  reg [11:0] r1,r2;
  reg ['NUM_STATE_BITS-1:0] present_state;

  always
    begin
      @(posedge sysclk) enter_new_state('IDLE);
        r1 <= @(posedge sysclk) x;
        ready = 1;
        if (pb)
          begin
            @(posedge sysclk) enter_new_state('INIT);
              r2 <= @(posedge sysclk) 0;
              while ((r1 >= y)|present_state !=='TEST)
                begin
                  @(posedge sysclk)enter_new_state('TEST);
                    if (r1 >= y)
                      begin
                        @(posedge sysclk enter_new_state('COMPUTE1);
                          r1 <= @(posedge sysclk) r1 - y;
                        @(posedge sysclk) enter_new_state('COMPUTE2);
                          r2 <= @(posedge sysclk) r2 + 1;
                      end
                end
          end
    end
task enter_new_state;
 input ['NUM_STATE_BITS-1:0] this_state;
 begin
   present_state = this_state;
   #1 ready=0;
 end
endtask
```

Continued

```
always @ (posedge sysclk) #20
  $display("%d ps=%b r1=%d r2=%d pb=%b ready=%b",
    $time, present_state,r1,r2, pb, ready);
endmodule
```

where the states are represented as:

```
'define    NUM_STATE_BITS 3
'define    IDLE        3'b000
'define    INIT        3'b001
'define    TEST        3'b010
'define    COMPUTE1    3'b011
'define    COMPUTE2    3'b100
```

The troublesome state here is state TEST. There is a Verilog while loop whose body includes state TEST and an if statement that includes the other states of the ASM loop. Three situations can occur with the Verilog while loop: It is possible that the while loop is being entered for the first time from state INIT, it is possible that the while loop is to be reexecuted from state COMPUTE2, and it is possible that the while loop is to exit from state TEST. In each of these three situations, the condition inside the Verilog while loop is evaluated. The only one of these three situations in which the Verilog loop body does not proceed to execute is when the ASM loop exits from state TEST. The other two situations (from state INIT and state COMPUTE2) are guaranteed to stay inside the Verilog while loop. Therefore, the present_state !== 'TEST condition makes sure that the next thing to execute in both of those two situations will be the algorithmic top of the Verilog while loop (state TEST).

In order to allow the Verilog while loop to exit at the identical $time that the ASM loop exits, there is a nested if inside the Verilog while loop, after state TEST. This if uses the same ASM condition ($r1 >= y$) that was also mentioned in the Verilog while loop. In the situation when this condition is false, no $time has elapsed before the Verilog while condition (($r1 >= y$) |present_state !== 'TEST) is re-evaluated. Since the present state is state TEST and ASM condition ($r1 >= y$) remains false since no $time has elasped, the Verilog while loop exits properly. Here is a simulation for $x=14$ and $y=7$:

```
9370    ps=000  r1=   13   r2=   1   pb=0   ready=1
9470    ps=000  r1=   14   r2=   1   pb=1   ready=1
9570    ps=001  r1=   14   r2=   1   pb=0   ready=0
9670    ps=010  r1=   14   r2=   0   pb=0   ready=0
9770    ps=011  r1=   14   r2=   0   pb=0   ready=0
9870    ps=100  r1=    7   r2=   0   pb=0   ready=0
9970    ps=010  r1=    7   r2=   1   pb=0   ready=0
10070   ps=011  r1=    7   r2=   1   pb=0   ready=0
10170   ps=100  r1=    0   r2=   1   pb=0   ready=0
10270   ps=010  r1=    0   r2=   2   pb=0   ready=0
10370   ps=000  r1=    0   r2=   2   pb=0   ready=1
```

5.4.3 Arbitrary gotos

It is poor style to use arbitrary gotos. Therefore such an example is not presented here. Nevertheless, regardless of how messy the ASM, some combination of ifs and whiles that use === and !== tests with the present_state can implement the ASM in Verilog.

5.5   Translating conditional command signals into Verilog

To translate a Mealy ASM, such as the one in section 5.1.1, that has conditional command signals (rather than conditional RTN), enter_new_state must include all the Mealy command signal outputs. In the example ASM of section 5.1.1, the task has four commands to initialize, each of which has a default of zero:

```
task enter_new_state;
   input ['NUM_STATE_BITS-1:0] this_state;
   begin
     present_state = this_state;
     #1 stop =0;
         speed = 0;
         stay = 0;
         leave = 0;
   end
 endtask
```

Initializing such conditional command signals is important because in many situations the Mealy command is explicitly mentioned only on certain paths through the ASM. By describing the default values for all outputs (whether they are Mealy or Moore) in enter_new_state, the behavioral Verilog will be a one-to-one mapping of the corresponding ASM chart.

The Mealy ASM chart of section 5.1.1 then can be translated into Verilog as shown below (using the !== technique described in section 5.4.1):

```
always
  begin
    @(posedge sysclk) enter_new_state('GREEN);
      speed = 3;
      while(count!=0 | present_state !== 'YELLOW)
        begin
          @(posedge sysclk) enter_new_state('YELLOW);
            stop = 1;
            speed = 1;
            count <= @(posedge sysclk) count + 1;
            if (count!=0)
                stay = 1;
          end
        leave = 1;
    @(posedge sysclk) enter_new_state('RED);
      stop = 1;
      count <= @(posedge sysclk) count + 2;
  end
```

The diamond and oval inside the loop simply translate into an if statement followed by the stay = 1 statement, with no intervening time control. Therefore, there is no time control between the return from enter_new_state and the execution of the if (and the possible consequent execution of stay=1.) Suppose that count is non-zero, which means stay becomes one at the same $time that speed and count become one. Since leave is not mentioned inside the loop, it retains its default value of zero.

On the other hand, suppose count is zero inside the loop. This means stay=1 does not execute, and so stay retains its default value (of zero) given to it by enter_new_state. No $time passes at the point where the while retests whether count !=0. Since count is zero, the while is guaranteed to exit, but still no $time has elasped. This means that the leave=1 statement executes at the same $time as the final call to enter_new_state('YELLOW) returns back to the loop body. Therefore the last cycle in which the machine is in state YELLOW will output leave as one, but stay as zero.

Since this is a correct translation of a bottom testing loop, the only way the machine exits from the while loop is from state YELLOW. (It is not possible to get directly from state GREEN to the exit of the while because of the present_state !== 'YELLOW.) Therefore, this Verilog is a one to one mapping of the ASM.

In the above Verilog, the states are represented as:

```
`define    NUM_STATE_BITS   2
`define    GREEN     2'b00
`define    YELLOW    2'b01
`define    RED       2'b10
```

and the following declarations occur at the beginning of the module:

```
reg stop;
reg [1:0] speed;
reg [2:0] count;
reg [`NUM_STATE_BITS-1:0] present_state;
reg stay,leave;
```

5.6   Single-state Mealy ASMs

As discussed in section 2.4.1, an ASM with four rectangles requires two bits to represent those states. An ASM with two rectangles requires only one bit to represent those states. What about an ASM with only one state? As explained in section 2.5, an ASM with only one rectangle represents pure combinational logic and therefore needs zero bits to represent the state. (The machine is always in that one state, and so it is not necessary to record which state the machine is in.)

The oval notation allows such an ASM to have an arbitrarily complex decision happening in that one state. For example, a decoder, whose input is a two-bit bus, inbus, and whose outputs are o0, o1, o2 and o3 can be described as:

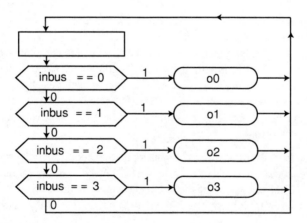

Figure 5-7. ASM for combinational logic (decoder).

Such an ASM bears a close resemblance to the way in which a combinational device is modeled in Verilog:

```
wire [1:0] inbus;
reg o0,o1,o2,o3;
always @(inbus)
  begin
    o0 = 0; o1 = 0; o2 = 0; o3 = 0;  //defaults
    case (inbus)
      0: o0 = 1;
      1: o1 = 1;
      2: o2 = 1;
      3: o3 = 1;
    endcase
  end
```

The only difference between the ASM and the Verilog is that the Verilog needs the proper time control for combinational logic, which is @ followed by a sensitivity list, rather than any mention of the system clock.

5.7 Conclusion

Moore machines have commands that occur when the machine is in a particular state. Mealy machines allow commands in a particular state to occur based on status. This chapter shows how Mealy ASMs allow a designer to express faster and better algorithms. Like Moore ASMs, Mealy ASMs have unconditional commands in rectangles. Unlike Moore ASMs, Mealy ASMs have conditional commands in ovals that follow diamonds. The conditional commands in the ovals happen at the same time as the unconditional commands in the rectangle and the decisions in the diamonds.

Translating a Mealy ASM into behavioral Verilog is usually simple, typically involving an if statement with no intervening time control. When the ASM involves command signals (rather than RTN), as would be the case at the mixed stage, the enter_new_state task must initialize the conditional commands. Some ASMs (both Moore and Mealy) cannot be expressed in the goto-less style with simple whiles and ifs. Such ASMs need to be translated into Verilog using !== tests of present_state. A common example of an ASM that must be translated into Verilog with a present_state !== test is a bottom testing loop. These techniques work only for simulation. See chapter 11 for synthesis techniques.

Single-state Mealy ASMs are a general notation to describe combinational logic in a behavioral fashion. As such, they are closely related to the behavioral Verilog description of combinational logic.

6. DESIGNING FOR SPEED AND COST

Chapter 2 uses ASM charts as a description of how to carry out each step of an algorithm in an arbitrarily chosen unit of physical time, known as the clock cycle. Chapter 3 introduces Verilog's $time variable, whose incrementation by the Verilog scheduler simulates the passage of physical time in the fabricated hardware. The example Verilog clock (section 3.7.1.3) used in chapters 3 through 5 has an arbitrarily chosen clock period of 100 units of $time. The exact amount of physical time that this 100 units of Verilog $time relates to is not specified nor is it of any concern in chapters 3 through 5. Up to this point, the emphasis has been on designing correct algorithms and implementing them properly in hardware. The only attempt to increase speed in chapters 2 through 5 is by doing more steps in parallel, thereby requiring fewer clock cycles. The number of clock cycles required by an ASM chart is a mathematical property of the algorithm and is separate from its physical implementation.

The total physical time required by a machine to compute an answer is the number of clock cycles required by the algorithm multiplied by the physical time of the clock period used in the hardware implementation. Physical clocks are often measured in frequency, rather than time. There is a reciprocal relationship between clock period and clock frequency. For example, what is the total time required to divide 14 by 7 using the machine described in section 2.2.7 when that machine is clocked at 200 MHz? According to the analysis there, the number of clock cycles required by the algorithm (including the time for two clock cycles in state IDLE) is 3+quotient=5 clock cycles. The clock period in this example is 5 ns, and so the total time is 25 ns.

In the first stages of design, algorithmic correctness and speed only in terms of clock cycles are the primary focus of the designer. The harsh physical reality of time should enter into the designer's thinking only after the design has been synthesized. This chapter shows how the $time features of a Verilog simulator allow a designer to experimentally determine the speed of a synthesized design without having to fabricate it. This chapter also illustrates three alternative design techniques that allow a trade-off between speed and cost should a synthesized design fail to meet its speed requirements.

6.1 Propagation delay

Every physical combinational logic device takes an amount of physical time, known as propagation delay, to compute its answer. Except in section 3.7.1.2, we have modeled combinational logic devices in Verilog as having no propagation delay. There are three reasons for modeling perfect (delayless) combinational logic. First, it is the only acceptable way to model combinational logic for synthesis. (The synthesis tool, rather than the designer, chooses the propagation delay of the synthesized hardware. Only after synthesis can the designer simulate the netlist to see how fast it is.) Second, the delayless style (section 3.7.2.1) is more efficient to simulate in the early stages of design, when the designer is more concerned with algorithmic correctness than with speed. Third, the delayless style is easier.

The purpose of top-down design is to defer (but not forget) as many details for as long as possible. Though it is simpler to ignore propagation delay, if we want to fabricate a practical machine, we must ultimately confront these details. Real-world machines must meet certain criteria of speed and cost. Even if a machine is algorithmically correct, if it is too slow or too expensive, it will not be practical to build. Running a machine at a higher clock frequency increases the speed of the machine, but there is a cost associated with operating at higher clock frequencies.

6.2 Factors that determine clock frequency

Synchronous devices (registers and the controller) are clocked by a clock signal of a particular frequency. If that frequency is low enough (the clock period is long enough), the machine will behave as predicted by the ASM chart and the Verilog simulations. For as long as the machine does not malfunction, increasing the clock frequency will proportionately increase the speed of the machine without having to redesign it. Unfortunately, there is a limit on how fast the clock can be. Therefore, to get the most out of the hardware, the designer wants to know what the maximum permissible clock frequency is.

The propagation delay of every combinational device and every gate in a machine potentially plays a role in determining the actual speed of the machine. There must be enough time in each clock cycle for every combinational logic path to stabilize on its correct value. A combinational device whose interconnection of gates is shallow can have its result ready to be clocked into a register faster than an equivalent device whose interconnection of gates is deep. Unfortunately, combinational devices whose internal interconnections are shallow tend to be more costly. Simulation and synthesis tools help the designer decide on a trade-off between speed and cost.

The factors that determine the maximum clock frequency include the delay of signals on each wire, the propagation delay of each gate, the way such gates and wires are interconnected to form building blocks (such as adders) and the way such building blocks are interconnected to form the problem-specific architecture and the controller. The designer does not have much influence on the first three factors. The implementation technology (not the designer or the synthesis tool) determines the delays of wires, gates and devices. (Of course, the designer can choose to use a more expensive technology to achieve higher speed. For example, by fabricating custom silicon with smaller chip dimensions, the propagation delays of each gate are correspondingly reduced.) At best, the designer can only give hints to the synthesis tool to favor either lower-cost building blocks or higher-speed building blocks. If speed is essential, the designer can manually create a netlist for a shallow (high-speed) building block, but this circumvents the advantages of the synthesis tool.

When using a synthesis tool, the only major factor to increase speed over which a designer has much control is the way building blocks are interconnected to form the architecture and the controller. The end of this chapter discusses various methods that the designer has at the architectural level to increase speed at minimum cost. But first, the next sections look at netlist level propagation delay, which is the underlying cause of this difficulty.

6.3   Example of netlist propagation delay

Putting propagation delays at the netlist level is easy. You simply instantiate the built-in gate with a parameter, which is the delay in units of $time. (This works only for built-in gates, since parameters in user-defined modules take on whatever meaning the user desires.)

Section 2.5 explains how a two-bit adder black box can be decomposed down to the gate level using hierarchical design. Sections 3.10.6 and 3.10.7 give the equivalent structural (hierarchical) Verilog modules for this adder assuming no propagation delay. As an example of modeling propagation delay, assume and and or gates have a delay of one unit of $time, and the more complicated xor gates have a delay of two units of $time. A slight change to the modules from sections 3.10.6 and 3.10.7 will provide a much more realistic model of what the actual hardware does:

```
module half_adder(c,s,a,b);
   input a,b;
   wire a,b;
   output c,s;
   wire c,s;
   xor #2 x1(s,a,b);
   and #1 a1(c,a,b);
endmodule

module full_adder(cout,s,a,b,cin);
   input a,b,cin;
   wire a,b,cin;
   output cout,s;
   wire cout,s;
   wire cout1,cout2,stemp;
   half_adder ha2(cout1,stemp,a,b);
   half_adder ha3(cout2,s,cin,stemp);
   or   #1      o1(cout,cout1,cout2);
endmodule

module adder(sum,a,b);
   input a,b;
   output sum;
   wire [1:0] a,b;
   wire [2:0] sum;
   wire c;
   half_adder ha1(c,sum[0],a[0],b[0]);
   full_adder fa1(sum[2],sum[1],a[1],b[1],c);
endmodule
```

Assuming this module is instantiated as before:

```
              adder adder1(sum,a,b);
```

the following diagram illustrates the interconnection of gates described by the above Verilog:

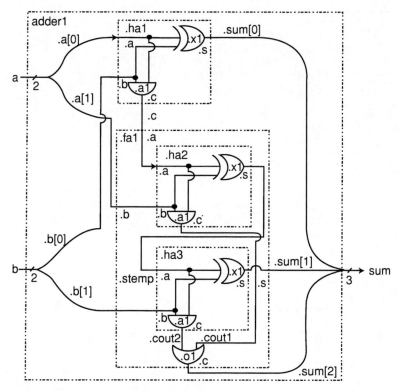

Figure 6-1. Adder with names used in structural Verilog.

6.3.1 A priori worst case timing analysis

Without using simulation, we could determine how long it takes in the worst case for the above modules to stabilize. There are many possible paths through the gates that have to be considered to determine the longest path. For example, there is a dependency of sum[2] on a[0]. A change in a[0] could cause an incorrect result for sum[2] until the effects of the change in a[0] can propagate through the path to sum[2]. The following shows where this change has to propagate, and how much $time is required:

Verilog name	total delay thus far
a[0]	0
adder1.a[0]	0
adder1.ha1.a	0
adder1.ha1.a1	1
adder1.ha1.c	1
adder1.c	1
adder1.fa1.a	1
adder1.fa1.ha2.a	1
adder1.fa1.ha2.a1	2
adder1.fa1.ha2.c	2
adder1.fa1.cout1	2
adder1.fa1.o1	3
adder1.fa1.c	3
adder1.sum[2]	3
sum[2]	3

There is also a dependency of sum[1] on a[0]. The following shows where the change in a[0] has to propagate, and how much $time is required:

Verilog name	total delay thus far
a[0]	0
adder1.a[0]	0
adder1.ha1.a	0
adder1.ha1.a1	1
adder1.ha1.c	1
adder1.c	1
adder1.fa1.a	1
adder1.fa1.ha2.a	1
adder1.fa1.ha2.x1	3
adder1.fa1.ha2.s	3
adder1.fa1.stemp	3
adder1.fa1.ha3.a	3
adder1.fa1.ha3.x1	5
adder1.fa1.ha3.s	5
adder1.sum[1]	5
sum[1]	5

There are other similar delay paths, but none of them are longer than five units of $time. Therefore, whatever code instantiates adder must wait more than five units of $time after changing a and b before using sum.

The test module in section 3.10.5 did not do this, so it would print out incorrect values for sum. If you simulate adder1 with the test code from section 3.10.5, most of the results printed out will be wrong. This is because the #1 in the test code is an inadequate amount of $time for adder1 to stabilize on the correct answer.

6.3.2 Simulation timing analysis

In circuits more complex than this trivial example, it is difficult to do a priori timing analysis, especially on a synthesized netlist. Also, since a priori analysis finds the worst case propagation delay, such analysis may be overly pessimistic. Therefore, designers often use post-synthesis simulation to obtain timing information about the design.

In this example, according to the a priori timing analysis, the test code for the adder module described in section 6.3.1 needs to wait longer than #5. If we use simulation to do the timing analysis, we would take a guess and check if the guess is an adequate amount of delay. Here we try #7:

```
module top;
```

```
   integer ia,ib;
   reg [1:0] a,b;
   wire [2:0] sum;
   reg [2:0] oldsum;

   adder adder1(sum,a,b);

   always #1
    begin
     #0 if (a+b==sum)
         $display("a=%d b=%d sum=%d CORRECT        $time=%d",
                   a,b,sum,$time);
        else
         if (sum==oldsum)
          $display("a=%d b=%d sum=%d WRONG LAG     $time=%d",
                   a,b,sum,$time);
         else
          $display("a=%d b=%d sum=%d WRONG GLITCH $time=%d",
                   a,b,sum,$time);
        oldsum = sum;
     end
   initial
```

Continued

```
  begin
    for (ia=0; ia<=3; ia = ia+1)
     begin
       a = ia;
       for (ib=0; ib<=3; ib = ib + 1)
        begin
           b = ib;
          #7 if (a + b === sum)
              $display("                      tested CORRECT");
              else
              $display("                      tested WRONG");
        end
     end
     $finish;
  end
endmodule
```

6.3.3 Hazards

In addition to the `initial` block in the test code of section 6.3.2, it is helpful to have an `always` block that monitors the change in sum at every unit of $time. At the end (#0) of each unit of $time, the `always` block checks if the current output of the adder (sum) is equal to a+b. If it is, it prints out the "CORRECT" message. If it is not, it prints out a message explaining the reason why. There are two possible explanations for a "WRONG" value of sum. The first is that the current value of sum is the same as what sum used to be at the previous unit of $time. In other words, sum is lagging behind the change in a or b. The other possible error is that sum has changed (due to the change in a or b) to an incorrect value. Such a momentary incorrect value from combinational logic with propagation delay is known as a *hazard* (also known as a *glitch*). Hazards occur when combinational logic internally has different path delays.

Here is a partial output of this simulation:

```
a=0  b=0  sum=X  WRONG  GLITCH  $time=          1
a=0  b=0  sum=X  WRONG  GLITCH  $time=          2
a=0  b=0  sum=X  WRONG  GLITCH  $time=          3
a=0  b=0  sum=0  CORRECT         $time=          4
a=0  b=0  sum=0  CORRECT         $time=          5
a=0  b=0  sum=0  CORRECT         $time=          6
                             tested CORRECT
a=0  b=1  sum=0  WRONG  LAG      $time=          7
a=0  b=1  sum=0  WRONG  LAG      $time=          8
a=0  b=1  sum=1  CORRECT         $time=          9
a=0  b=1  sum=1  CORRECT         $time=         10
a=0  b=1  sum=1  CORRECT         $time=         11
a=0  b=1  sum=1  CORRECT         $time=         12
a=0  b=1  sum=1  CORRECT         $time=         13
                             tested CORRECT
. . .
a=2  b=2  sum=3  WRONG  LAG      $time=         70
a=2  b=2  sum=3  WRONG  LAG      $time=         71
a=2  b=2  sum=6  WRONG  GLITCH  $time=         72
a=2  b=2  sum=6  WRONG  LAG      $time=         73
a=2  b=2  sum=4  CORRECT         $time=         74
a=2  b=2  sum=4  CORRECT         $time=         75
a=2  b=2  sum=4  CORRECT         $time=         76
                             tested CORRECT
. . .
```

In the above, for cases such as a=0 b=1, the output of sum simply retains its old value
until sum makes a single change to the correct value. In essence, in these cases, it is
like describing the adder with the following behavioral block:

```
module adder(sum,a,b);
parameter DELAY=1;
output sum;
input a,b;
reg [2:0] sum;
wire [1:0] a,b;
  always (a or b)
    # DELAY sum=a+b;
endmodule
```

where DELAY is an integer propagation delay. Although the above is an attractive way
of viewing propagation delay, it does not describe the more complex behavior that
occurs in other cases. For example, in the simulation of the adder given in section 6.3,
for cases such as a=2 b=3, at first (like the other cases) the output makes no change

(since the input change has not yet propagated to the output). Later, the output changes to an incorrect result that is different from the earlier value of `sum`. Finally, the output stabilizes on the correct result.

Although the a priori analysis using the circuit diagram indicates that more than #5 would always be safe, we could use simulation to see if #4 would be enough. Here is a partial output of this simulation:

```
a=0 b=0 sum=X WRONG GLITCH $time=              1
a=0 b=0 sum=X WRONG GLITCH $time=              2
a=0 b=0 sum=X WRONG GLITCH $time=              3
                              tested WRONG
a=0 b=1 sum=0 WRONG GLITCH $time=              4
a=0 b=1 sum=0 WRONG LAG    $time=              5
a=0 b=1 sum=1 CORRECT      $time=              6
a=0 b=1 sum=1 CORRECT      $time=              7
                              tested CORRECT
. . .
a=2 b=2 sum=3 WRONG LAG    $time=             40
a=2 b=2 sum=3 WRONG LAG    $time=             41
a=2 b=2 sum=6 WRONG GLITCH $time=             42
a=2 b=2 sum=6 WRONG LAG    $time=             43
                              tested WRONG
. . .
```

Although four units of `$time` was enough for the adder to stabilize in many cases, it failed to stabilize for all of them. Therefore, a longer period is required for completely correct behavior for any possible input. On the other hand, if we knew before we build our machine that the inputs would *always* be among those cases where the combinational logic stabilizes early, we could run the machine faster. For an adder, such a situation is unlikely, but for other kinds of combinational logic, we might be able to use Verilog simulation to determine that our machine can run faster than a priori worst case analysis would predict.

6.3.4 Advanced gate-level modeling

Verilog provides three additional features for modeling gate-level delays: rising/falling delays, minimum/typical/maximum delays and `specify` blocks. The first of these allows us to model the fact that, for many electronic gate technologies, the time for a gate to change its output to one is not the same as the time for the gate to change its output to zero. For example, suppose the `xor` in the half-adder of section 6.3 takes two units of `$time` when its output changes to a one, but three units of `$time` when its output changes to a zero:

```
xor #(2,3) x1(s,a,b);
```

Note that the xor in section 6.3 could have been described as:

```
xor #(2,2) x1(s,a,b);
```

The second of these advanced gate delay features allows us to model that there are certain variations in the fabrication process that affect the speed of supposedly identical gates at the time of physical manufacturing. Even though the gates are supposed to be identical, these minor variations mean some of the gates will be slower than others. The maximum speed possible would be obtained if there were no variation during fabrication. Using statistical quality control methods, manufacturers determine the typical speed expected given random variations and determine a minimum acceptable speed by discarding parts that do not obtain this speed. Many Verilog simulators allows resimulation at each of these three speeds without recompilation by specifying these three delays separated by colons. For example, the following:

```
xor #(1:2:3) x1(s,a,b);
```

indicates a minimum delay of 1, a typical delay of 2 and a maximum delay of 3. The way in which this is used depends on a particular simulator, and not all simulators implement this feature.

The third of these advanced gate delay features, known as the specify block, allows us to model delays within modules without having to indicate delays on individual gates. For example, the following module is equivalent to the one given in section 6.3:

```
module half_adder(c,s,a,b);
   input a,b;
   wire a,b;
   output c,s;
   wire c,s;

   specify
     (a >= c) = 1;
     (a >= s) = 2;
     (b >= c) = 1;
     (b >= s) = 2;
   endspecify
```

Continued

```
  xor x1(s,a,b);
  and a1(c,a,b);
endmodule
```

Not all simulators support `specify` blocks. For more information, check the documentation for your simulator, or see the book by Palnitkar mentioned at the end of this chapter.

6.4   Abstracting propagation delay

As the previous sections illustrate, once a design has been synthesized down to the gate level, Verilog can provide a fairly accurate model of propagation delay. A problem arises if one wishes to estimate propagation delay before synthesis. For a given technology, manufacturers usually publish a priori estimates of worst case propagation delays for bus-width building blocks (such as adders). We would like to be able to use such worst case estimates to simulate the propagation delay of an architecture when it is still at the mixed stage (block diagram). The problem is that the propagation delay of a physical bus-width device exhibits itself only as specific hazards (like those illustrated in section 6.3.3) that require a synthesized netlist to be simulated.

This section illustrates how Verilog can be used to model abstractly the propagation delay of a bus-width device. The correct Verilog code for doing this uses some relatively advanced features of Verilog. To motivate the need for these features, we will first consider some incorrect attempts at modeling propagation delay.

6.4.1   Inadequate models for propagation delay

The simplest Verilog code for a bus-width device that includes some notation of propagation delay is similar to the code given in 6.3.3, except that the port sizes are defined by the first parameter, and the propagation delay is defined by the second parameter:

```
module adder(s,a,b);
parameter SIZE = 1, DELAY = 0;
output s;
input a,b;
reg [SIZE-1:0] s;
wire [SIZE-1:0] a,b;
  always @(a or b)
    # DELAY s=a+b;
endmodule
```

As explained in section 6.3.3, this code is deficient because it does not model cases where there is a hazard but instead always models the error as a lag.

How should a hazard be represented abstractly? The specific value that presents itself when a hazard occurs can only be predicted from the synthesized netlist. Instead, at the abstract level, we will use ′bx to represent the hazard.

```
module adder(s,a,b);
parameter SIZE = 1, DELAY = 0;
output s;
input a,b;
reg [SIZE-1:0] s;
wire [SIZE-1:0] a,b;
  always @(a or b)
    begin
      s = 'bx;
      # DELAY s=a+b;
    end
endmodule
```

Although this is an improvement, the above still has a flaw. To see why it is deficient, consider the following design which instantiates the above adder twice:

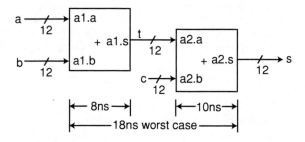

Figure 6-2. Two instantiations of adder.

The following test code gives an example how the last definition of adder gives a misleading result from simulation:

```
module test;
   reg [11:0] a,b,c;
   wire [11:0] t,s;
   adder #(12,8)   a1(t,a,b);
   adder #(12,10) a2(s,t,c);
   initial
     begin
            a = 0;
            b = 0;
            c = 0;
        #30 a = 100;
            b = 20;
            c = 3;
        #40 $finish;
     end
   always @(s)
     $display("$time=%d s=%d",$time,s);
   always @(t)
     $display("$time=%d t=%d",$time,t);
endmodule
```

This is illustrated by the simulation produced from the above test code:

```
         $time=            0 t=x
         $time=            0 s=x
         $time=            8 t=   0
         $time=           10 s=   0
         $time=           30 t=x
         $time=           30 s=x
         $time=           38 t= 120
         $time=           40 s= 123
```

The test code instantiates two adders. The first instance, a1, has a propagation delay of 8, and the second instance, a2, has a delay of 10. At $time 30, the test code causes a, b and c to change. A worst case analysis indicates that it should take 18 additional $time units ($time=48) to produce the sum of 100+20+3; however the simulation shows the correct sum in only 10 $time units ($time=40).

This flaw exists because the always block for a2 is still delaying (#10) when the change in t (also known as a2.a) occurs at $time=38. Rather than delaying an additional 10 units of $time from $time=38, Verilog simply returns to the @ time control at $time 40. Since a2.a is stable at $time 40, the algorithmically correct answer 123 is available too soon. This Verilog model allows us to conclude that the machine could run faster than is physically possible.

6.4.2 Event variables in Verilog

Verilog provides special variables, known as event variables, that are helpful in overcoming the problems shown in the last section. Variables declared as events are used only in two places: First, in a triggering statement:

```
event e;
...
->e;
```

and, second, in time control:

```
always @ e
  ...
```

Note: There are no parentheses around the variable in the @ time control. The -> triggers the corresponding @ to be scheduled. For example, the following prints "10" and "30":

```
event e;
 initial
   begin
     #10;
     -> e;
     #20
     -> e;
   end
always @ e
   $display($time);
```

Here is an example of how an event could be used to model the adder:

```
module adder(s,a,b);
  parameter SIZE = 1, DELAY = 0;
  output s;
  input a,b;
  reg [SIZE-1:0] s;
  wire [SIZE-1:0] a,b;
  event change;

  always @(a or b)
    start_change;

  always @change
    # DELAY s=a+b;

  task start_change;
    begin
      s = 'bx;
      -> change;
    end
  endtask
endmodule
```

Unfortunately, the above produces the same incorrect model of the adder (the correct result 123 is available too soon) for the same reasons discussed in the previous section.

6.4.3   The `disable` statement

This statement causes Verilog to cease execution of a labeled block. For `always` blocks, it causes them to restart at the top. (It has another use, explained in section 7.5.) An optional label is given after a colon on a `begin` statement. Here is how the `disable` statement overcomes the problem shown in the previous section:

```
module adder(s,a,b);
  parameter SIZE = 1, DELAY = 0;
  output s;
  input a,b;
  reg [SIZE-1:0] s;
  wire [SIZE-1:0] a,b;
  event change;

  always @(a or b)
        start_change;
```

Continued

```
    always @change
        begin : change_block
          # DELAY s=a+b;
        end

    task start_change;
        begin
          s = 'bx;
          disable change_block;
          #0;
          -> change;
        end
      endtask
endmodule
```

The task start_change can be called many times from the first always block without $time advancing. This way, every change in the inputs will be noticed by the Verilog scheduler. The only # control in start_change is #0. This is required so the disable statement can take effect. After change_block has been disabled, the change event is retriggered. This, in turn, causes the full # DELAY before the output changes from 'bx.

The #DELAY is in the block (change_block) which can be disabled. There is no way that changes that occur in the middle of a #DELAY will be missed. Therefore, instantiating a series of these adders will produce a correct model of the propagation delay. For example, here is the simulation using the same test code as section 6.4.1:

```
                    $time=          0  t=x
                    $time=          0  s=x
                    $time=          8  t=    0
                    $time=         18  s=    0
                    $time=         30  t=x
                    $time=         30  s=x
                    $time=         38  t=  120
                    $time=         48  s=  123
```

Note that s is 'bx from $time 30 until $time 48, as is predicted by worst case timing analysis.

6.4.4 A clock with a PERIOD parameter

The reason we wish to simulate propagation delays is ultimately to design faster machines that are clocked at higher frequencies. Therefore, it is desirable to have a clock with PERIOD as a parameter:

```
module cl(clk);
    parameter TIME_LIMIT = 110000,
              PERIOD = 100;
    output clk;
    reg clk;
    initial
      clk = 0;
    always
      begin
        #(PERIOD/2) clk = ~clk;
        #(PERIOD-PERIOD/2) clk = ~clk;
      end
    always @(posedge clk)
      if ($time > TIME_LIMIT) #(PERIOD-1) $stop;
endmodule
```

Note that if PERIOD is omitted in the instantiation, it will default to 100, as has been the situation in all earlier simulations.

6.4.5 Propagation delay in the division machine

Suppose the propagation delays are 70 for the ALU, 25 for the mux, 20 for the comparator and 10 for the inverter. To backannotate this in the original code of section 4.4.5, simply include the propagation delay parameter with the instantiation:

```
module slow_div_arch(aluctrl,muxctrl,ldr1,clrr2,
                     incr2,ldr3,r1gey,x,y,r3bus,sysclk);
  input aluctrl,muxctrl,ldr1,clrr2,incr2,ldr3,x,y,sysclk;
  output r1gey,r3bus;
  wire [5:0] aluctrl;
  wire muxctrl,ldr1,clrr2,incr2,ldr3,r1gey,sysclk;
  wire [11:0] x,y,r3bus;
  wire [11:0] muxbus,alubus,r1bus,r2bus;

  enabled_register #12   r1(alubus,r1bus,ldr1,sysclk);
  mux2             #(12,25)  mx(x,y,muxctrl,muxbus);
  alu181           #(12,70)  alu(r1bus,muxbus,aluctrl[5:2],
                               aluctrl[1],aluctrl[0],,alubus,);
  comparator       #(12,20)  cmp(r1lty,,,r1bus,y);
  not              #10       inv(r1gey,r1lty);
```

```
    counter_register #12   r2(,r2bus,,1'b0,incr2,clrr2,sysclk);
    enabled_register #12   r3(r2bus, r3bus,ldr3,sysclk);
    ...
endmodule
```

When this is simulated with a clock period of 100, it works:

```
cl #(20000,100) clock(sysclk);
slow_div_system   slow_div_machine(pb,ready,x,y,
            quotient,reset,sysclk);
```

as is illustrated by the following timing diagram:

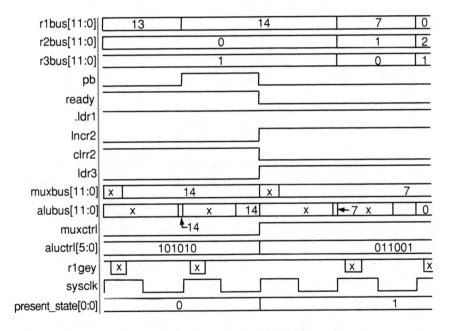

Figure 6-3. Timing diagram for division machine with abstract propagation delay.

A designer might want to experiment to see if the clock can be speeded up to have a period of, say, 90:

```
cl #(20000,90) clock(sysclk);
slow_div_system  slow_div_machine(pb,ready,
    x,y,quotient,reset,sysclk);
```

The test code will detect an error:

```
    6365 r1=x r2=   2 r3=   1 pb=0 ready=1
            1    10       0   muxbus=x alubus=x  x= 14 r1gey=x
                             muxctrl=0   aluctrl=101010
error x=  14 y=   7 x/y=   2 quotient=   1
```

because the ALU will not have had a chance to stabilize by the time of the rising edge
of the clock.

6.5   Single cycle, multi-cycle and pipeline

The solution to many problems involves performing the same kind of computation on
large amounts of independent data values. The term *independent* means that the result
of doing computation on one data value does not affect nor is affected by doing the
computation on any of the other data values. For example, three-dimensional computer
graphics (such as occur in virtual reality systems) usually require evaluating the same
formulae at millions of points. The order in which such points are processed does not
matter. You will get the same answer if you start processing on the lower-left-hand side
of the screen as you will get if you start processing on the upper-right-hand of the
screen, or if you process in any other order that you might choose. In problems like this
that have complete data independence, many possible hardware solutions exist. All
such hardware solutions are correct in that they all eventually arrive at the desired
result. These hardware solutions differ in terms of their speed and cost.

If cost were not a constraint, problems with totally independent data values could be
solved by building one combinational logic machine for each data value to be processed. Each such machine could compute its answer in parallel to all the other machines. Although this kind of massively parallel approach is sometimes used, it is not
practical in many situations due to cost constraints.

Because practical problems with perfectly independent data are commonplace where
cost is as or more important than speed, three standard techniques have been developed
that allow the designer to choose the trade-off between speed and cost. These three
techniques are known as the single-cycle, pipelined, and multi-cycle approaches. What
these three techniques share in common is that no more than one complete result is
produced per clock cycle.

At one extreme is the single-cycle approach. With the *single-cycle* approach, the result for one of the independent values is completely computed (start to finish) in a single clock cycle. The single-cycle approach is perhaps the most natural approach to think about, but it is usually not the most efficient.

At the opposite extreme is the multi-cycle approach. With the *multi-cycle* approach, the result for one of the independent values requires several clock cycles to be computed. Thinking about the multi-cycle approach is quite analogous to thinking about the software paradigm (section 2.2.2). The multi-cycle approach is usually the slowest, but it often requires minimum hardware because it can be implemented with a central ALU (section 2.3.4).

In between the single-cycle approach and the multi-cycle approach is the pipelined approach. The *pipelined* approach usually requires more hardware than the other approaches but often is the fastest and most efficient. In order to understand the pipelined approach, it is necessary to investigate the two other approaches first.

As discussed earlier in this chapter, the total time required by a machine is the number of clock cycles multiplied by the clock period. The three approaches discussed in this section differ both in terms of the number of clock cycles required and the clock period. We can understand the algorithmic distinctions among these three approaches at the behavioral stage and even predict the number of clock cycles required at the behavioral stage; however, we cannot predict which approach will be fastest at the behavioral stage. This is because the clock period is determined by the propagation delay in the architecture, which we cannot predict until the mixed stage, or when the hardware has been synthesized.

6.5.1   Quadratic polynomial evaluator example

The quadratic polynomial `a*x*x + b*x + c` is a simple example of a formula that a machine might evaluate many times with different values of x, but the same values of a, b and c (which remain unchanged for a suitable period before, during and after the quadratic evaluations). For each unique x value, the computation of the quadratic formula is independent of the computation for other values of x. Although the formulae used for practical problems, such as computer graphics, are more complex than this familiar old quadratic, the nature of the formulae in such practical problems is very similar to this quadratic.

Although a practical machine would probably store the x values in a synchronous memory, for the sake of simplicity in this example, assume the values of x are contained in a ROM. The goal of the machine is to evaluate the quadratic polynomial for each of these x values and store the corresponding y values into a synchronous memory (see section 8.2.2.3.1 for how a synchronous memory can be implemented). The address where each y value is stored should be the same as the address in the ROM from which the corresponding x value was fetched.

Suppose a is 1, b is 2 and c is 3. If the contents of x are as shown on the left in decimal, when the machine is done, the contents of the y memory will be as shown on the right:

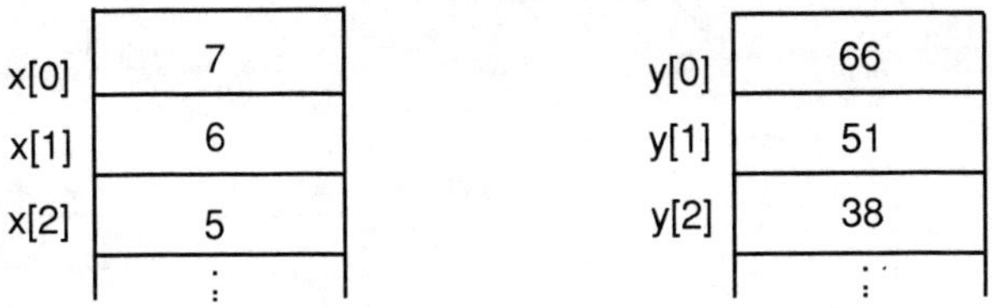

x[0]	7
x[1]	6
x[2]	5
	⋮

y[0]	66
y[1]	51
y[2]	38
	⋮

Figure 6-4. Example contents of x and y for quadratic machine.

For example, $y[2] = a*x[2]*x[2] + b*x[2] + c = 25+10+3 = 38$.

The machine will use a push button interface, similar to the one described in section 2.2.1. The machine will wait in state IDLE until pb is pressed. While the machine is in state IDLE, it leaves the contents of y alone. Some time after the machine leaves state IDLE it will begin to fill y with the correct results.

The following sections will look at behavioral ASMs to illustrate how this machine can be implemented with the single-cycle, multi-cycle and pipelined approaches. Several incorrect versions of pipelining will be presented before the final correct pipelined solution is shown in section 6.5.7.

6.5.2   Behavioral single cycle

The ASM chart for the single-cycle approach is quite simple and obvious. The machine only needs to have a memory address (ma) register. The ma register in the single- cycle approach provides the address used by both the x and y during each clock cycle. In each clock cycle, the content of the ROM is fetched from the address indicated by ma, the quadratic is evaluated using the value of x fetched from the ROM and the result is stored into y at the same address indicated by ma.

Suppose the maximum memory address is the constant MAXMA. The first of the following two ASM charts (figure 6-5) describes the single-cycle approach in the simplest possible fashion. The computation of the quadratic actually involves several multiplications and additions. The second (equivalent) ASM chart (figure 6-6) makes this clearer:

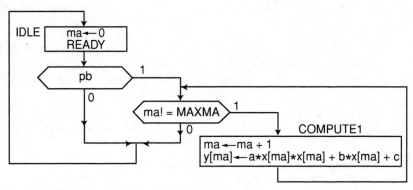

Figure 6-5. Behavioral single cycle ASM with only ←.

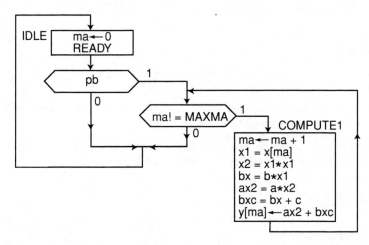

Figure 6-6. Equivalent single cycle ASM with = for combinational logic.

Note the use of = rather than ← for the intermediate results (x1, x2, bx, ax2 and bxc). As discussed in chapter 2, the = means that combinational logic computes all of these values in one clock cycle. Note that x2 and bx are dependent on x1; ax2 is dependent on x2; and bxc is dependent on bx. This means that the minimum clock period for the single-cycle approach must allow enough time for the computations of all of these intermediate results to stabilize. The amount of time it takes for the combinational logic to finish computing these intermediate values is not something we can predict at the behavioral stage. Up until this chapter, we have neglected such propagation delays, but later in this chapter, we will estimate what these delays will be.

Since we do not know how fast the machine can be clocked, let us assume that the clock period is 100 units of Verilog $time for the purpose of the following and later simulations. Also, for reasons to be explained later, we will assume that each word in y is initialized to 'bz prior to execution of this ASM. In the following partial simulation output, the $time and registers are printed on one line with the contents of y on the following line:

```
 349 ps=000 ma= 0 x1=x    x2=x  bx=x   ax2=x  bxc=x
       z    z    z    z    z    z    z    z
 449 ps=001 ma= 0 x1=   7 x2=49 bx=14 ax2=49 bxc=17
       z    z    z    z    z    z    z    z
 549 ps=001 ma= 1 x1=   6 x2=36 bx=12 ax2=36 bxc=15
      66    z    z    z    z    z    z    z
  . . .
1149 ps=001 ma= 7 x1=   0 x2= 0 bx= 0 ax2= 0 bxc= 3
      66   51   38   27   18   11    6    z
1249 ps=000 ma= 8 x1=x    x2=x  bx=x   ax2=x  bxc=x
      66   51   38   27   18   11    6    3
```

In state COMPUTE1 at $time 449, the machine obtains the value (7) of x1 from the ROM. After getting this from the ROM, but during the same clock cycle, the machine computes the square (49), and the product (49) of a and the square. Also in this clock cycle, the machine computes the product (bx=14) of b and x1. After computing bx, but before the end of the clock cycle, the machine computes the sum (bxc=17) and the sum (ax2+bxc). This final result (66) is scheduled to be stored in the memory. This value appears at the correct place (y[0]) by $time 549.

The number of clock periods required for this single cycle ASM to complete is MAXMA+1, because one result is produced each clock cycle.

Here is the behavioral Verilog code used to produce the above simulation:

```
`define NUM_STATE_BITS 3
`define IDLE          3'b000
`define COMPUTE1      3'b001
module poly_system(pb,a,b,c,ready,sysclk);
 input pb,a,b,c,sysclk;
 output ready;
 wire pb,sysclk;
 wire [11:0] a,b,c;
 reg [11:0] x['MAXMA:0],y['MAXMA:0];
 reg ready;
 reg [11:0] ma;
```

Continued

```
reg [11:0] x1,x2,bx,ax2,bxc;
reg ['NUM_STATE_BITS-1:0] present_state;
integer i;
initial
 begin
  for (i=0;i<='MAXMA;i=i+1)
   begin
    x[i]='MAXMA-i;
    y[i]='bz;
   end
 end

always
 begin
  @(posedge sysclk) enter_new_state('IDLE);
   ma <= @(posedge sysclk) 0;
   ready = 1;
   if (pb)
    begin
     while (ma != 'MAXMA)
      begin
       @(posedge sysclk) enter_new_state('COMPUTE1);
        ma <= @(posedge sysclk) ma + 1;
        x1 = x[ma];
        x2 = x1*x1;
        bx = b*x1;
        ax2 = a*x2;
        bxc = bx + c;
        y[ma] <= @(negedge sysclk) ax2 + bxc;
      end
    end
 end
endmodule
```

Note that the order of the intermediate computations (=) matters in Verilog.

The non-blocking assignment to the memory location, `y[ma]`, uses `@(negedge sysclk)` rather than the `@(posedge sysclk)` typical for non-blocking assignment to ordinary registers (see section 3.8.2). The problem here arises because new values are stored into *distinct* elements of y during *every clock cycle*. Some simulators will do the proper thing in a situation like this even if you were to use:

```
ma  <=  @(posedge sysclk)  ma + 1;

. . .

 y[ma]  <=  @(posedge sysclk) ax2 + bxc;  // not portable
```

However, @(negedge sysclk) is necessary to produce the correct result on other Verilog simulators. To understand why, remember that there are two separate concepts in Verilog: sequence and $time. A non-blocking assignment by itself will cause the new value to be stored as the last event at the specified $time. All Verilog simulators save the right-hand expressions (ax2 + bxc and ma + 1 in this example) until the specified clock edge. For left-hand values that are not arrays, this is sufficient to model the behavior of synchronous registers. But when you are dealing with a synchronous memory (y[ma]), a Verilog simulator must also save the address (ma) for the left-hand memory until the next clock edge. The sequence in which a simulator will update ma and y[ma] is not defined. The Verilog standard is ambiguous on this issue, and some vendors, for reasons of efficiency, have chosen not to save the address.

To overcome this problem at the pure behavioral stage, we need to remember that the statements in a particular state will execute one unit of $time after the rising edge. The falling edge of the system clock will occur prior to the next rising edge, but after all non-blocking assignments of this state have been scheduled. Therefore, ma will still have the correct value, but changing y[ma] at the falling edge will not disturb some dependent computations[1] within this state.

This is another illustration of choosing an appropriate level of abstraction (see section C.1). In the pure behavioral stage, we are only interested in the values of registers and memories at the moment of the rising edge. The negedge memory approach simulates the values at the rising edge properly on *all* Verilog simulators. How a particular simulator arranges to simulate between the rising edges is irrelevant as the designer explores different algorithmic possibilities. Of course, in the actual architecture, a synchronous memory will change its values only at the rising edge, *synchronously* to the ordinary registers in the design. At the mixed or later stages of design, such details are significant to determine the proper clock frequency, but since the mixed stage removes all <=, there is no problem. Before the mixed stage, the designer should not worry about clock frequency; thus the negedge memory approach is perfectly acceptable.

---

[1] There are no dependent computations in this example.

6.5.3 Behavioral multi-cycle

The single-cycle approach discussed in the previous section does all of the intermediate computations in one clock cycle. The multi-cycle approach, on the other hand, does each intermediate computation in a separate clock cycle. This of course means it takes several clock cycles to produce one result. In the multi-cycle approach, each intermediate result is stored in a register, thus the ASM uses the ← notation. The multi-cycle approach is like the software paradigm (section 2.2.2), where each intermediate computation occurs in a separate rectangle by itself. Here is an ASM chart for the multi-cycle version of the quadratic evaluator:

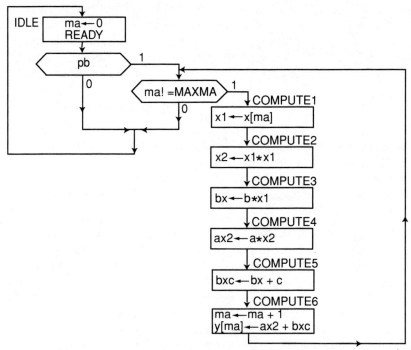

Figure 6-7. Behavioral multi-cycle ASM.

This machine has six registers (`ma`, `x1`, `x2`, `bx`, `ax2` and `bxc`), and has six states inside the loop. Here is a partial simulation, again assuming a clock period of 100 (which may be much longer than is actually required):

```
 349 ps=000 ma= 0 x1=x   x2=x   bx=x   ax2=x   bxc=x
        z    z    z    z    z    z    z    z
 449 ps=001 ma= 0 x1=x   x2=x   bx=x   ax2=x   bxc=x
        z    z    z    z    z    z    z    z
 549 ps=010 ma= 0 x1= 7 x2=x   bx=x   ax2=x   bxc=x
        z    z    z    z    z    z    z    z
 649 ps=011 ma= 0 x1= 7 x2=49 bx=x   ax2=x   bxc=x
        z    z    z    z    z    z    z    z
 749 ps=100 ma= 0 x1= 7 x2=49 bx=14 ax2=x   bxc=x
        z    z    z    z    z    z    z    z
 849 ps=101 ma= 0 x1= 7 x2=49 bx=14 ax2=49 bxc=x
        z    z    z    z    z    z    z    z
 949 ps=110 ma= 0 x1= 7 x2=49 bx=14 ax2=49 bxc=17
        z    z    z    z    z    z    z    z
1049 ps=001 ma= 1 x1= 7 x2=49 bx=14 ax2=49 bxc=17
       66    z    z    z    z    z    z    z

 . . .
5249 ps=000 ma= 8 x1= 0 x2= 0 bx= 0 ax2= 0 bxc= 3
       66   51   38   27   18   11    6    3
```

In state COMPUTE1 around $time 449, x1 is scheduled to be loaded with
x[ma]=x[0]=7. This change in x1 shows up by $time 549. In state COMPUTE2
around $time 549, the square (49) of this value is scheduled to be loaded into x2,
which shows up by $time 649. In state COMPUTE3 around $time 649, the product
(14) of x1 and b is scheduled to be loaded into bx, which shows up by $time 749. In
state COMPUTE4 around $time 749, the product (49) of a and x2 is scheduled to be
loaded into ax2, which shows up by $time 849. In state COMPUTE5 around $time
849, the sum of bx and c (17) is scheduled to be loaded into bxc, which also shows up
by $time 949. Finally, in state COMPUTE6 around $time 949, the sum of ax2 and
bxc is scheduled to be loaded into y, and this shows up by $time 1049.

The number of clock periods required for this multi-cycle ASM to complete is
6*(MAXMA+1). Although at first glance, this appears much slower than the single-
cycle approach of section 6.5.2, it need not be that much slower. Later we will be able
to predict the propagation delay of the architecture for the multi-cycle approach (which
determines the maximum clock frequency). Since there is less computation being done
in each clock cycle, it should be possible to clock the multi-cycle machine faster than
the single-cycle machine. The relative performance of these two machines is something we can only predict given a structural architecture.

Here is the behavioral Verilog code used to produce the above simulation:

```
`define NUM_STATE_BITS 3
`define IDLE          3'b000
`define COMPUTE1      3'b001
`define COMPUTE2      3'b010
`define COMPUTE3      3'b011
`define COMPUTE4      3'b100
`define COMPUTE5      3'b101
`define COMPUTE6      3'b110
...
 always
  begin
   @(posedge sysclk) enter_new_state(`IDLE);
    ma <= @(posedge sysclk) 0;
    ready = 1;
    if (pb)
     begin
      while (ma != `MAXMA)
       begin
        @(posedge sysclk) enter_new_state(`COMPUTE1);
         x1 <= @(posedge sysclk) x[ma];
        @(posedge sysclk) enter_new_state(`COMPUTE2);
         x2  <= @(posedge sysclk) x1*x1;
        @(posedge sysclk) enter_new_state(`COMPUTE3);
         bx  <= @(posedge sysclk) b*x1;
        @(posedge sysclk) enter_new_state(`COMPUTE4);
         ax2 <= @(posedge sysclk) a*x2;
        @(posedge sysclk) enter_new_state(`COMPUTE5);
         bxc <= @(posedge sysclk) bx + c;
        @(posedge sysclk) enter_new_state(`COMPUTE6);
         ma <= @(posedge sysclk) ma + 1;
         y[ma] <= @(negedge sysclk) ax2 + bxc;
       end
     end
  end
```

6.5.4 First attempt at pipelining

The single-cycle approach puts all the computation steps into one clock cycle but uses
= (corresponding only to combinational logic) for the intermediate results. The multi-
cycle approach spreads the computation steps across separate clock cycles, but uses ←
(corresponding to registers) for the intermediate results. The pipelined approach is half-
way between these two approaches.

The pipelined approach puts all the computation steps into one clock cycle and uses ←
(corresponding to registers) for the intermediate results. This means, unlike the other
two approaches, each intermediate computation in the pipelined approach occurs in
parallel to the other intermediate computations. The only reason that the intermediate
computations can occur in parallel is that a machine like this is processing a large
amount of independent data in an identical fashion.

A pipelined machine is very much like a factory assembly line. Factories are efficient
because they mass produce many copies of an identical item. At each point in time,
each worker in the factory does one thing to a partially assembled item on the production line. For example:

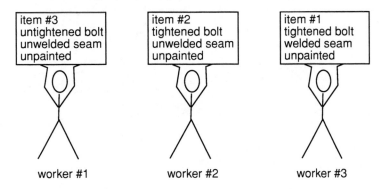

Figure 6-8. Analogy to factory.

Worker #1 might tighten a bolt, worker #2 might weld a seam and worker #3 might
paint the item. Each worker acts in parallel to the other workers. In the above picture,
worker #1 is tightening the bolt on item #3 at the same time that worker #2 is welding
item #2 (which already has its bolt tightened) and that worker #3 is painting item #1
(which has it bolt tightened and which has been welded). Each item has experienced
the correct sequence in order (tightening, welding and painting), but the tightening,
welding and painting that happens at any given instance occurs to independent items.

With this analogy in mind, we can understand that each of the intermediate computations (←) in the pipelined quadratic evaluator produces an intermediate result derived
from a different x value. Here is a first (somewhat flawed) attempt to describe the
factory-like operation of this pipelined system:

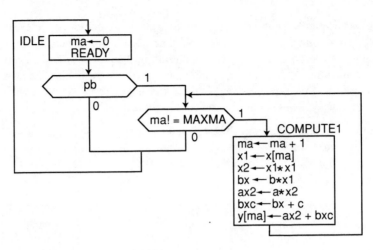

Figure 6-9. Incorrect pipelined ASM.

Here is a simulation showing how this ASM malfunctions:

```
 349 ps=000 ma= 0 x1=x   x2=x   bx=x      ax2=x      bxc=x
       z    z    z    z    z    z    z    z
 449 ps=001 ma= 0 x1=x   x2=x   bx=x      ax2=x      bxc=x
       z    z    z    z    z    z    z    z
 549 ps=001 ma= 1 x1= 7 x2=X   bx=4094 ax2=4095 bxc=x
       x    z    z    z    z    z    z    z
 649 ps=001 ma= 2 x1= 6 x2=49 bx=   14 ax2=   1 bxc= 1
       x    x    z    z    z    z    z    z
 749 ps=001 ma= 3 x1= 5 x2=36 bx=   12 ax2=  49 bxc=17
       x    x    2    z    z    z    z    z
 849 ps=001 ma= 4 x1= 4 x2=25 bx=   10 ax2=  36 bxc=15
       x    x    2   66    z    z    z    z
 949 ps=001 ma= 5 x1= 3 x2=16 bx=    8 ax2=  25 bxc=13
       x    x    2   66   51    z    z    z
1049 ps=001 ma= 6 x1= 2 x2= 9 bx=    6 ax2=  16 bxc=11
       x    x    2   66   51   38    z    z
1149 ps=001 ma= 7 x1= 1 x2= 4 bx=    4 ax2=   9 bxc= 9
       x    x    2   66   51   38   27    z
1249 ps=000 ma= 8 x1= 0 x2= 1 bx=    2 ax2=   4 bxc= 7
       x    x    2   66   51   38   27   18
```

There are several problems with this, but before discussing what is wrong, let us consider what is almost right. In state COMPUTE1 around $time 449, x1 is scheduled to be loaded with x[ma]=x[0]=7. This change in x1 shows up by $time 549. Around $time 549, the square (49) of this value is scheduled to be loaded into x2, which shows up by $time 649. In parallel, the product (14) of x1 and b is scheduled to be loaded into bx, which also shows up by $time 649. Around $time 649, the product (49) of a and x2 is scheduled to be loaded into ax2, which shows up by $time 749. In parallel, the sum of bx and c (17) is scheduled to be loaded into bxc, which also shows up by $time 749. Finally, around $time 749, the sum of ax2 and bxc is scheduled to be loaded into y, and this shows up by $time 849.

The problem at $time 849 is that although 66 is the correct value, it shows up at the wrong address for y. This is because ma has necessarily been incremented in each of the clock cycles. This machine has forgotten that the 66 is supposed to stored at y[0]. Instead it stores it at y[3]. Aside from the fact that the addresses to y are offset by three, the machine continues to compute a correct result each clock cycle. By $time 949, 51 is stored into y. By $time 1049, 38 is stored into y, and so forth.

In addition to storing the correct results at the wrong addresses, this machine also has another flaw. It does not finish the complete job (storing 11, 6 and 3). The intermediate results needed to produce 11, 6 and 3 are left frozen in the pipeline when the machine returns to state IDLE.

Another, less obvious, flaw is that garbage values ('bx, 'bx and 2) are stored into the memory during the first clock cycles (449, 549 and 649). Initializing y to 'bz rather than 'bx to highlights this flaw.

6.5.5   Pipelining the ma

The major problem with the ASM chart of section 6.5.4 is that the memory address used to store into y does not correspond to the value being stored. To overcome this problem, we can introduce three additional registers, ma1, ma2 and ma3, that will save the memory addresses from the previous three clock cycles. In a given clock cycle, ma1 is the value of ma one clock cycle ago, ma2 is the value of ma two clock cycles ago, and ma3 is the value of ma three clock cycles ago.

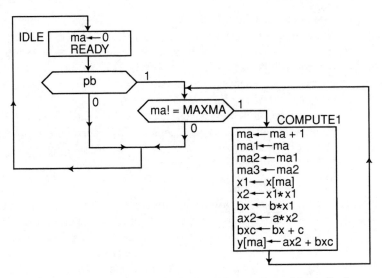

Figure 6-10. Pipelined ASM with multiple addresses but without flush.

Here is a simulation that shows how addresses flow through the ma pipeline:

```
349 ps=000 ma=    0 x1=x x2=x bx=x
           ma1=    0 ax2=x bxc=x ma2=    0 ma3=    0
     z    z    z    z    z    z    z    z
449 ps=001 ma=    0 x1=x x2=x bx=x
           ma1=    0 ax2=x bxc=x ma2=    0 ma3=    0
     z    z    z    z    z    z    z    z
549 ps=001 ma=    1 x1=    7 x2=X bx=4094
           ma1=    0 ax2=4095 bxc=x ma2=    0 ma3=    0
     x    z    z    z    z    z    z    z
649 ps=001 ma=    2 x1=    6 x2=   49 bx=   14
           ma1=    1 ax2=    1 bxc=    1 ma2=    0 ma3=    0
     x    z    z    z    z    z    z    z
749 ps=001 ma=    3 x1=    5 x2=   36 bx=   12
           ma1=    2 ax2=   49 bxc=   17 ma2=    1 ma3=    0
     2    z    z    z    z    z    z    z
849 ps=001 ma=    4 x1=    4 x2=   25 bx=   10
           ma1=    3 ax2=   36 bxc=   15 ma2=    2 ma3=    1
     66    z    z    z    z    z    z    z
949 ps=001 ma=    5 x1=    3 x2=   16 bx=    8
           ma1=    4 ax2=   25 bxc=   13 ma2=    3 ma3=    2
     66   51    z    z    z    z    z    z
```

Continued.

```
1049 ps=001 ma=    6 x1=    2 x2=    9 bx=    6
            ma1=    5 ax2=   16 bxc=   11 ma2=    4 ma3=    3
      66   51   38   z    z    z    z    z
1149 ps=001 ma=    7 x1=    1 x2=    4 bx=    4
            ma1=    6 ax2=    9 bxc=    9 ma2=    5 ma3=    4
      66   51   38   27   z    z    z    z
1249 ps=000 ma=    8 x1=    0 x2=    1 bx=    2
            ma1=    7 ax2=    4 bxc=    7 ma2=    6 ma3=    5
      66   51   38   27   18   z    z    z
```

Although the addition of `ma1`, `ma2` and `ma3` solves the addressing problem, the revised ASM still does not finish the complete job (storing 11, 6, and 3). As was the case in the ASM of section 6.5.4, the intermediate results needed to produce 11, 6 and 3 are left frozen in the pipeline when the machine returns to state IDLE. Also, garbage values (`'bx`, `'bx` and 2) are still stored into the memory during the first clock cycles, although now they are stored in `y[0]` each time.

6.5.6 Flushing the pipeline

In order to prevent the final values from being frozen in the pipeline, there need to be some additional clock cycles spent "flushing" those values out of the pipeline. Returning to the factory analogy, when the factory is about to cease production of a particular model item, worker #1 can stop work earliest, but the other workers must finish their tasks on the last item worker #1 tightened. Similarly, worker #2 can stop before worker #3. So it is with flushing the pipeline.

The ASM needs three states, FLUSH1, FLUSH2 and FLUSH3, that perform the required computations on the valid data in the pipeline. For those registers that have valid data, the computations are identical to those in state COMPUTE1. At each successive flushing state, there are fewer registers in the pipeline that contain valid data; thus each successive state has fewer computations to perform.

6.5.7 Filling the pipeline

The reason that garbage values have been stored by all the previous pipeline attempts is because of the assignment to `y[ma3]` in state COMPUTE1. During the first clock cycles when state COMPUTE1 executes, `ma3`, `ax2` and `bxc` do not have legitimate values. Therefore, to store `ax2+bxc` at address `ma3` is illegitimate. The situation is the opposite problem from flushing the pipeline. In this case, the pipeline must be filled prior to storing the first result in y. The following is a completely correct pipelined ASM that accomplishes this operation by introducing states FILL1, FILL2 and FILL3:

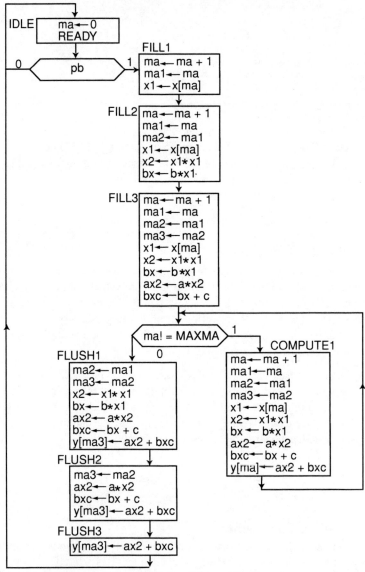

Figure 6-11. Correct pipelined ASM that fills and flushes.

Here is a simulation that shows the proper values filling the pipeline:

```
349 ps=000 ma=    0 x1=x x2=x bx=x
            ma1=    0 ax2=x bxc=x ma2=    0 ma3=    0
      z   z   z   z   z   z   z   z
449 ps=010 ma=    0 x1=x x2=x bx=x
            ma1=    0 ax2=x bxc=x ma2=    0 ma3=    0
      z   z   z   z   z   z   z   z
549 ps=011 ma=    1 x1=    7 x2=x bx=x
            ma1=    0 ax2=x bxc=x ma2= 0 ma3= 0
      z   z   z   z   z   z   z
649 ps=100 ma=    2 x1=    6 x2=49 bx=14
            ma1=    1 ax2=x bxc=x ma2= 0 ma3= 0
      z   z   z   z   z   z   z   z
749 ps=001 ma=    3 x1=    5 x2=   36 bx=   12
            ma1=    2 ax2=49 bxc=17 ma2= 1 ma3= 0
      z   z   z   z   z   z   z   z
849 ps=001 ma=    4 x1=    4 x2=   25 bx=   10
            ma1=    3 ax2=36 bxc=15 ma2= 2 ma3= 1
      66  z   z   z   z   z   z   z
      ...
```

The rest of the simulation is similar to the previous example. Here is the behavioral
Verilog code used to produce the simulation of the correct pipelined machine:

```verilog
`define NUM_STATE_BITS 3
`define IDLE           3'b000
`define COMPUTE1       3'b001
`define FLUSH1         3'b101
`define FLUSH2         3'b110
`define FLUSH3         3'b111
`define FILL1          3'b010
`define FILL2          3'b011
`define FILL3          3'b100
...
always
 begin
  @(posedge sysclk) enter_new_state(`IDLE);
    ma <= @(posedge sysclk) 0;
    ready = 1;
    if (pb)
     begin
      @(posedge sysclk) enter_new_state(`FILL1);
        ma <= @(posedge sysclk) ma + 1;
        ma1 <= @(posedge sysclk) ma;
        x1  <= @(posedge sysclk) x[ma];
      @(posedge sysclk) enter_new_state(`FILL2);
        ma <= @(posedge sysclk) ma + 1;
```

Continued.

```
     ma1 <= @(posedge sysclk) ma;
     ma2 <= @(posedge sysclk) ma1;
     x1  <= @(posedge sysclk) x[ma];
     x2  <= @(posedge sysclk) x1*x1;
     bx  <= @(posedge sysclk) b*x1;
   @(posedge sysclk) enter_new_state('FILL3);
    ma <= @(posedge sysclk) ma + 1;
    ma1 <= @(posedge sysclk) ma;
    ma2 <= @(posedge sysclk) ma1;
    ma3 <= @(posedge sysclk) ma2;
    x1  <= @(posedge sysclk) x[ma];
    x2  <= @(posedge sysclk) x1*x1;
    bx  <= @(posedge sysclk) b*x1;
    ax2 <= @(posedge sysclk) a*x2;
    bxc <= @(posedge sysclk) bx + c;
    while (ma != 'MAXMA)
      begin
       @(posedge sysclk) enter_new_state('COMPUTE1);
        ma <= @(posedge sysclk) ma + 1;
        ma1 <= @(posedge sysclk) ma;
        ma2 <= @(posedge sysclk) ma1;
        ma3 <= @(posedge sysclk) ma2;
        x1  <= @(posedge sysclk) x[ma];
        x2  <= @(posedge sysclk) x1*x1;
        bx  <= @(posedge sysclk) b*x1;
        ax2 <= @(posedge sysclk) a*x2;
        bxc <= @(posedge sysclk) bx + c;
        y[ma3] <= @(negedge sysclk) ax2 + bxc;
      end
   @(posedge sysclk) enter_new_state('FLUSH1);
    ma2 <= @(posedge sysclk) ma1;
    ma3 <= @(posedge sysclk) ma2;
    x2  <= @(posedge sysclk) x1*x1;
    bx  <= @(posedge sysclk) b*x1;
    ax2 <= @(posedge sysclk) a*x2;
    bxc <= @(posedge sysclk) bx + c;
    y[ma3] <= @(negedge sysclk) ax2 + bxc;
   @(posedge sysclk) enter_new_state('FLUSH2);
    ma3 <= @(posedge sysclk) ma2;
    ax2 <= @(posedge sysclk) a*x2;
    bxc <= @(posedge sysclk) bx + c;
    y[ma3] <= @(negedge sysclk) ax2 + bxc;
   @(posedge sysclk) enter_new_state('FLUSH3);
    y[ma3] <= @(negedge sysclk) ax2 + bxc;
   end
end
```

As described in section 6.5.2, the use of `negedge` memory modeling is necessary with many Verilog simulators.

6.5.8 Architectures for the quadratic evaluator

Only by proceeding to the mixed stage of the top-down design process can the maximum speed of the quadratic evaluator be estimated for the single-cycle, multi-cycle and pipelined versions. The mixed stage for each version requires choosing an architecture appropriate for each algorithm. Since the computations (to be performed by combinational logic) in each version are identical, the combinational logic devices (adders and multipliers) in these three versions will be identical. The three versions differ only with respect to whether and when intermediate computations are saved in registers. The following sections describe the architectures for these three versions.

6.5.8.1 Single-cycle architecture

The behavioral single-cycle ASM charts in section 6.5.2 describe all of the computations that must occur in one clock cycle before a result can be loaded into y. The following architecture implements these computations:

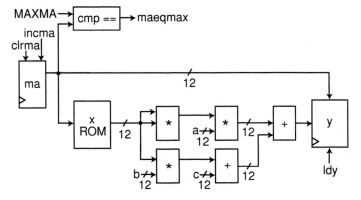

Figure 6-12. Single-cycle architecture.

The ma register provides the same address to x and y. The mabus is also compared against MAXMA to produce the status signal maeqmax.

There are three multipliers in the architecture. The ma register selects a particular word from the ROM. This value is fed to the first two multipliers. One of these multipliers produces the square, and the other multiplies this value by b. There is a third multiplier that multiplies the square by a.

There are also two adders in the architecture. The first adder produces the sum of bx and c. The other adder produces the final result. The final result is loaded into y[ma] when the ldy signal is asserted.

The following Verilog code shows the definition of this single-cycle architecture, along with the corresponding mixed controller:

```verilog
module poly_ctrl(pb,maeqmax,ldy,incma,clrma,ready,sysclk);
 input pb,maeqmax,sysclk;
 output ldy,incma,clrma,ready;
 wire pb,maeqmax,sysclk;
 reg ldy,incma,clrma,ready;
 reg ['NUM_STATE_BITS-1:0] present_state;
 always
  begin
   @(posedge sysclk) enter_new_state('IDLE);
    //ma <= @(posedge sysclk) 0;
    ready = 1;
    clrma = 1;
    if (pb)
     begin
      while (~maeqmax)
       begin
        @(posedge sysclk) enter_new_state('COMPUTE1);
         //ma <= @(posedge sysclk) ma + 1;
         //x2 = x[ma]*x[ma];
         //bx = b*x[ma];
         //ax2 = a*x2;
         //bxc = bx + c;
         //y[ma] <= @(negedge sysclk) ax2 + bxc;
         incma = 1;
         ldy = 1;
       end
     end
  end
...
endmodule


module poly_arch(maeqmax,ldy,incma,clrma,a,b,c,sysclk);
 output maeqmax;
 input ldy,incma,clrma,a,b,c,sysclk;
 wire maeqmax,ldy,incma,clrma,sysclk;
 wire [11:0] a,b,c;
 wire [11:0] x2,bx,ax2,bxc,xbus,ydibus,mabus;
```

Continued

```
counter_register #12   ma(,mabus,,1'b0,incma,clrma,sysclk);
comparator        #12   cmp(,maeqmax,,mabus,'MAXMA);
rom               #(12,23)  x(mabus,xbus);

multiplier        #(12,24)  m1(x2,xbus,xbus);
multiplier        #(12,24)  m2(bx,b,xbus);
multiplier        #(12,24)  m3(ax2,a,x2);
adder             #(12,25)  a1(bxc,bx,c);
adder             #(12,25)  a2(ydibus,ax2,bxc);

ram               #12   y(ldy,mabus,ydibus,,sysclk);
endmodule
```

In the above, it is assumed that the propagation delays of the ROM, multipliers and adders are 23, 24 and 25 units of $time, respectively. 'CLOCK_PERIOD is 100, which is just barely long enough for all the combinational logic to stabilize before a result is clocked into y, as illustrated by the following timing diagram produced by a Verilog simulator:

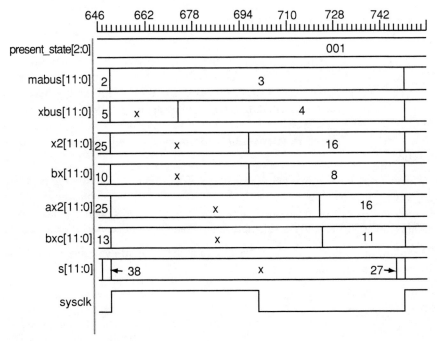

Figure 6-13. Timing diagram for single-cycle ASM.

6.5.8.2 Multi-cycle architecture

The behavioral multi-cycle ASM chart in section 6.5.3 can be implemented by many different possible architectures. Some of these possible architectures could be considerably cheaper than the architecture presented in this section; however, the architecture in this section was chosen for its consistency with the architecture in the previous section:

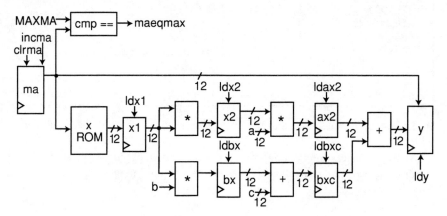

Figure 6-14. Multi-cycle architecture.

The only difference between this architecture and the previous architecture is the insertion of registers for x1, x2, bx, ax2 and bxc. As indicated by the ASM chart, it takes six clock cycles for each computation to travel through this architecture. In the first cycle, only ldx1 is asserted. In the second cycle, only ldx2 is asserted. In the third cycle, only ldbx is asserted. In the fourth cycle, only ldax2 is asserted. In the fifth cycle, only ldbxc is asserted. In the sixth cycle, finally ldy and incma are asserted.

As mentioned above, this is not a particularly efficient architecture for the multi-cycle approach because in any given clock cycle, five-sixths of the architecture is not performing any useful computation. Nevertheless, the insertion of the registers allows this architecture to be clocked considerably faster than the architecture in section 6.5.8.1.

The following Verilog code shows the definition of this multi-cycle architecture, along with the corresponding mixed controller:

```
always
 begin
  @(posedge sysclk) enter_new_state('IDLE);
   //ma <= @(posedge sysclk) 0;
   clrma = 1;
   ready = 1;
   if (pb)
    begin
     while (~maeqmax)
      begin
       @(posedge sysclk) enter_new_state('COMPUTE1);
        //x1 <= @(posedge sysclk) x[ma];
        ldx1 = 1;
       @(posedge sysclk) enter_new_state('COMPUTE2);
        //x2  <= @(posedge sysclk) x1*x1;
        ldx2 = 1;
       @(posedge sysclk) enter_new_state('COMPUTE3);
        //bx  <= @(posedge sysclk) b*x1;
        ldbx = 1;
       @(posedge sysclk) enter_new_state('COMPUTE4);
        //ax2 <= @(posedge sysclk) a*x2;
        ldax2 = 1;
       @(posedge sysclk) enter_new_state('COMPUTE5);
        //bxc <= @(posedge sysclk) bx + c;
        ldbxc = 1;
       @(posedge sysclk) enter_new_state('COMPUTE6);
        //ma <= @(posedge sysclk) ma + 1;
        //y[ma] <= @(negedge sysclk) ax2 + bxc;
        incma = 1;
        ldy = 1;
      end
    end
 end
...
counter_register #12   ma(,mabus,,1'b0,incma,clrma,sysclk);
comparator       #12   cmp(,maeqmax,,mabus,'MAXMA);
rom              #(12,23)   x(mabus,xbus);
enabled_register #12            x1(xbus,x1bus,ldx1,sysclk);
multiplier       #(12,24)   m1(x2dibus,x1bus,x1bus);
enabled_register #12            x2(x2dibus,x2dobus,ldx2,sysclk);
multiplier       #(12,24)   m2(bxdibus,b,x1bus);
enabled_register #12            bx(bxdibus,bxdobus,ldbx,sysclk);
multiplier       #(12,24)   m3(ax2dibus,a,x2dobus);
enabled_register #12            ax2(ax2dibus,ax2dobus
                                       ldax2,sysclk);
adder            #(12,25)   a1(bxcdibus,bxdobus,c);
```

Continued

```
enabled_register #12          bxc(bxcdibus,bxcdobus
                              ldbxc,sysclk);
adder             #(12,25)    a2(ydibus,ax2dobus,bxcdobus);
ram               #12   y(ldy,mabus,ydibus,,sysclk);
```

In the above, it is assumed that the propagation delays are the same as in the single-cycle approach of section 6.5.8.1. With the multi-cycle approach, 'CLOCK_PERIOD can now be 26 in this example, which is nearly four times faster than the single-cycle approach. The faster clock is possible because there is less logic that has to stabilize before each intermediate result is clocked into one of the registers. The following timing diagram illustrates this:

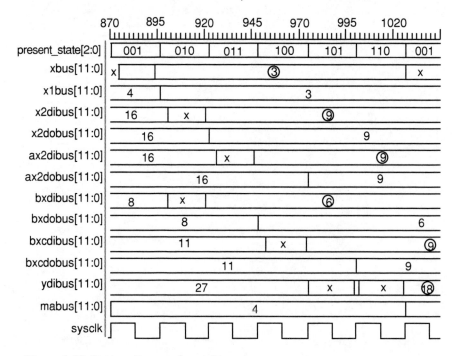

Figure 6-15. Timing diagram for multi-cycle.

6.5.8.3 Pipelined architecture

The correct behavioral ASM for the pipelined method given in section 6.5.7 requires three additional registers: `ma1`, `ma2` and `ma3`. In a pipelined design, all of the registers inserted in the single cycle architecture to make it a pipelined architecture are referred to as *pipeline registers*. In this architecture, every register, except `ma`, is a pipeline register.

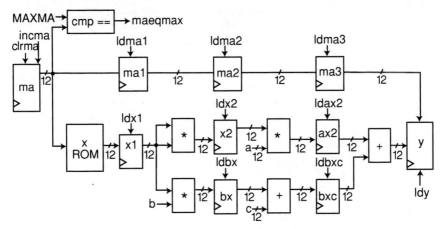

Figure 6-16. Pipelined architecture.

Notice that the pipeline registers are drawn in columns. The `ma1` and `x1` registers are drawn in the leftmost pipeline register column. The `ma2`, `x2` and `bx` registers are in the next pipeline register column. The `ma3`, `ax2` and `bxc` registers are in the third pipeline register column. In between each pipeline register column are buses and combinational logic only. Such a column of combinational logic to the left of a pipeline register is known as a *pipeline stage*. For example, the first pipeline stage is the ROM. The second pipeline stage involves two multipliers and a bus (that passes `ma1`). The third stage involves an adder and a multiplier. The fourth stage involves just an adder. (This architecture assumes that a value can be clocked into the appropriate word of `y` as though it just a clocked register. See section 8.2.2.3.1 for details.)

The following Verilog code shows the definition of this pipelined architecture, along with the corresponding mixed controller:

```verilog
always
 begin
  @(posedge sysclk) enter_new_state('IDLE);
   //ma <= @(posedge sysclk) 0;
   ready = 1;
   clrma = 1;
   if (pb)
    begin
     @(posedge sysclk) enter_new_state('FILL1);
      //ma  <= @(posedge sysclk) ma + 1;
      //ma1 <= @(posedge sysclk) ma;
      //x1  <= @(posedge sysclk) x[ma];
      ldx1 = 1;
      incma = 1;
      ldma1 = 1;
     @(posedge sysclk) enter_new_state('FILL2);
      //ma  <= @(posedge sysclk) ma + 1;
      //ma1 <= @(posedge sysclk) ma;
      //ma2 <= @(posedge sysclk) ma1;
      //x1  <= @(posedge sysclk) x[ma];
      //x2  <= @(posedge sysclk) x1*x1;
      //bx  <= @(posedge sysclk) b*x1;
      ldx1 = 1;
      ldx2 = 1;
      ldbx = 1;
      incma = 1;
      ldma1 = 1;
      ldma2 = 1;
     @(posedge sysclk) enter_new_state('FILL3);
      //ma  <= @(posedge sysclk) ma + 1;
      //ma1 <= @(posedge sysclk) ma;
      //ma2 <= @(posedge sysclk) ma1;
      //ma3 <= @(posedge sysclk) ma2;
      //x1  <= @(posedge sysclk) x[ma];
      //x2  <= @(posedge sysclk) x1*x1;
      //bx  <- @(posedge sysclk) b*x1;
      //ax2 <= @(posedge sysclk) a*x2;
      //bxc <= @(posedge sysclk) bx + c;
      ldx1 = 1;
      ldx2 = 1;
      ldbx = 1;
      ldax2 = 1;
      ldbxc = 1;
      incma = 1;
      ldma1 = 1;
      ldma2 = 1;
      ldma3 = 1;
```

Continued

```
   while (~maeqmax)
    begin
     @(posedge sysclk) enter_new_state('COMPUTE1);
      //ma  <= @(posedge sysclk) ma + 1;
      //ma1 <= @(posedge sysclk) ma;
      //ma2 <= @(posedge sysclk) ma1;
      //ma3 <= @(posedge sysclk) ma2;
      //x1  <= @(posedge sysclk) x[ma];
      //x2  <= @(posedge sysclk) x1*x1;
      //bx  <= @(posedge sysclk) b*x1;
      //ax2 <= @(posedge sysclk) a*x2;
      //bxc <= @(posedge sysclk) bx + c;
      //y[ma3] <= @(negedge sysclk) ax2 + bxc;
      ldx1 = 1;
      ldx2 = 1;
      ldbx = 1;
      ldax2 = 1;
      ldbxc = 1;
      incma = 1;
      ldma1 = 1;
      ldma2 = 1;
      ldma3 = 1;
      ldy = 1;
    end
   @(posedge sysclk) enter_new_state('FLUSH1);
    //ma2 <= @(posedge sysclk) ma1;
    //ma3 <= @(posedge sysclk) ma2;
    //x2  <= @(posedge sysclk) x1*x1;
    //bx  <= @(posedge sysclk) b*x1;
    //ax2 <= @(posedge sysclk) a*x2;
    //bxc <= @(posedge sysclk) bx + c;
    //y[ma3] <= @(negedge sysclk) ax2 + bxc;
    ldx2 = 1;
    ldbx = 1;
    ldax2 = 1;
    ldbxc = 1;
    ldma2 = 1;
    ldma3 = 1;
    ldy = 1;
   @(posedge sysclk) enter_new_state('FLUSH2);
    //ma3 <= @(posedge sysclk) ma2;
    //ax2 <= @(posedge sysclk) a*x2;
    //bxc <= @(posedge sysclk) bx + c;
```

Continued

```
       //y[ma3] <= @(negedge sysclk) ax2 + bxc;
       ldax2 = 1;
       ldbxc = 1;
       ldma3 = 1;
       ldy = 1;
     @(posedge sysclk) enter_new_state('FLUSH3);
       //y[ma3] <= @(negedge sysclk) ax2 + bxc;
       ldy = 1;
   end
 end
. . .
counter_register  #12   ma(,mabus,,1'b0,incma,clrma,sysclk);
enabled_register  #12   ma1(mabus,ma1bus,ldma1,sysclk);
enabled_register  #12   ma2(ma1bus,ma2bus,ldma2,sysclk);
enabled_register  #12   ma3(ma2bus,ma3bus,ldma3,sysclk);
comparator        #12   cmp(,maeqmax,,mabus,'MAXMA);
rom               #(12,23)   x(mabus,xbus);
enabled_register  #12       x1(xbus,x1bus,ldx1,sysclk);
multiplier        #(12,24)  m1(x2dibus,x1bus,x1bus);
enabled_register  #12       x2(x2dibus,x2dobus,ldx2,sysclk);
multiplier        #(12,24)  m2(bxdibus,b,x1bus);
enabled_register  #12       bx(bxdibus,bxdobus,ldbx,sysclk);
multiplier        #(12,24)  m3(ax2dibus,a,x2dobus);
enabled_register  #12       ax2(ax2dibus,ax2dobus
       ldax2,sysclk);
adder             #(12,25)  a1(bxcdibus,bxdobus,c);
enabled_register  #12       bxc(bxcdibus,bxcdobus
       ldbxc,sysclk);
adder             #(12,25)  a2(ydibus,ax2dobus,bxcdobus);
ram               #12   y(ldy,ma3bus,ydibus,,sysclk);
```

In the above, it is assumed that the propagation delays are the same as in the single- and multi-cycle approaches. With the pipelined approach, 'CLOCK_PERIOD can usually be about as fast as in the multi-cycle approach (26 in this example), but the number of such clock cycles is nearly the same as in the single-cycle approach. Unlike the multi-cycle approach, where five sixths of the combinational logic is unproductive during each clock cycle, when the pipeline is full, all of the combinational logic is doing productive work. The following timing diagram illustrates how the pipelined approach gets the complete job done faster than the single-cycle or multi-cycle approach:

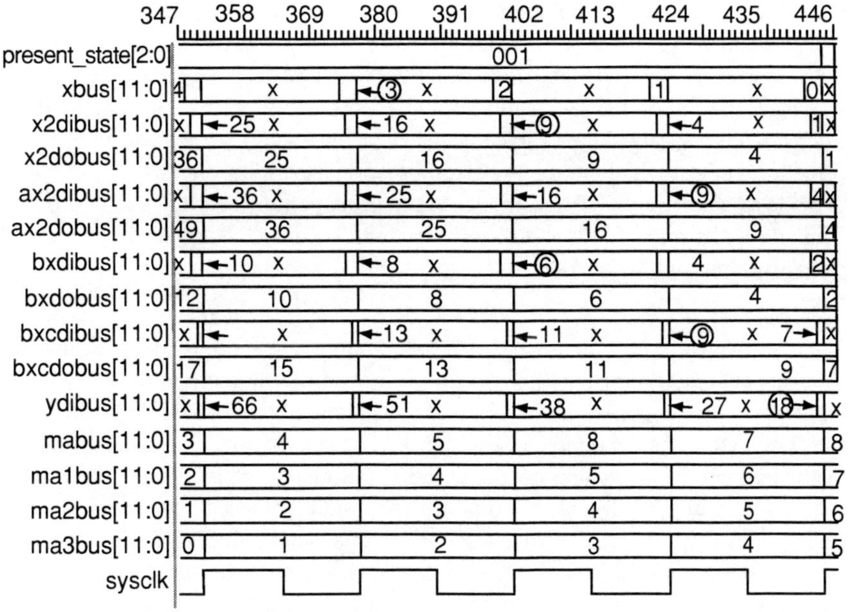

347   358   369   380   391   402   413   424   435   446

present_state[2:0] 001

xbus[11:0] 4   x   ←③   x   2   x   1   x   0 x

x2dibus[11:0] x   ←25 x   ←16 x   ←⑨   x   ←4   x   1 x

x2dobus[11:0] 36   25   16   9   4   1

ax2dibus[11:0] x   ←36 x   ←25 x   ←16 x   ←⑨   x   4 x

ax2dobus[11:0] 49   36   25   16   9   4

bxdibus[11:0] x   ←10 x   ←8   x   ←⑥   x   4   x   2 x

bxdobus[11:0] 12   10   8   6   4   2

bxcdibus[11:0] x   ←   x   ←13 x   ←11 x   ←⑨   x   7→ x

bxcdobus[11:0] 17   15   13   11   9   7

ydibus[11:0] x   ←66 x   ←51 x   ←38 x   ←27 x   ⑱→ x

mabus[11:0] 3   4   5   8   7   8

ma1bus[11:0] 2   3   4   5   6   7

ma2bus[11:0] 1   2   3   4   5   6

ma3bus[11:0] 0   1   2   3   4   5

sysclk

Figure 6-17. Timing diagram for pipelined ASM.

6.6   Conclusion

The first duty of a designer is to produce a correct design. The top-down design process explained in earlier chapters helps organize a designer's thinking to achieve this goal. Often, in addition to being algorithmically correct, a practical design must meet the criteria of speed and cost. This chapter explains how Verilog can help a designer determine if a design meets its speed goals. This chapter also explains different design alternatives that allow a designer to trade off speed and cost.

The speed of an algorithm implemented in hardware depends on two factors: the clock period and the number of clock cycles. The algorithm itself determines how many clock cycles are required, but the limiting factor on how fast the clock period can be is gate-level propagation delay. Synthesizable Verilog cannot have propagation delay, but once a design is synthesized, it is easy to annotate the built-in gates of the netlist with propagation delays. (Some synthesis tools automatically backannotate the netlist for post-synthesis simulation.) Gates with delays create the possibility of spurious wrong outputs, known as hazards.

There are many ways that a building block, such as an adder, can be synthesized into a gate level netlist. Each such unique netlist may give rise to unique patterns of hazards that can only be simulated in detail after synthesis. Despite the fact that we cannot know the details of a hazard prior to synthesis, it is possible to abstractly model a hazardous period of a signal using Verilog events, the `disable` statement and `'bx`. Such models are not synthesizable but instead provide more accurate timing simulation prior to synthesis. Instantiations of combinational building blocks defined this way provide an accurate worst case model of the propagation delay for bus-width devices. From the `$time` the inputs to any of the instantiated devices change until such changes propagate through all of the instantiated devices, the final output of the collection is `'bx`.

Many problems require that the same computation be performed on large amounts of independent data. There are three common algorithmic alternatives to solve such problems, known as single-cycle, multi-cycle, and pipelined. In the single-cycle approach, the computation on each independent piece of data is begun and finished in one clock cycle. In the multi-cycle approach, the computation on each independent piece of data requires several clock cycles to complete before another piece of data can be processed. The pipelined approach, like an assembly line in a factory, does different aspects of the computation with different pieces of independent data at the same time. Although it still takes several clock cycles to complete the computation for a particular piece of data, once the pipeline is filled, it produces one result per clock cycle.

What the multi-cycle and pipelined approaches have in common is that they both can be clocked by "faster" clocks, determined by the worst case delay of a single building block. The single-cycle approach demands a "slower" clock, determined by the delay path through several devices. The single-cycle approach produces exactly one result per clock cycle, and the pipelined approach usually produces almost one result per clock cycle (because the time to fill the pipeline is usually negligible compared to how much independent data is to be processed). The multi-cycle approach needs several clock cycles to produce each result. Therefore, the pipelined approach is usually fastest and most efficient because it can be clocked fast and it produces nearly one result per clock cycle.

Although in recent years pipelining has become important in the design (and marketing) of personal computers, pipelining is not a new concept. It has been used since the 1960s to design general-purpose computers (chapter 9), but its use with special-purpose computers has a much longer history. Pipelining is one of those algorithmic concepts that endure. It was first applied to computer design by Babbage in the 1820s. On Babbage's machine, the clock cycle was generated when one turned a crank. To avoid muscle strain, Babbage designed for speed and cost and chose a pipelined design. Despite almost unimaginable technological change in two centuries, many designers since then have followed in Babbage's algorithmic footsteps and have chosen pipelined designs. One does not have to be a genius like Babbage to understand pipelining today, because modern tools like Verilog simulators make these intricate `$time` related concepts much easier to understand.

6.7 Further reading

GAJSKI, DANIEL D., *Principles of Digital Design*, Prentice Hall, Upper Saddle River, NJ, 1997. Chapter 8 discusses how pipelining can be applied to both the controller and the architecture (datapath).

PALNITKAR, S., *Verilog HDL: A Guide to Digital Design and Synthesis,* Prentice Hall, PTR,Upper Saddle River, NJ, 1996. Chapters 5 and 10 explain about sophisticated gate-level delay modeling in Verilog.

PATTERSON, DAVID A. and JOHN L. HENNESSY, *Computer Organization and Design: The Hardware/Software Interface,* Morgan Kaufmann, San Mateo, CA, 1994. Chapters 5 and 6 discuss the trade-offs of the single-cycle, multi-cycle and pipelined approaches.

6.8 Exercises

6-1. A complex number, X, can be represented inside a machine as two integers: the real part, `xr`, and the imaginary part, `xi`. Mathematicians say that X = `xr+i*xi`, where `i` is the square root of minus one. (Some electrical engineers use the symbol `j` instead of `i`.) To add two complex numbers, X and Y, simply requires adding the real and imaginary parts separately. To multiply two complex numbers, X and Y, requires computing `xr*yr-xi*yi` and `xr*yi+xi*yr`. Suppose that a machine has four ROMs: `xr[ma]`, `xi[ma]`, `yr[ma]` and `yi[ma]`. Design a multi-cycle behavioral ASM suitable for a central ALU architecture that computes the sum of the products of the complex values in X and Y ROMs. This computation has many practical applications in the field of digital signal processing, such as filtering out unwanted noise in a telephone conversation. Note that there is no need for a memory in this problem because the desired answer is a single complex sum composed of `sumr` and `sumi`. You may assume the ALU can do either an integer addition, subtraction or multiplication in a single cycle.

6-2. Implement a pure behavioral Verilog simulation and test code that verifies your design in problem 6-1.

6-3. Draw a block diagram for the architecture of problem 6-1. Assume the propagation delays (in nanoseconds) of building blocks are the same as in section 6.5.8.1. How many seconds will it take for your machine to compute the sum assuming there are ten million words in each ROM?

6-4. Implement a mixed Verilog simulation that verifies your architecture for problem 6-3. Assume the propagation delays of building blocks are the same as in section 6.5.8.1.

6-5. Design a single-cycle behavioral ASM for problem 6-1. The architecture that will eventually implement the register transfers of this single-cycle machine may have as many integer adders, subtractors and multipliers as necessary.

6-6. Implement a pure behavioral Verilog simulation and test code that verifies your design in problem 6-5.

6-7. Draw a block diagram for the architecture of problem 6-5. Assume the propagation delays (in nanoseconds) of building blocks are the same as in section 6.5.8.1. How many seconds will it take for your machine to compute the sum assuming there are ten million words in each ROM?

6-8. Implement a mixed Verilog simulation that verifies your architecture for problem 6-7. Assume the propagation delays of building blocks are the same as in section 6.5.8.1.

6-9. Design a pipelined behavioral ASM for problem 6-1. The architecture that will eventually implement the register transfers of this pipelined machine may have as many pipeline stages and as many integer adders, subtractors and multipliers as necessary.

6-10. Implement a pure behavioral Verilog simulation and test code that verifies your design in problem 6-9.

6-11. Draw a block diagram for the architecture of problem 6-9. Assume the propagation delays (in nanoseconds) of building blocks are the same as in section 6.5.8.1. How many seconds will it take for your machine to compute the sum assuming there are ten million words in each ROM?

6-12. Implement a mixed Verilog simulation that verifies your architecture for problem 6-11. Assume the propagation delays of building blocks are the same as in section 6.5.8.1.

7. ONE HOT DESIGNS

The manual process of translating an ASM (or the equivalent Verilog) into hardware is quite involved. The final step of creating a structural controller is tedious because we have to determine what the next state is in every situation. There is an alternative way to create a structural controller directly from the behavioral Verilog or from the mixed ASM without the need to consider the next state logic and the present state register. The final hardware structure that is created by this alternative method is slightly more expensive than what is created by the process described in chapter 4, but it is much easier to understand.

This approach is known as a *one hot* method. The one hot controller uses as many flip flops as there are states in the ASM. As described in appendix D, a D-type flip flop is a one-bit register, whose output Q is simply its input D, delayed by one clock cycle.

The reason the one hot method is preferred is that the translation from behavior to structure is much more straightforward than the process described in chapter 4. Given a behavioral ASM or the equivalent behavioral Verilog, there is a one-to-one mapping to the circuit diagram for the one hot controller. (There is a tool described in appendix F that automates this process.)

7.1 Moore ASM to one hot

As explained in chapter 2, a Moore ASM is composed of three symbols: rectangles describing states, diamonds describing decisions and arrows describing control flow. With the one hot technique, each of these behavioral symbols (or the corresponding Verilog) translates directly into a physical piece of hardware.

7.1.1 Rectangle/flip flop

A rectangle in an ASM [or the equivalent @ (posedge sysclk) in Verilog] translates into a flip flop. This technique is known as one hot because it is assumed that only one of the flip flops is hot (i.e., contains a one) in any clock cycle. The rest of the flip flops will be cold (contain 0). The rules of the one hot technique ensure that if this assumption is true shortly after $time 0, the one hot interconnections will guarantee this one hot property will remain in effect from then on.

7.1.2 Arrow/wire

An arrow in an ASM (or the implicit flow of control in Verilog) corresponds to a physical wire in a one hot controller. When that wire is hot it means that the corresponding statement in Verilog is active during the current clock cycle. Several statements (that execute in parallel) might be hot in a particular clock cycle, but only one flip flop (corresponding to a state) is hot in that clock cycle.

7.1.3 Joining together/OR gate

It is common for two or more arrows to join together in an ASM. This joining together occurs because in different clock cycles there are different paths to arrive at the same next state. Because they could fight (and produce a 1'bx in Verilog simulation), it is illegal to tie together two wires that are each connected to an output. Therefore, when arrows in an ASM join together, the corresponding physical wires in the one hot design must be ORed together.

7.1.4 Decision/demux

A decision (diamond in an ASM or an equivalent `if` or `while` in Verilog) translates into a one bit wide demux. Recall from appendix C that the combinational logic for a demux is very different from that of a mux. The following truth table describes the outputs (`out0` and `out1`) of the demux, given its two inputs (`in` and `cond`):

in	cond	out1	out0
0	0	0	0
0	1	0	0
1	0	0	1
1	1	1	0

Notice when `in` is cold (i.e., 0), both outputs are cold. When `in` is hot (i.e., 1), only one of its two outputs is hot; hence it preserves the one hot property.

7.1.5 Summary of the one hot method

The following diagram illustrates the above four rules for translating a Moore ASM into a one hot design:

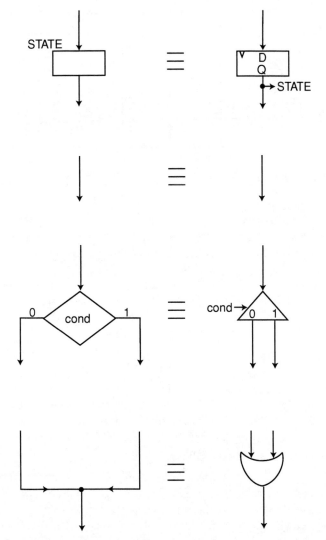

Figure 7-1. Moore ASMs and corresponding components of one hot controllers.

The one hot method uses more flip flops than the method shown in chapter 4. The number of flip flops of the present state register in chapter 4 is approximately the base two logarithm of the number of states. The following table shows how many flip flops are required in each method:

number of states	Ch 4	one hot	one hot + power on
2	1	2	3
3	2	3	4
4	2	4	5
5	3	5	6
6	3	6	7
7	3	7	8
8	3	8	9

The need for the extra flip flop for "power on" will be explained in the next section.

7.1.6 First example

Consider the first ASM given in section 2.2.2 (and the mixed ASM in section 2.3.3), which has five states. The diagram below shows how to translate the ASM from section 2.2.2 into one hot hardware. There are five states in this ASM; therefore there are five flip flops in the corresponding one hot controller. For clarity, the wires connected to the Q outputs of these five flip flops are labeled with states that correspond to them. Later we will use a numeric labeling scheme, but for now using the state names as wire names will emphasize how the one hot method works.

Coming out of state IDLE, there is a decision based on pb. If pb is false, there is a path that eventually leads back to state IDLE. If pb is true, there is a path that leads to state INIT. In the one hot controller, this decision corresponds to a demux whose input is the Q output of the flip flop for IDLE. The out0 output of the demux goes to OR gates that in turn provide the D input for the flip flop for IDLE. The out1 output of the demux connects to the D input for the flip flop for INIT. In other words, the ASM chart and the circuit diagram have an identical structure.

In the ASM, there are two paths that lead to state TEST. One comes from state INIT and the other comes from state COMPUTE2. In the one hot controller, this corresponds to another OR gate. In state TEST there is a decision based on r1gey (which tells if r1 >= y). If this condition is true, the ASM proceeds to state COMPUTE1; otherwise the ASM proceeds to state IDLE. This decision corresponds to a demux whose input is the Q output of the flip flop for state TEST. The cond input for this demux is r1gey. The out1 output of this demux connects to the D input of the flip flop for state COM-PUTE1. The out0 output of this demux connects to the OR gate that leads back to the flip flop for state IDLE.

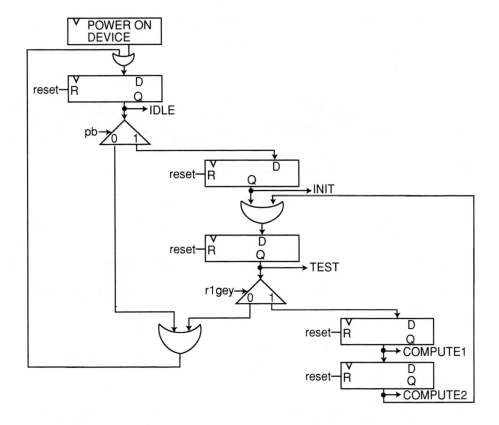

Figure 7-2. One hot controller for ASMs of sections 2.2.2 and 2.3.3.

In order to guarantee that the one hot property holds at $time 0, all of the flip flops are connected to an asynchronous reset signal. In physical hardware, shortly after $time 0 is when this reset signal ceases to be active. It is never used again.

By itself, just resetting the flip flops that correspond to the states in a particular ASM would have the effect of making all those flip flops cold at $time 0. Exactly one of these flip flops (the one for state IDLE in this example) needs to become hot at the first rising edge of the clock. To accomplish this, we need to OR an additional wire on the path to the flip flop for state IDLE.

This extra wire will be the output of a power-on device that will be hot only between $time 0 until the first rising edge of the clock. After the first rising edge of the clock, this power on device will will be cold thereafter.

This power-on device is constructed as a D flip flop with its D input tied to a one. The Q output of this flip flop is complemented to form the signal that "ignites" the first flip flop of the actual controller.

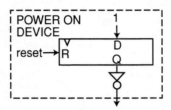

Figure 7-3. Power-on device for one hot controllers.

The next diagram shows the translation of the ASM from sections 2.2.7 and 2.3.1 (or the corresponding Verilog from section 4.1.5) into a one hot controller. Like any other controller, this also needs to generate the command outputs required for the architecture. This is very similar to the netlist given in section 4.4.2, except instead of `present_state[0]` and `~present_state[0]`, we have IDLE and COMPUTE as wires.

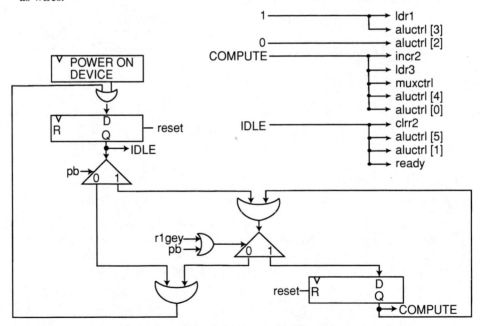

Figure 7-4. One hot controller for ASMs of sections 2.2.7 and 2.3.1.

7.2 Verilog to one hot

It usually takes a little more hardware to implement a machine with the one hot method, so why is this method worthwhile? The one hot method has a tremendous advantage: not only is there a one-to-one mapping of ASMs into one hot designs—there is also a similar correspondence between an implicit style behavioral Verilog block and the one hot circuit. Because of this direct translation process, it is relatively simple to write software that translates such Verilog into a one hot design without the need for the designer to go through the lengthy process described in chapter 4.

VITO is an automated preprocessor tool that performs the translation from implicit style behavioral Verilog into a one hot design. How to use VITO is described in appendix F. In order to understand the approach used in VITO, it is necessary to appreciate the one-to-one mapping between Verilog and the one hot circuit. Such an appreciation is best developed by working through a few examples. The following examples describe manual translation of implicit style Verilog directly into a one hot circuit. Rather than instantiating built-in gates, the following examples will use what is called *continuous assignment*.

7.2.1 Continuous assignment

Continuous assignment is a shorthand way of describing combinational logic. It is equivalent to defining a module without having to declare the ports and so forth that would otherwise be required. Let's consider an example of a continuous assignment:

```
module test;
   reg ff_1,ff_2;
   ...   /code that deals with ff_1 and ff_2

   wire s_3;
   assign s_3 = ff_1 | ff_2;
endmodule
```

There also is an additional shorthand for continuous assignment that allows the wire declaration to occur on the same line. For example, the above is equivalent to:

```
wire s_3 = ff_1 | ff_2;
```

Note that continuous assign is not a behavioral statement. The left-hand side is a `wire`, not a `reg`. Clearly, using continuous assign shortens the code considerably compared to the "`hidden_module`" shown below:

```
module test;
  reg ff_1,ff_2;
  ...  /code that deals with ff_1 and ff_2

  wire s_3;
  hidden_module h1(s_3, ff_1, ff_2);
endmodule
```

This "hidden_module" defines combinational logic in the usual way with a reg for the output port and a sensitivity list for the input wires:

```
module hidden_module(s_3, ff_1, ff_2);
  output s_3;
  input ff_1, ff_2;
  reg s_3;
  wire ff_1, ff_2;
  always @(ff_1 or ff_2)
    s_3 = ff_1 | ff_2;
endmodule
```

The computation, ff_1 | ff_2, is the same as given in the continuous assign. The power of the continuous assign is that it allows arbitrarily complicated expressions (of arbitrary width) to be evaluated. For example, the following:

```
module test;
  reg [11:0] a,b;
  ...  /code that deals with a and b

  wire [11:0] sum;

  assign sum = a + b;
endmodule
```

is equivalent to instantiating a hidden_adder:

```
module test;
  reg [11:0] a,b;
  ...  /code that deals with a and b
```

Continued

```
    wire [11:0] sum;

    hidden_adder h2(sum, a, b);
  endmodule
  module hidden_adder(sum, a, b);
    input a,b;
    output sum;
    wire [11:0] a,b;
    reg [11:0] sum;
    always @(a or b)
      sum = a + b;
  endmodule
```

An advantage of the continuous assignment is that you do not have to specify the widths of sum, a and b multiple times—their previous declarations are sufficient. In contrast, with the hidden adder approach you have to duplicate the declaration of their widths inside the declaration of hidden_adder.

Also, continuous assignment allows the use of the conditional operator (? :). For example, the following:

```
  module test;
    reg [11:0] a,b;
    reg sel;
    ...  /code that deals with a,b,sel

    wire [11:0] muxout;

    assign muxout =  sel ? b : a;
  endmodule
```

is equivalent to instantiating a hidden instance of a mux2, whose portlist is given in section 4.2.1.5.

```
      mux2 #(12) h3(a, b, sel, muxout);
```

7.2.2 One hot using continuous assignment

The wires that implement the combinational logic of a one hot circuit can be described with continuous assignment. This is done as a notational convenience because continuous assignments are equivalent to structural instances but are much more concise. For example, the adder and mux in the last section could have been of any width, but the syntax of the actual continuous assignment would have been the same. Synthesis tools available from many different vendors are able to translate continuous assignments into the structural instances (netlist) required to fabricate hardware. Each flip flop required by the one hot circuit will be described by a separate one-bit `reg`. Such `reg`s are also synthesizable. The names of these `wire`s and `reg`s will relate to the statement numbers of the lines in the Verilog `always` block from which they derive.

7.2.2.1 One hot with `if else`

The following example Verilog (taken from section 3.8.2.3.3) illustrates implicit style behavioral Verilog with an `if else` statement. In this example `@(posedge sysclk)` `#1` occurs on lines 3, 5, 9, 14 and 17, so the names of the five flip flops for the one hot controller will be `ff_3`, `ff_5`, `ff_9`, `ff_14` and `ff_17`:

```
 1:always
 2: begin
 3:  @(posedge sysclk) #1;    //FIRST   is ff_3
 4:    a <= @(posedge sysclk) 1;
 5:  @(posedge sysclk) #1;    //SECOND is ff_5
 6:    b <= @(posedge sysclk) a;
 7:    if (a == 1)
 8:      begin
 9:       @(posedge sysclk) #1;//THIRD   is ff_9
10:         a <= @(posedge sysclk) b;
11:      end
12:    else
13:      begin
14:       @(posedge sysclk) #1;//FOURTH is ff_14
15:         b <= @(posedge sysclk) 4;
16:      end
17:  @(posedge sysclk) #1;    //FIFTH   is ff_17
18:    a <= @(posedge sysclk) 5;
19: end
```

It is easier to give each flip flop a name that relates to what statement number the `@(posedge sysclk)` `#1` occurs on than to use the name from the original ASM. The reason that we do not use the names FIRST, SECOND, THIRD, FOURTH and

FIFTH for the flip flops is that those names were inside comments, which are ignored by Verilog. [The reason we do not use the `enter_new_state` task (sections 3.9.1.2 and chapter 4] to give each state a name is that the VITO preprocessor does not support tasks.) The example in section 7.1 of translating from an ASM to a one hot circuit used the state names given in the ASMs as the names of the flip flops only for the purpose of illustrating the nature of the one hot method. Since this translation will now be automated, the names do not matter. In the automated process, the designer will seldom notice what name is given to each `wire`.

Every statement also has a `wire` associated with it that is active when the corresponding statement executes. (In the earlier example, these names were also the original ASM state names. In general this need not be the case, and so separate names are appropriate for an automated tool.) In this example, there are nineteen `wires` (s_1 through s_19) that correspond to statements in the original implicit style Verilog code. Of these `wires`, five act as command signals to the architecture:

```
     wire          action in architecture when wire is active
     s_4                   a <= @(posedge sysclk) 1;
     s_6                   b <= @(posedge sysclk) a;
     s_10                  a <= @(posedge sysclk) b;
     s_15                  b <= @(posedge sysclk) 4;
     s_18                  a <= @(posedge sysclk) 5;
```

The other wires (s_1, s_2, s_3, s_5, s_7, s_8, s_9, s_11, s_12, s_13, s_14, s_16, s_17 and s_19) are used to define the rest of the one hot controller. Some of those `wires` are synonymous with each other. For example s_11 (an end statement) is synonymous with the s_10 wire that precedes it.

Although there are nineteen `wire` names in the one-hot controller, only the above five are sent to the architecture. Using the methodical approach (such as in sections 2.3.1 and 8.4.1) for designing the architecture, we sort the above list according to the left-hand side of the <=, and separate them according to these destinations:

```
     s_4              a <= @(posedge sysclk) 1;
     s_10             a <= @(posedge sysclk) b;
     s_18             a <= @(posedge sysclk) 5;

     s_6              b <= @(posedge sysclk) a;
     s_15             b <= @(posedge sysclk) 4;
```

Of course, there are many architectures that could implement the above register transfers. In earlier chapters we have focused on using standard building blocks, such as enabled registers. In the approach of this chapter, we instead choose an architecture that is easier to describe using continuous assignment. For this reason, we will use simple (non-enabled) D-type registers that have no command inputs whatsoever. In other words, such registers can be described simply as:

```
reg [11:0] a,b;
wire [11:0] new_a,new_b;
always @ (posedge sysclk)
  a = new_a;
always @ (posedge sysclk)
  b = new_b;
```

All of the actions normally encapsulated inside a register building block (of the kind described in appendix D) now have to be given with the combinational logic that computes new_a and new_b. From the sorted list above, one approach would be to use three muxes for computing new_a and two muxes for computing new_b:

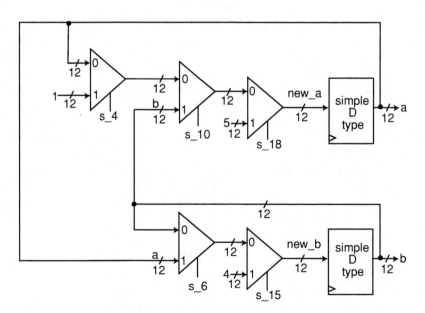

Figure 7-5. Architecture generated from implicit Verilog of sections 7.2.2.1 and 3.8.2.3.3.

The combinational logic in the above diagram can be expressed as two continuous assignments:

```
assign new_a =
   (s_18)  ? 5 :
   (s_10)  ? b :
   (s_4)   ? 1 :
               a;

assign new_b =
   (s_15)  ? 4 :
   (s_6)   ? a :
               b;
```

Because of the nature of correct one hot designs, it is guaranteed that s_18+s_10+s_4 <= 1 and s_6+s_15 <= 1. In other words, within each group (dealing with the same destination register), no more than one of the command signals will be active in any clock cycle. This means there are several permutations of the muxes that would also suffice, such as:

```
assign new_a =
   (s_4)   ? 1 :
   (s_10)  ? b :
   (s_18)  ? 5 :
               a;

assign new_b =
   (s_6)   ? a :
   (s_15)  ? 4 :
               b;
```

Notice that the architecture was created by a *textual transformation* of the original Verilog. The block diagram given above was shown only as an aid to understand how the Verilog continuous assignment works. The preprocessor produces similar Verilog by just rearranging the original text of the Verilog.

Having defined the architecture, all that remains is to define the controller. The following circuit diagram shows how each implicit style behavioral statement translates into hardware:

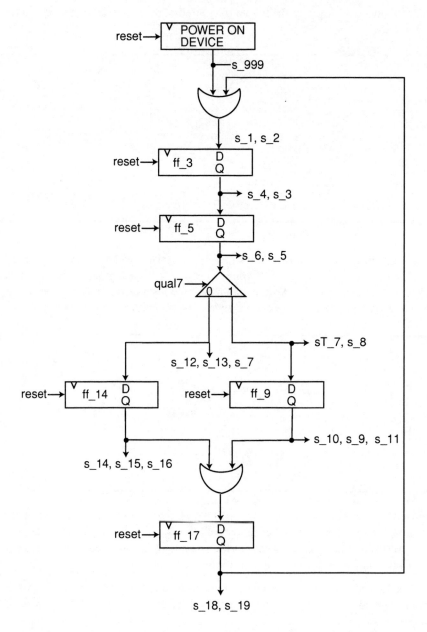

Figure 7-6. Controller generated from implicit Verilog of sections 7.2.2.1 and 3.8.2.3.3.

Again, VITO does not draw such a circuit diagram. The above diagram is provided only to help explain the textual transformations that the preprocessor performs. Starting from the original Verilog, the preprocessor generates the many continuous assignments required to describe the above circuit.

7.2.2.2 One hot with *if*

The following example is taken from the implicit style example in section 3.8.2.3.4. It illustrates a slightly different one hot controller than the last example because the original Verilog uses an if without an else:

```
 1:always
 2: begin
 3:   @(posedge sysclk) #1;    //FIRST    is ff_3
 4:     a <= @(posedge sysclk) 1;
 5:   @(posedge sysclk) #1;    //SECOND   is ff_5
 6:     b <= @(posedge sysclk) a;
 7:     if (a == 1)
 8:       begin
 9:       @(posedge sysclk) #1;//THIRD    is ff_9
10:         a <= @(posedge sysclk) b;
11:       @(posedge sysclk) #1;//FOURTH   is ff_11
12:         b <= @(posedge sysclk) 4;
13:       end
14:   @(posedge sysclk) #1;    //FIFTH    is ff_14
15:     a <= @(posedge sysclk) 5;
16: end
```

The if statement translates to a demux whose input, qual7, comes from the comparator for statement 7 that implements the condition a == 1. The 1 output, sT_7, corresponds to when this condition is true at the $time the if executes. The 0 output, s_7 corresponds to when this condition is false at the $time the if executes.

The preprocessor generates the following one hot controller. In the following, only some of the wire names are shown:

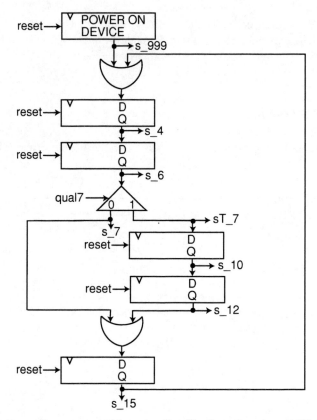

Figure 7-7. Controller generated from implicit Verilog of sections 7.2.2.2 and 3.8.2.3.4.

7.2.2.3   *One hot with* `while`

The following example is taken from the implicit style example in section 3.8.2.3.5. It is similar to the last example, except it uses a `while` loop (and a corresponding OR gate in the one hot controller to indicate the two paths that lead to the top of the `while`):

```
1: always
2: begin
3:   @ (posedge sysclk) #1;    //FIRST  is ff_3
4:     a <= @ (posedge sysclk) 1;
5:   @ (posedge sysclk) #1;    //SECOND is ff_5
6:     b <= @ (posedge sysclk) a;
7:     while (a == 1)
8:       begin
9:         @ (posedge sysclk) #1;//THIRD  is ff_9
```

Continued

```
10:        a <= @(posedge sysclk) b;
11:      @(posedge sysclk) #1;//FOURTH is ff_11
12:        b <= @(posedge sysclk) 4;
13:     end
14:  @(posedge sysclk) #1;   //FIFTH  is ff_14
15:    a <= @(posedge sysclk) 5;
16: end
```

The preprocessor generates the following one hot controller:

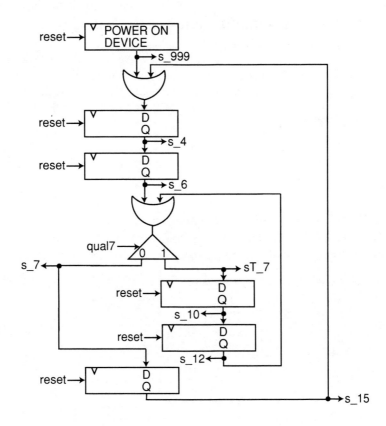

Figure 7-8. Controller generated from implicit Verilog of sections 7.2.2.3 and 3.8.2.3.5.

7.3   Mealy commands in a one hot machine

Chapter 5 describes Mealy machines, where a command can be conditional. Conditional commands, shown as ovals in ASM charts, are not guaranteed to execute simply because the machine is in a particular state. In order for a Mealy command to execute, some specified condition must also be true. In a one hot controller, this condition corresponds to an output `wire` from the proper demux. In a Mealy one hot controller, some of the wires associated with the statements that compose a state may not necessarily be active when the machine is in that state. This is why the preprocessor creates a `wire` for every statement: any statement in a Mealy machine might be conditional.

7.4   Moore command signals with Mealy <=

The VITO preprocessor only permits <=. At first glance, this might appear to prevent implementation of mixed ASMs or of ASMs that have external command outputs, such as `ready` in the ASM of section 2.2.7. In fact, as long as such command signals are unconditional (Moore), they can be described using Mealy <=. By doing so, the cost of the architecture will increase by some extra flip flop(s); however this is usually a small fraction of the total cost.

In addition, using the technique described below ensures that the command signal is hazard-free,[1] which is necessary in certain situations, such as interfacing to asynchronous memories (section 8.2.2.3.2).

7.4.1   Example to illustrate the technique

As an example to explain this technique, consider the following machine that asserts an external command signal, `comm`, when the machine is in state BOT:

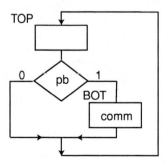

Figure 7-9. Example with Moore command signal.

---

[1] The physical cause for hazards is explained in section 6.3.3.

This can be described with the following implicit style behavioral Verilog:

```
always
   begin
     @(posedge sysclk) #1;        //TOP
       comm = 0;
       if (pb)
         begin
           @(posedge sysclk) #1;//BOT
             comm = 1;
         end
   end
```

As explained in section 3.9.1.2, the comm = 0 statement can been hidden inside the enter_new_state task so that 0 is the default value for comm:

```
always
   begin
     @(posedge sysclk) enter_new_state('TOP);
       if (pb)
         begin
           @(posedge sysclk) enter_new_state('BOT);
             comm = 1;
         end
   end

task enter_new_state;
  input ['NUM_STATE_BITS-1:0] this_state;
  begin
    present_state = this_state;
    #1 comm =0;
  end
endtask
```

Since the VITO preprocessor only allows <=, we need to describe a machine without using = whose behavior will be identical to the above after the first clock cycle.

One of the essential ideas used throughout this entire book is the meaning of the non-blocking assignment. It computes a value now but assigns that value to a register at **the next rising edge of the clock**. Since the above Verilog is a Moore machine, the command is synonymous with the machine being in a particular state. As described in sections 2.4 and 4.4.5, such Moore commands can be generated by combinational logic that is part of the next state logic:

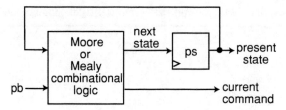

Figure 7-10. Current command approach suitable for Moore or Mealy controller.

Although Mealy machines must be defined using the above, we can look at a Moore machine such as this example in a different way. We know what the next state is going to be, and we know that there is a command synonymous with being in that next state. Instead of using combinational logic for the current command as we have done in previous chapters, we can instead use a register that will contain the *next command:*

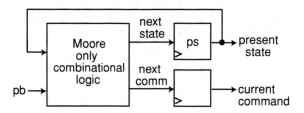

Figure 7-11. Next command approach suitable only for Moore controller.

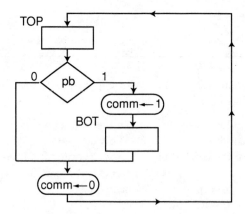

Figure 7-12. Behavioral ASM with ← for next command approach.

In the implicit style behavioral Verilog corresponding to figure 7-12, we use `<=` to assign values to the next command (`comm`) register:

```
always
  begin
    @(posedge sysclk) #1;       //TOP
      if (pb)
        begin
            comm <= @(posedge sysclk) 1;
          @(posedge sysclk) #1;//BOT
        end
      comm <= @(posedge sysclk) 0;
  end
```

Except for the time prior to the second full clock cycle (during which time `comm` is $1'bx$), the above machine has the same behavior as the original Verilog code that used `=`. Note that the non-blocking assignments are pure behavioral Mealy commands, and so the assignment is conditional. The machine only schedules the assignment of 1 to `comm` when the machine is already on the path where the next state will be state BOT. In other words, we only know that the next command will be 1 when we already know that the next state will be state BOT. The only way to get to state BOT is conditionally in state TOP when `pb == 1`.

Likewise, the machine only schedules the assignment of 0 to `comm` when the machine is already on the path where the next state will be state TOP. The next command will be 0 only when we know that the next state will be state TOP. There are of course two ways that we could know that the next state is state TOP: conditionally in state TOP when `pb == 0`, and unconditionally in state BOT. The non-blocking assignment, `comm <= 0`, only has to be described once because it was given after the `if` statement.

By rearranging the `comm <= @(posedge sysclk) 0` to the top of the `always` loop, the following is identical to the original Moore ASM, including the first *full* clock cycle:

```
always
  begin
      comm <= @(posedge sysclk) 0;
    @(posedge sysclk) #1;       //TOP
      if (pb)
        begin
            comm <= @(posedge sysclk) 1;
          @(posedge sysclk) #1;//BOT
        end
  end
```

This works on the assumption[2] that the power on circuit can reliably provide `s_999` as the command signal that clears `comm` prior to that first full clock cycle.[3] When this assumption holds, the above example illustrates a simple rule for converting a Moore machine with external commands into a Mealy machine using <=: put the <= for the next commands just prior to the @ (`posedge sysclk`) that marks the beginning of the corresponding next state. All commands must be described, including those to take on default values.

7.4.2   Pure behavioral two-state division example

Section 4.1.5 shows how to translate the two-state ASM chart for the childish division algorithm (section 2.2.7) into implicit style behavioral Verilog in a way that is suitable for simulation only (using the `enter_new_state` task). With the technique illustrated in the last section, it is very easy to modify this source code so that you can synthesize physical hardware directly from the implicit style (pure behavioral) Verilog without having to go through the tedious "mixed" and "pure structural" stages (sections 4.2 and 4.3). The boldface in the following shows the changes to the code from section 4.1.5:

```
module slow_div_system(pb,ready,x,y,r3,sysclk);
 input pb,x,y,sysclk;
 output ready,r3;
 wire pb;
 wire [11:0] x,y;
 reg ready;
 reg [11:0] r1,r2,r3;
 //reg ['NUM_STATE_BITS-1:0] present_state;
 wire sysclk;
 always
  begin
    ready <= @(posedge sysclk) 1;
   @(posedge sysclk) #1;                 //IDLE
    r1 <= @(posedge sysclk) x;
    r2 <= @(posedge sysclk) 0;
    //ready = 1;
```

---

[2] Some synthesis tools might not produce a reliable circuit under these circumstances, and so the former method (assigning 0 to `comm` at the bottom of the `always`) might be preferred. The latter Verilog code is logically correct, but its physical implementation may be unreliable, depending on the clock frequency and the physical properties of the technology.

[3] Actually, the signal that clears the register is `s_3`, which is the OR of `s_999` and the `wire` from the bottom of the `always` loop.

Continued

```
    if (pb)
      begin
        while (r1 >= y | pb)
          begin
            ready <= @(posedge sysclk) 0;
            @(posedge sysclk) #1;        //COMPUTE
            r1 <= @(posedge sysclk) r1 - y;
            r2 <= @(posedge sysclk) r2 + 1;
            r3 <= @(posedge sysclk) r2;
          end
      end
  end
  // task enter_new_state; ...
endmodule
```

7.4.3   Mixed two-state division example

Although the code in section 7.4.2 alone is enough to create physical hardware using appropriate synthesis tools (perhaps with the help of the VITO preprocessor described in appendix F), sometimes (for reasons of availability, speed or cost), the designer may want to create the architecture manually, as described in sections 2.3.1 and 4.2.3. The reader of chapter 4 may be left with the impression that in such a case, the designer must also create the controller manually. In fact, as long as the controller is a Moore machine, and the designer is willing to expend a few extra flip flops for the controller, it is possible to go straight from the mixed stage to physical hardware, without going through the tedious details of the "pure" structural stage.

Section 4.2.4 shows how to translate the mixed Moore ASM of the two-state division machine, (which generates command signals for a specific architecture that the designer has selected) into mixed Verilog. Using the techniques described in the preceding sections, here is equivalent implicit style Verilog acceptable to the synthesis preprocessor (bold indicates differences from section 4.2.4):

```
module slow_div_ctrl(pb,ready,aluctrl,muxctrl,ldr1,
              clrr2,incr2,ldr3,r1gey,sysclk);
  input pb,r1gey,sysclk;
  output ready,aluctrl,muxctrl,ldr1,clrr2,incr2,ldr3;

  //reg [`NUM_STATE_BITS-1:0] present_state;
  wire pb;
  reg ready;
  reg [5:0] aluctrl;
  reg muxctrl,ldr1,clrr2,incr2,ldr3;
  wire r1gey,sysclk;
```

Continued

```
always
 begin
    ready <= @(posedge sysclk) 1;
    aluctrl <= @(posedge sysclk) 'PASSB;
    muxctrl <= @(posedge sysclk) 0;
    ldr1 <= @(posedge sysclk) 1;
    clrr2 <= @(posedge sysclk) 1;
    incr2 <= @(posedge sysclk) 0;
    ldr3 <= @(posedge sysclk) 0;
   @(posedge sysclk) #1;              //IDLE
    //r1 <= @(posedge sysclk) x;
    //r2 <= @(posedge sysclk) 0;
    //ready = 1;
    //aluctrl = 'PASSB;
    //muxctrl = 0;
    //ldr1 = 1;
    //clrr2 = 1;
    if (pb)
     begin
       while (r1gey | pb)
        begin
           ready <= @(posedge sysclk) 0;
           aluctrl <=@(posedge sysclk) 'DIFFERENCE;
           muxctrl <= @(posedge sysclk) 1;
           ldr1 <= @(posedge sysclk) 1;
           clrr2 <= @(posedge sysclk) 0;
           incr2 <= @(posedge sysclk) 1;
           ldr3 <= @(posedge sysclk) 1;
          @(posedge sysclk) #1;     //COMPUTE
           //ready = 0;
           //r1 <= @(posedge sysclk) r1 - y;
           //r2 <= @(posedge sysclk) r2 + 1;
           //r3 <= @(posedge sysclk) r2;
           //aluctrl = 'DIFFERENCE;
           //muxctrl = 1;
           //ldr1 = 1;
           //incr2 = 1;
           //ldr3 = 1;
         end
      end
  end
endmodule
```

7.5 Bottom testing loops with `disable` inside `forever`

Although `while` loops are sufficient to implement any algorithm, and are preferable for many mathematical problems, there are situations when a bottom testing loop is more convenient than a `while` loop. Chapter 5 describes how the `enter_new_state` approach allows bottom testing loops to be simulated. The problem is that the technique used in chapter 5 is not acceptable to synthesis tools including the VITO preprocessor. There is another approach for bottom testing loops, involving the `disable` statement inside a `forever` loop that the preprocessor accepts.

The `disable` statement in Verilog has two main uses: stopping execution of a parallel block in a simulator (explained in section 6.4.3) and implementing bottom testing loops for synthesis (explained below).

As an example of a bottom testing loop, consider the following ASM. It is supposed to go through the bottom testing loop (consisting of states TOP and BOT) five times before returning to state OUTSIDE:

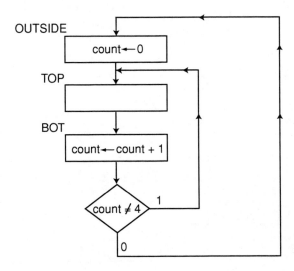

Figure 7-13. Example bottom testing loop.

Using the simulation technique of section 5.4.1, this could be translated to Verilog as follows:

```
always
 begin
  @(posedge sysclk) enter_new_state('OUTSIDE);
   count <= @(posedge sysclk) 0;
   while (count!=4 & present_state !== 'BOT);
    begin
     @(posedge sysclk) enter_new_state('TOP);
     @(posedge sysclk) enter_new_state('BOT);
      count <= @(posedge sysclk) count + 1;
    end
 end
```

The above simulates correctly. However, the above cannot be synthesized into a one hot machine using the VITO preprocessor.

An alternative way to implement a bottom testing loop is to use a `forever` statement with a `disable` statement inside. Using a `disable` statement requires an extra block that has a label to surround the `forever`. The `forever` by itself would never exit, and so the `disable` statement causes a `goto` the `end` that matches the labeled `begin`. For example, the above ASM could be translated into:

```
always
 begin
  @(posedge sysclk) #1;            //OUTSIDE
   count <= @(posedge sysclk) 0;
   begin : looplab
    forever
     begin
      @(posedge sysclk) #1;       //TOP
      @(posedge sysclk) #1;       //BOT
       count <= @(posedge sysclk) count + 1;
       if(count==4)
        begin
         disable looplab;
        end
     end
   end
 end
```

The above works correctly with the VITO preprocessor. However, this will not simulate properly on most Verilog simulators because the `disable` statement will also disable the non-blocking assignment. Putting #1 in front of the `disable` may help on some simulators, but on many simulators there seems to be no way to use `disable` in this way properly. Therefore, the Verilog you choose for a bottom testing loop will be quite different if you want to simulate than if you want to synthesize. This book has avoided using bottom testing loops in most examples in order that simulation may agree with synthesis, but there are situations where hardware designers prefer bottom testing loops.

7.6 Conclusion

One hot encoding provides a more natural way of translating complex algorithms into hardware than the binary encoded approach described in earlier chapters. Because of this, a preprocessor tool is available that directly translates an algorithm written in implicit style behavioral Verilog into a one hot circuit. There is a one to one mapping between the Verilog (or the equivalent ASM) and the one hot controller.

There are *graphical* software tools that can automatically translate an ASM chart into Verilog, but the use of such tools is often ill advised. The use of such tools locks the designer into a proprietary file format. Although manually drawn ASM charts are useful to a designer in the early stages of design, they lack the expressive power of Verilog to hide the details of a design with good notation. Instead, this book uses graphical **ASM charts** only as the **master plan** for the design. The **details** of the design occur in *textual* form as **implicit style** behavioral **Verilog**. With one of several commercial synthesis tools and perhaps the VITO synthesis preprocessor described in appendix F (that uses the one-hot techniques given in this chapter), implicit style Verilog alone is often enough to create operational hardware.

The central concept of this book is that algorithms can be described using pure behavioral ASM charts (with RTN) or the equivalent pure behavioral Verilog (with implicit style `whiles` and `ifs` together with the non-blocking assignment). This approach is different than traditional software because of the potential for parallel processing and because of the idea of the system clock. This approach is different than traditional hardware because of the emphasis on algorithms and behavior. Such implicit style behavioral Verilog algorithms (or their equivalent ASM charts) describe in an abstract fashion the operations carried out by some specific synchronous architecture. Chapters 4 and 5 show how you can manually design such architectures using Verilog, but the Verilog to one hot preprocessor (explained in this chapter) eliminates the need for such manual translation.

7.7 Further reading

ARNOLD, MARK G. and JAMES D. SHULER, "A Preprocessor that Converts Implicit Style Verilog into One-hot Designs," *6th International Verilog HDL Conference*, Santa Clara, CA, March 31-April 3, 1997, pp. 38-45. Gives more information about the VITO preprocessor.

PROSSER, FRANKLIN P. and DAVID E. WINKEL, *The Art of Digital Design: An Introduction to Top Down Design*, Prentice Hall PTR, Englewood Cliffs, NJ, 2nd ed., 1987. Chapter 5 gives examples of the one hot technique.

7.8 Exercises

7-1. Draw a circuit diagram for a one hot controller corresponding to the Moore ASM given in section 2.2.4. Label the output of each flip flop with the name of the state. Assume the command and status signals of the architecture are the same as in sections 2.3.1 and 4.2.3.

7-2. Draw a circuit diagram for a one hot controller corresponding to the Mealy ASM given in section 5.2.4. Label the output of each flip flop with the name of the state. Assume the command and status signals of the architecture are the same as in sections 2.3.1 and 4.2.3.

7-3. Draw a block diagram using muxes and combinational logic which is equivalent to the following continuous assignment (assume that the 12-bit a and b are defined elsewhere):

```
wire [11:0] new_a;
assign new_a =
  (s_10) ? a+b :
  (s_20) ? 2*a-b :  a;
```

7-4. Rewrite the pure behavioral Verilog of section 4.1.3 into the implicit style form suitable for the VITO preprocessor. (Eliminate the enter_new_state task and convert ready to <= as described in section 7.4.2.) Use the preprocessor to produce the continuous assignments that are equivalent to the one hot design. Draw a circuit diagram for the one hot controller labeled with the names used in the output of the preprocessor. Also draw a block diagram for the architecture constructed only with combinational logic, muxes and simple (non-enabled) D-type registers corresponding to the ? : in the output of the preprocessor.

7-5. Rewrite the pure behavioral Verilog of section 5.4.2 into the implicit style form suitable for the VITO preprocessor. (Eliminate the enter_new_state task and convert ready to <= as described in section 7.4.2. Also, use the disable statement in a different way than was described in this chapter.) Use the preprocessor to produce the continuous assignments that are equivalent to the one hot design. Draw a circuit diagram for the one hot controller labeled with the names used in the output of the preprocessor. Also draw a block diagram for the architecture constructed only with combinational logic, muxes and simple (non-enabled) D-type registers corresponding to the ? : in the output of the preprocessor.

8. GENERAL-PURPOSE COMPUTERS

8.1 Introduction and history

The machines described earlier in this book have each implemented a single algorithm that solves a specific problem, such as the childish division algorithm given in chapter 2. We use the term *special-purpose computer* to describe such machines, which are designed to solve only one problem. The designer of a special-purpose computer transforms the algorithm which solves just that one problem into the hardware structure that directly performs the computations required by that specific algorithm. The history of automation is filled with examples of such machines. Prior to the electronic age, Blaise Pascal's 1642 calculator, Jacquard's automated loom, Charles Babbage's 1823 difference engine, Herman Hollerith's[1] 1887 electromechanical punch card counter (which tabulated the 1890 U.S. census), Leonardo Torres y Quevedo's 1911 electromechanical chess playing machine as well as all of the calculators and business equipment of the early twentieth century are illustrations of special-purpose computers.

It is not surprising that the first "computers"[2] implemented with electronic (vacuum tube) technologies were also special-purpose machines. C. E. Wynn-Williams' 1932 binary counter (for nuclear particles), John V. Atanasoff and Clifford Berry's 1938 computer (later dubbed the ABC machine) at Iowa State University for solving simultaneous equations and the British Colossus machines of World War II (that cracked German coded messages) are all illustrations of successful vacuum tube implementations of special-purpose machines that followed specific algorithms.

In contrast to such specialized machines, a *general-purpose computer* is designed to solve any problem, limited only by the size of the machine. The idea of a general-purpose computer is independent of any technology. Babbage and Augusta Ada (Lady Lovelace) envisioned a machine that could create its own programs. Alan Turing published a theoretical paper in 1936 which proved there are mathematical functions that cannot be computed mechanically. To do this, he envisioned a (technologically inefficient but plausible) machine programmed via a "tape" that could be both read and written. The theoretical machines envisioned by Babbage and Turing were "universal" because they would have the capability of self-modification.

---

[1] Hollerith started a company that later became IBM.

[2] The term "computer" did not develop its current meaning as a machine that processes information until the 1950s. Previously, a "computer" was a person hired by a scientist to carry out an algorithm manually.

Governments on both sides during World War II focused more resources on the design of computers than had ever occurred before. Although at first many such machines were justified because they solved some important special-purpose problem, such as ballistics, the huge expense required to build and maintain such machines motivated several independent groups to design machines that could be reused to solve different problems. These wartime machines were not fully general-purpose in the modern sense, but were programmable via punched tape. The tape moved in one direction past a reader, and the holes told the machine what to do. Although on most such machines looping was not possible (because the tape moved in only one direction) and self-modification was not possible (because once a tape was punched, it could not be repunched), such machines made it easy to change the program by changing the tape.

Konrad Zuse filed a patent in Germany in 1936 on such a tape-controlled machine and built several versions of this machine, the first of which became operational in 1941. Colossus, in fact, was tape-controlled due to the flexibility required by British mathematicians (including Turing) who sought to break ever-changing German codes. George R. Stibitz and others at Bell Labs built several tape-controlled relay computers, some of which remained in use for over a decade after the war. In 1943 Howard Aiken and others at Harvard, with the help of engineers from IBM, built the tape-controlled Harvard Mark I, which was used by U.S. Navy personnel (including the later to become famous Admiral Grace Hopper). Near the end of the war, IBM started to build the SSEC, which was unique among the tape-controlled computers of the war because it had some limited ability for the type of self-modification alluded to by Babbage and Turing (and was therefore almost a true general-purpose computer).

John P. Eckert, John W. Mauchly and others in the Moore School at the University of Pennsylvania built ENIAC from 1943 to 1945 for ballistic computations required by the U. S. Army. It was the largest computer built during the war, constructed with an order of magnitude more vacuum tubes (nearly 20,000) than any of the other machines. Unlike other machines of the era, it was not programmed via a tape, but instead it had to be rewired (via a plugboard) to solve a different problem. (Designing a "program" for the ENIAC was similar to designing the controller and architecture as illustrated in chapter 2). This made ENIAC far more specialized and inconvenient than the tape-controlled machines. Recognizing this inconvenience, people at the Moore School (notably John von Neumann) proposed building EDVAC, which would represent programs in the same memory as data, rather than on tape or with a plugboard.

Although EDVAC was not the first general-purpose computer to become operational, von Neumann's 1945 proposal was profoundly influential. To this day, his name is synonymous with general-purpose computers that store their programs in the same memory as their data and that use what we now call the fetch/execute algorithm. The hardware implementation of the actual EDVAC machine did not become operational until 1951, in part because von Neumann left the Moore School to join Princeton's Institute for Advanced Studies where he designed the IAS machine.

The first general-purpose computer to become operational was a prototype (known as "Baby Mark I") built by Frederic C. Williams and Tom Kilburn at the University of Manchester in England. Although small (less than one thousand vacuum tubes) and limited (only 32 words of memory and no hardware support for division), it ran its first software program (division using the childish algorithm described in chapter 2) on June 21, 1948. A later version of this machine, the Manchester Mark I, became operational in 1949 with 128 words of memory. A commercial version of the Mark I was produced in Britain a few years later by Ferranti. Also in Britain during 1949, Maurice Wilkes and others at Cambridge completed the EDSAC, which had 512 words of memory.

In the U.S., the first operational general-purpose computer was BINAC with 512 words of memory, built in 1949 by Eckert and Mauchly after they left the Moore School to start their own company. Their company later produced the UNIVAC, which was the first general-purpose computer that was commercially successful in the U.S. (more than 20 were sold).

Since the early 1950s, thousands of slightly different implementations of general-purpose computers have proliferated worldwide. Although they differ quite significantly in the details, all of them implement essentially the same algorithm in hardware: fetch/execute, which is the subject of this chapter.

8.2   Structure of the machine

Since the Manchester Mark I, almost all general-purpose machines have had the same overall top-level structure, illustrated by the following diagram:

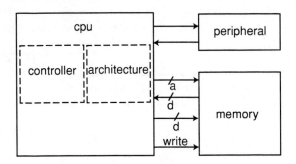

Figure 8-1. Block diagram of typical general-purpose computer.

8.2.1 CPU, peripheral and memory

There are typically three components of a general-purpose machine:

1. The Central Processing Unit (CPU) is composed of a controller and an architecture. As discussed in chapter 2, for special-purpose computers, the architecture is where the ALU(s), register(s) and other computational elements reside.

2. The peripheral(s) are distinct special-purpose computer(s) that interface the CPU to outside actors, such as a keyboard. Each interface typically has its own controller and architecture, including synchronizers as required.

3. Memory is a special kind of device that stores the bits that can represent both programs and data.

Sections 8.3 and 8.4 of this chapter will illustrate how the same techniques used for special-purpose computers in chapter 2 can implement the CPU of a general-purpose computer. The next section describes memory.

8.2.2 Memory: data and the stored program

The one component that has not been discussed in detail previously is memory.[3] From a behavioral standpoint, a *memory* is simply an array of words. The subscript to this array is known as an *address*. We will refer to the number of bits required for the address as a. The designer can choose how many bits (d) are in each word, and also how many words (2^a) are in the memory. For example, Williams and Kilburn's first machine had a word size of 32 bits, and there were 32 such words in the memory (five address bits). Therefore, their memory had $32*2^5 = 1024$ bits total. Later in this chapter we will design a machine with 4096 words (12-bit address), and a wordsize of 12-bits. This machine will require a memory with $12*2^{12} = 49,152$ bits. Most machines typically have memories that hold billions of bits, but that is a detail that is irrelevant to learning the essential ideas of this chapter.

In this chapter and in appendices A and B, we will indicate both address and contents of memory in octal. As an abbreviation, the address will be shown on the left, and the contents will be shown on the right, with a slash separating the address from the contents. For example,

```
0123/4567
```

---

[3] The memory we are talking about can be used both to store and retreive bits. It is refered to as "RAM" by some people.

means the address is 0123_8 == 0000010100011_2 == 83 and the contents is 4567_8 == 100101110111_2 == 2423, or more succinctly in array notation, memory[83] == 2423. We will sometimes abbreviate even further to say m[83] == 2423.

There are five independent issues that can be used to categorize memory: unidirectional versus bidirectional (section 8.2.2.1), deterministic versus non-deterministic access time (section 8.2.2.2), synchronous versus asynchronous (section 8.2.2.3), static versus dynamic (section 8.2.2.4) and volatile versus non-volatile (section 8.2.2.5).[4]

8.2.2.1 *Unidirectional buses versus a bidirectional bus*

There are two common variations on how a memory device connects to the rest of the system. One approach uses simple unidirectional buses of the type seen in chapter 2, and the other combines two data buses together into what is known as a bidirectional bus.

The simplest form of memory has two input buses and one output bus. This simple type of memory with only unidirectional buses is what we will use in this chapter. In this type of memory, the d-bit-wide din bus is an input to the memory device, and the d-bit-wide dout bus is the output of the memory device. Also, the a-bit wide addr bus is another input to the memory device. There must be additional input(s) to the memory device, which are described in later sections.

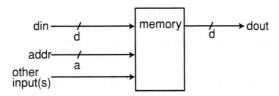

Figure 8-2. Symbol for memory with unidirectional data buses.

A *bidirectional* bus is one that is used to send information two ways. In the following diagram, a bidirectional bus is indicated by an arrow that points both ways. In this case, bidirectionality allows combining the din and dout buses, into a single data bus as illustrated in the following:

---

[4] A different issue related to memory (not discussed in this chapter) is how many ports the memory has. Multi-ported memory is discussed in section 9.8, but in this chapter all memory is assumed to have only one read port and one write port, as illustrated in the following sections.

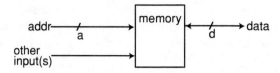

Figure 8-3. Symbol for memory with bidirectional data bus.

The advantage of bidirectionality is that there are fewer wires connecting the memory device to the rest of the system, however, interfacing to such a device is more complicated. This requires the use of tri-state buffers. Except for such tri-state buffers, the internal structure of this memory is identical to that of a memory with separate `din` and `dout` buses. We will not consider memory devices with bidirectional buses in this chapter.

8.2.2.2 Deterministic versus non-deterministic access time

Like all other physical devices, the actual memory hardware is not instantaneous. Although at the early behavioral stages of the design we prefer to ignore the time it takes to use the memory (known as the *access time*), ultimately, the speed of the memory will have a major influence on the speed of a general-purpose computer (since both the program and the data have to be obtained from the same memory). In the final stages of design, the designer must consider memory timing.

When the access time is *deterministic*, the time to obtain an arbitrary bit from memory does not vary significantly as a function of the address.

Almost all primary memories used today have deterministic access times. Almost all secondary memories (e.g., disk drives) have non-deterministic access times.

8.2.2.3 Synchronous versus asynchronous memory

There are two different ways in which the memory timing can occur: synchronous and asynchronous. The difference is whether or not the memory uses the system clock.

8.2.2.3.1 Synchronous memory

This kind of memory is the fastest, most expensive and simplest for the designer to incorporate. This kind of memory is commonly used where speed is important, such as in pipelined systems (section 6.5.8.3) or RISC computers (chapter 10). Because of its cost, it has not generally been used for the primary memory of a stored program computer, although recently, as clock speeds have increased, synchronous primary memories have become more commonplace.

In addition to the address and data buses, there must be a command input, ldmem, that tells the memory what to do. In order for the memory to be synchronous with the rest of the machine, it also needs a clock input. The top-level structure of a synchronous memory with separate din and dout is shown below:

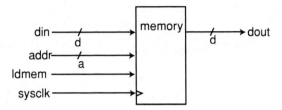

Figure 8-4. Symbol for synchronous memory.

There are two things that the device does. Given enough time, the output of the device reflects the contents of the memory at the word indicated by the address bus. In other words, neglecting the propagation delay (i.e., neglecting the access time),

$$dout = memory[addr]$$

The second thing that the memory can do is based on the ldmem command signal. On the next rising edge after ldmem becomes true, the word in memory indicated by the address bus changes to become the value of the data input bus,

$$memory[addr] \leftarrow din$$

Note that almost instantly after this change takes effect, dout will also change. At most one word in memory can be changed in one clock cycle when a memory is single-ported.[5] If ldmem is not true, memory remains unchanged.

A synchronous memory device can be thought of as a bank of registers. Although it is not usually the most efficient way to build a memory, the following diagram shows a structure that achieves this goal:

---

[5] Chapters 9 and 10 describe multi-ported memories that allow more than one memory operation per clock cycle.

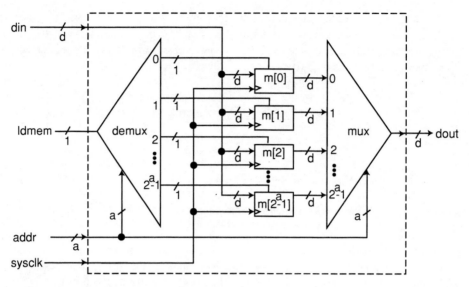

Figure 8-5. Implementation of synchronous memory.

This diagram uses 2^a enabled registers, each containing d bits. It also has a $1*2^a$ demux and a $d*2^a$ mux.

Let's ignore the left-hand side of this diagram (the demux that connects to the ldmem signal) for the moment. For example, assume the proper data has already been placed in each register, and the user wishes to obtain the contents of one of the memory locations. The user provides the address of the desired memory cell on the addr bus. The mux selects the output of the corresponding register, and outputs that on the dout bus.

To understand how a user is able to write new data into memory, you need to recall what the combinational logic of a demux does. When ldmem is 0, the demux will produce 0s on all of its 2^a outputs, so none of the registers would change. When ldmem is 1, the demux will output a 1 on **exactly** one of its outputs and 0s on the others. The output that is 1 will be determined by the value of the addr bus. Therefore if ldmem is 1, only the contents of the register corresponding to the current addr bus will change at the next rising edge of the clock.

Notice that the above implementation has a deterministic access time (essentially the propagation delay of the mux). It is possible to build a synchronous clocked memory based on shift registers, where the access time varies depending on how many clock cycles are required to shift the desired bit out. However, there is seldom any advantage to such a memory.

8.2.2.3.2 Asynchronous memory

A significant portion of the cost of the memory shown in 8.2.2.3.1 is due to the clock being provided to each register. All of the cheaper memory technologies invented have been asynchronous, which means they do not use the system clock. If the designer can cope with a memory that is asynchronous, the cost of memory can be reduced.

The block diagram for such an asynchronous memory (with separate `din` and `dout` ports) is:

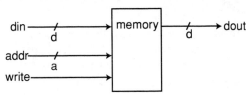

Figure 8-6. Symbol for asynchronous memory.

Here the `write` signal combines the roles of the `ldmem` signal and the `sysclk` signal. Asynchronous memory may also have a bidirectional data bus instead of two unidirectional buses.

In general, asynchronous design is highly unsafe, and should only be attempted by expert designers. Proper asynchronous design involves consideration of much lower (electronic) details than is the case for synchronous design. With the introduction of HDLs, the vast majority of design (such as CPUs) in industry is synchronous because synchronous designs are much more likely to be synthesized correctly. Asynchronous design is beyond the scope of this text, and so we will not consider the internal structure that implements this memory (although it is similar in concept to the diagram in 8.2.2.3.1).

Fortunately, since memory is such an important commodity, electronic experts have hidden most of the asynchronous ugliness inside commonly available memory chips. To cope safely with such devices, there are only three extra things that the designer has to do:

1. Choose a conservative clock speed for the rest of the system relative to the access time of the memory. In other words, the access time of the memory should be a small fraction of the clock period. Some memories have a different time for read and write, and so you should choose the larger of these.

2. Hold `addr` and `din` constant for at least one clock cycle before and during the cycle `write` is active. This means both `addr` and `din` should come from registers in the architecture that are not changed during this time.

3. Make sure there are no hazards in `write` (see section 6.3.3). As long as the controller is a Moore machine, this is easy to guarantee (section 7.4).

Sometime after the memory receives the `write` pulse but before the access time has elapsed, the asynchronous memory will latch the new value into memory. In this way, the asynchronous `write` pulse combines the role of the system clock and the `ldmem` signal.

8.2.2.4 Static versus dynamic memory

Static memory retains its contents regardless of whether it is used or not. The only limitation on using static memory is that the clock speed can be no faster than the access time. (More precisely, the maximum clock speed needs to consider all propagation delays, including access time. Examples of calculating the maximum possible clock speed are given in chapter 6.)

Dynamic memory places an additional requirement on the designer. Not only is there a maximum possible clock speed, but there is also a minimum clock speed. This is because every word in dynamic memory needs to be *refreshed*. Dynamic memory technology is usually based on very cheap electrical devices known as capacitors, which store charge. Over time, the charge leaks away. Unless the capacitors are refreshed, the information will disappear.

Dynamic memory is the cheapest kind of fast memory that is currently available. For problems where the machine will continually use all memory addresses over and over, there is no extra inconvenience to use dynamic memory. For most other problems, where this cannot be guaranteed, it is best to use a *dynamic memory controller* between the dynamic memory and the rest of the system.

8.2.2.5 Volatile versus non-volatile

A memory device is *volatile* if it loses its contents when the power is turned off. A memory device is *non-volatile* if it can retain its contents without consuming any power. A memory technology that is inherently non-volatile would be desirable because it allows a program to remain in memory when the power is turned off; however with most technologies, it is more cost-effective to provide a backup battery to preserve memory contents when the power to the rest of the system is removed.

8.2.3 History of memory technology

One of the recurring themes of this book is that *technology changes, but algorithms endure*. This means the time you spend honing your problem solving skills will benefit you throughout your career because such skills do not become outdated. Learning about the fetch/execute algorithm in particular is important because it is the algorithm which makes your career possible.[6]

From the historical account in section 8.1, you might wonder why it took so long for the general-purpose computer to be realized. The reason is a corollary to the italicized phrase above: some algorithms have to wait to be implemented until technology changes enough so the hardware is affordable and practical. To see how this postponed implementation of general-purpose computers until after World War II, consider the three components of a general-purpose machine: CPU, peripherals and memory.

As the ABC, Colossus and ENIAC illustrate, vacuum tube technology was available by the start of World War II to implement CPUs and peripherals. The technological problem from the late 1930s until the early 1950s was memory. Although the pioneers were aware of techniques like section 8.2.2.1.1 and used small memories of this sort for data access, the cost of storing programs in such memory was prohibitive. Currently, memories of this kind are commonly used, but not as the primary memory for stored program computers. It takes about six switching devices (relays, vacuum tubes, transistors or whatever technology is in vogue) to construct an enabled flip flop, so it would take $6*d*2^a$ switching devices to build the registers. It takes approximately $a*2^a$ switching devices to construct the demux, and $d*a*2^a$ switching devices for the mux. This makes the total about $((a+6)*d + a)*2^a$ switching devices to construct a working memory unit along the lines shown in section 8.2.2.3.1.

For Williams and Kilburn's 32 word memory (which, even in 1948, was considered too small for practical programming), this would require $((5+6)*32 + 5)*32 = 11,424$ vacuum tubes, which is more than an order of magnitude more tubes than was required for their entire CPU. (The ENIAC used about 20,000 vacuum tubes because it stored all its **data** in vacuum tubes. Also, storing a **program** in vacuum tubes would have been unrealistic.)

In order to build their machines, the pioneers had to invent technologies for memory that were more efficient and reliable than simple vacuum tubes. Zuse invented a binary mechanical memory. Atanasoff invented a rotating drum using capacitors (which is conceptually similar to the dynamic memory chips in widespread use today). Although creative, neither of these technologies would have been reasonable for a general-purpose computer in the 1940s.

The breakthrough came when Williams and Kilburn invented a TV-like tube for storing bits in the Mark I. Using the terminology defined above, the Williams tube was bidirectional, asynchronous, dynamic and volatile. Most importantly, the Williams tube was the first affordable technology that had the same kind of deterministic access time

---

[6] For example, modern Verilog simulators and synthesis tools are possible only because of large and fast general-purpose computers.

provided by the mux in section 8.2.2.3.1. The Williams tube also had the desirable side effect that a programmer could actually see every bit in memory (since they were actually stored as glowing dots of electric charge on the phosphor screen).

The only other cost-effective memory technology available before 1953 was the delay line, which had been used during the war for (analog) radar signals. These were also bidirectional, asynchronous, dynamic and volatile. The problem with delay lines is that they do not have deterministic access times. A delay line memory recirculates the same data over and over again. One has to wait until the desired data comes out of the delay line before it can be used. Therefore, there is a range of possible access times, the longest of which is quite slow. Wilkes as well as Eckert and Mauchly used such delay lines for the memory on their first general-purpose machines. Most computers of the early 1950s used delay lines, including IBM's first general-purpose machine, the model 701.

In 1953, Jay W. Forrester at MIT invented a new memory technology that stores each bit on a small doughnut-shaped magnetic *core*. Wires were woven through the cores to interface them to the system. By selecting which wire electric current flows through, the system could selectively magnetize or demagnetize each core (corresponding to storing a 1 or a 0). Core memory is fast compared to earlier technologies. It is also nearly ideal from a designer's viewpoint. It is unidirectional, deterministic, asynchronous, static and non-volatile. It is far less expensive to construct a core memory of a given size than to construct a comparable memory (of the kind outlined in sections 8.2.2.3.1 and 8.2.2.3.2) using vacuum tubes or transistors. Core memory dominated the computer industry until the late 1970s so much so that the term "core" became synonymous with the primary memory of a general-purpose computer. The first practical application of core memory was in Forrester's general-purpose Whirlwind computers, used by the U. S. Air Force for strategic defense for decades. Although it is not as cheap as the technology that replaced it, the military continues to use core memory because it is non-volatile, and it can retain its contents better than any other technologies when in close proximity to a nuclear explosion.

The final technological change for memory occurred at the end of the 1960s when Robert Noyce, Gordon Moore and others at Intel developed semiconductor integrated circuit memories that had all of the hardware for a memory similar to the one described in 8.2.2.3.1 on a single chip of silicon.

Integrated circuit memories come in many varieties. Today there are many competing manufacturers worldwide of interchangeable memory chips. A designer is able to choose from many different chips in a trade-off between speed, cost and ease of design.

Some chips are unidirectional, but others are bidirectional. Since the number of pins on an integrated circuit tends to be more of a constraint than the number of devices that can fit on the chip, larger integrated circuit memories tend to be bidirectional.

Almost all integrated circuit memories have deterministic access times. In the 1970s, research occurred in magnetic "bubble" memories with non-deterministic access times, however these memories did not succeed in the marketplace.

Some integrated circuit memories are synchronous, but many others are asynchronous. Larger memories have tended to be asynchronous because that enables more bits to be packed onto a chip of comparable area. The difference is between about six switching devices per bit versus only two switching devices per bit.

Many integrated circuit memories are static, which allows designers to observe them operate slowly enough for the details to be intelligible. Such memories are ideal for a computer design lab. Also, static memories tend to be faster than comparable dynamic memories. Despite these desirable properties, larger memories are dynamic. The difference is between two switching devices per bit versus one switching device and a capacitor per bit. Synchronous dynamic memories offer high speed at low cost.

Almost all existing integrated circuit memories are volatile. Successful research into ferrous semiconductors that would in essence put core memory on a chip occurred in the early 1990s. Whether such memories will be successfully commercialized remains to be seen at the time of this writing.

One principal limitation of integrated circuit memories is the number of pins that connect the memory to the CPU. In the mid 1990s, attempts were made by Cray Computer and others to overcome this restriction by fabricating multiple CPUs on the same chip as memory. Whether such "smart memories" will be successfully commercialized remains to be seen at the time of this writing.

For the first quarter of a century of the computer age, the physical appearance of memory devices changed radically as technology improved. For the second quarter of a century, memory looked basically the same: a silicon chip. As bit densities increased, packaging changed to hold more bits, but the semiconductor electronics that store each bit have remained essentially the same. By the mid 1990s, Single In-line Memory Modules (SIMMs) and Dual In-line Memory Modules (DIMMs) that can fit in the palm of your hand and that can contain billions of bits became a common way to package several dynamic memory chips.

What the preferred memory technology will look like by the end of the 21st century is, of course, unclear. Although visibility of bits was simply a side effect of the properties of the phosphor in the Williams tube, the idea of using light to store information has not gone away. In the late 1990s, prototype photochemical and holographic memories accessed using lasers were demonstrated that have the potential of storing orders of magnitude more bits than semiconductor memories. Daydreaming farther into the future, perhaps some nanomechanical computer designer in the 21st century might even pursue Zuse's memory designs at the atomic scale!

What is certain is that the cost, speed and capacity of integrated circuit memory has improved radically in the last quarter century. It is likely these improvements will continue well into the 21st century. That these technological factors have improved exponentially is in large part responsible for the success of the general-purpose computer, which needs to store both its programs and its data in memory.

8.3 Behavioral fetch/execute

The three components of a general-purpose computer described in section 8.2 (CPU, peripherals and memory) act together as a unified system that implements the fetch/execute algorithm. This section describes how to model the behavior of this unified system with an ASM chart, without regard to the structural interconnection of the hardware. This section explains what is referred to in chapter 2 as the "pure" behavioral stage of the top-down design process. Later, in section 8.4, the "mixed" stage of the top-down design process shows some of the structural interconnections for the CPU and memory.

This section focuses on the algorithm that makes the general-purpose possible: fetch/execute. Although the details of the fetch/execute algorithm vary widely among the thousands of general-purpose machines designed and built since 1948, the fundamental operations of the fetch/execute algorithm have remained essentially the same:

1. Fetch the current instruction from memory
2. If needed, fetch data from memory
3. Prepare for fetching the next instruction
4. Decode and execute the current instruction
    a) Interpret what the current instruction means
    b) Carry out the operation asked for by the current instruction, possibly modifying memory

Steps 2 and 4 have details that are machine specific. It may be possible to rearrange the order in which steps 2, 3 and 4 occur, depending on these machine-specific details.

A general-purpose computer can modify its instructions without programmer intervention because it uses the same memory to store instructions as it uses to store data. In other words, it can treat instructions as though they were data. This characteristic of universal machines, known as self-modification, is difficult for programmers to use effectively. However, this capability is the key to the success of the general-purpose computer. The ability for self-modification allows software (known as compilers and assemblers) to translate programs automatically from an easy to understand high-level language (C, Java, Pascal, Verilog, etc.) to the much more tedious machine language that is specific to the hardware.

For readers without intimate experience with low-level programming, appendix A gives a short introduction to machine and assembly language (and how they relate to high-level language) using an example of adding three numbers. This example will also be used in later sections of this chapter.

8.3.1 Limited instruction set

Although the fetch/execute algorithm is similar on all general-purpose computers, the machine-specific details depend on the instruction set being implemented. The *instruction set* is the set of machine language bit patterns that the hardware can interpret. All software on a particular machine is eventually translated to such instructions. The hardware is only capable of executing instructions that are in its unique instruction set. Although conceptually similar, a different model computer probably has an entirely different instruction set.

8.3.1.1 The PDP-8

We need a simple yet concrete example of an instruction set so that we can go through the stages of the top-down process, starting at the abstract algorithm for fetch/execute (which has remained essentially unchanged for half a century) and concluding with a unique hardware structure that implements those instructions. The instruction set that we will use as an example in this chapter is a subset of the PDP-8's instruction set. The PDP-8 is a classic example of what is called *a single-accumulator, one-address instruction set*. (All early stored program machines, including the Manchester Mark I, had this simple kind of instruction set.)

8.3.1.2 History of the PDP-8

The PDP-8, which was designed by C. Gordon Bell and Ed DeCastro at DEC in 1965, is pivotal in the history of general-purpose computers. It was the first computer that cost only a few thousand rather than hundreds of thousands of dollars. Bell was able to achieve this with core memory and transistor technology by keeping the instruction set simple and the memory small. The PDP-8 continued to be manufactured (with improved technologies) into the 1990s due to the simplicity and elegance of its instruction set. These attributes also make it an ideal example of the fetch/execute algorithm.

8.3.1.3 Instruction subset

The complete PDP-8 instruction set, which is described in appendix B, has about thirty instructions. Even though the PDP-8 has one of the simplest and smallest instruction sets ever designed, attempting to concentrate on all thirty of these instructions at once would distract from our primary goal: understanding the enduring fetch/execute algorithm. Therefore, for the example in this section, we will implement a machine that executes only the following four PDP-8 instructions (that are also used in the example of summing three numbers in appendix A):

mnemonic	octal machine language	what the mnemonic stands for
TAD	1xxx	add memory to accumulator (Twos complement ADd)
DCA	3xxx	Deposit accumulator in memory and Clear Accumulator
CLA	7200	CLear Accumulator
HLT	7402	HaLT

This subset does not contain enough of the PDP-8's instruction set for practical programming, but it provides a good introduction to fetch/execute. The capitalized letters explain what the mnemonic stands for. The first two instructions (TAD and DCA) are memory reference instructions, which require the machine to calculate an *effective address* (the address of the data in memory that the instruction is going to reference). Although, like most other instruction sets, the PDP-8 has several variations (known as *addressing modes*) on how to calculate the effective address, we will only consider the simplest one of these, known as *direct* page zero addressing. (Two bits in the instruction register determine which addressing mode the machine uses.) With direct page zero addressing mode, the effective address is simply the low-order seven bits of the instruction, denoted as xxx in the octal machine language above. The reason the PDP-8 is known as a one-address machine is because each instruction uses at most a single effective address.

The next instruction (CLA) manipulates the accumulator register without referencing memory; therefore it does not need an effective address. The final instruction (HLT) causes the machine to stop executing a program and instead proceed to a special state where the machine waits until an external signal tells it to run another program.

8.3.1.4   Registers needed for fetch/execute
The pure behavioral ASM for a general-purpose computer uses register transfer notation, similar to that of a special-purpose computer, as explained in chapter 2. Therefore, we need to determine what registers will be manipulated in the behavioral fetch/execute ASM.

Some of the registers are specified by the specific instruction set. The details of these registers are machine dependent. In the case of the PDP-8, the 12-bit accumulator, `ac`, is the primary register that the machine language programmer uses. (There are a few other registers, such as the `link`, that are specific to the PDP-8. As was done in appendix A, we will ignore these for the moment in order to keep this example simple.) Other machines, such as the Pentium, have different registers that the programmer can manipulate. We refer to the registers that are visible to the programmer as the *programmer's model*. Some people refer to these as the *computer architecture*; however we do not use this term since the registers in the programmer's model are not everything contained in the internal architecture of the CPU.

In addition to the registers required to implement a specific machine language, the fetch/execute algorithm requires the hardware to have two registers: the program counter, `pc`, and the instruction register, `ir`. Typically, the `pc` contains the address in memory of the next instruction to execute, and the `ir` contains the current instruction which is about to execute. If the machine did not have an HLT instruction, the machine would simply loop forever doing the four phases of the algorithm:

1. fetch the instruction from `m[pc]` into `ir`
2. calculate the effective address
3. increment the `pc` (prepare for fetch of *next* instruction)
4. decode and execute the instruction in the `ir`

where `m` refers to memory array. Most machines, including the PDP-8, have some form of HLT instruction. In order to keep track of whether the machine has halted or not, there needs to be an additional one-bit register, `halt`. When the machine has not executed an HLT instruction, `halt` is 0. When the machine has just executed an HLT instruction, `halt` becomes 1. The fetch/execute algorithm proceeds only when `halt` is 0.

The machine needs a register to hold the effective address of data in memory to be manipulated by an instruction. This register, which may be used for other purposes at different times, is known as the memory address register, `ma`. It will be convenient to have an additional register, known as the memory buffer register, `mb`, to contain the data that was in memory at the effective address prior to the execution of the instruction.

In later stages of the top-down design process, it will be convenient to have `ma` as the sole source providing the `addr` input to the memory device. At the "pure" behavioral stage, we can ensure this is possible by restricting the use of the memory array. All references to memory must be `m[ma]`, rather than the somewhat more natural references, `m[pc]`. Also it will be convenient to have `mb` as the sole source providing the `din` input to the memory device. In the restricted behavioral ASM, the only way to store something into memory is by saying `m[ma]` ← `mb`. This will require that the behavioral ASM have states that initialize `mb` properly.

8.3.1.5 ASM for fetch/execute

The following is an ASM that implements the fetch/execute algorithm for the tiny
instruction set described in section 8.3.1.3:

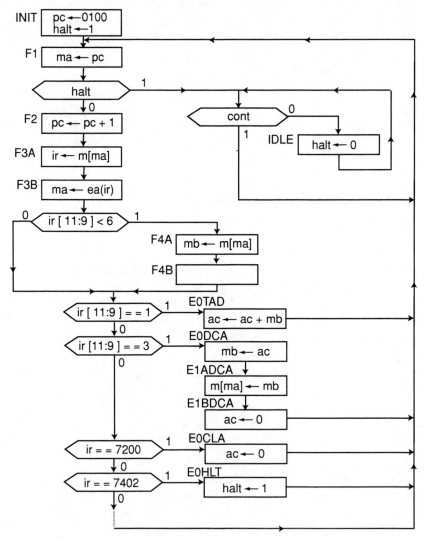

Figure 8-7. ASM implementing four instructions of PDP-8.

For ease of notation, the four-digit constants in the above ASM (0100, 7200 and 7402) are in octal. Smaller constants are shown in decimal. Although the names are arbitrary, the first letter in the names chosen for most of the states indicates the role of the state in the fetch/execute algorithm. States whose names begin with the letter "F" primarily have to do with the fetch aspect of the algorithm. States whose names begin with the letter "E" have to do with execution of machine language instructions. States whose names begin with other letters (e.g., "I") have to do with aspects of the machine besides the fetch/execute algorithm, such as interfacing to the programmer.

When the machine is first powered on, it goes through the INIT state once, where halt is initially set to 1, and pc is set to an arbitrary address (0100 octal in this example) where we assume the program resides. (Later we will make the location of the program more flexible, but for this limited example, assume the program always starts at 0100.)

When halt is 1, the machine proceeds to the IDLE state. When the machine is in the IDLE state, it waits for an external signal, known on the PDP-8 as cont ("continue"), that tells it to run another program. When cont is true in state IDLE, the machine proceeds to the beginning of the actual fetch/execute algorithm, which starts at state F1. Since state IDLE clears the halt register, the machine will not return to state IDLE until the fetch/execute algorithm (F1, F2, F3A, F3B, ...) has executed an HLT instruction.

State F1 is necessary because of the restriction on the use of memory described at the end of section 8.3.1.4. Since pc was set up to contain the address of the next instruction that should be executed (either by state INIT for the first time through the fetch/execute algorithm, or by some state inside the algorithm for later instructions), it would be natural to say something like ir ← m[pc]; however this violates the aforementioned restriction. At the time the machine enters state F1, the ma contains no useful information, and so it is possible to copy the program counter into ma. This will be used by a later state (F3A) to actually fetch the current instruction into the instruction register.

State F1 also needs to check the halt register to see if the previous instruction that just finished executing was an HLT instruction or if the machine has just been powered up (the machine just came from state INIT). If halt is true, the machine proceeds to state IDLE as explained above.

If halt is not true, the machine proceeds to state F2. This state prepares the machine for executing the next instruction after the current one by incrementing the program counter. The placement of state F2 is somewhat arbitrary since a copy of the original program counter has been preserved for the moment in the memory address register. (For example, state F2 could be placed after state F3B, which would more closely match the description given in appendix A.)

The next state after state F2 is state F3A. State F3A fetches the instruction stored in memory pointed to by the original program counter, which is now in the memory address register. In the behavioral ASM, we use the same register transfer notation for dealing with memory, `ir←m[ma]`, as is used for dealing with other registers. In later stages of the design, the timing of memory may be somewhat more difficult than that of the internal CPU registers, but at this early stage, we will ignore these details.

The next state after state F3A is state F3B. State F3B calculates the effective address using the information in the instruction register. This calculation is denoted by a function referred to as `ea(ir)`. For example, if the `ir` is 1107_8, `ea(ir)` is 0107_8. If the `ir` is 3111_8, `ea(ir)` is 0111_8. In the later stages of the top-down design process, the `ea` function will be realized using combinational logic. Appendix A assumes there is an additional register for the effective address, but this is not necessary here since state F3B uses the existing `ma` register to hold the effective address. (To implement the complete instruction set of the PDP-8 described in appendix B, more complicated combinational logic is required for `ea`. This is because the `ea` function for some of the addressing modes not implemented in this chapter needs an additional argument.)

State F3B has a decision to determine what state occurs next. If the instruction is a Memory Reference Instruction (MRI), the next state after F3B will be F4A. If the instruction is not an MRI, the next state after F3A will be one that implements the operation requested by the instruction ("E"xecute it). Even though in this section the only MRIs in our PDP-8 subset start with 1 and 3, we will describe how to test for any MRI. In the complete PDP-8 instruction set (given in appendix B), an instruction is MRI if the high order octal digit (three bits) of the instruction is between 0 through 5 inclusive. There are several ways one could write this test. It could be written as `ir < 6000`, however, this does not emphasize that only the high-order three bits of the instruction register determine the outcome. The test could be written as `ir/1000₈ < 6` or `ir>>9 < 6` to emphasize that the outcome is based on the high-order three bits, but neither one of these tests is the most succinct way to express this. We need a notation that clearly says "just look at these bits." Although the material in this section does not depend on any knowledge of Verilog, Verilog does indeed have such a bit selection notation: `ir[11:9]` says form a three-bit value using bits 11 through 9 of `ir`, which is roughly equivalent to `(ir >> 9) & 7`, which in this case is equivalent to `ir >> 9` since `ir` is 12-bits. For example, if `ir` is 1107_8, `ir[11:9]` is 1. If `ir` is 3111_8, `ir[11:9]` is 3. We will use this aspect of Verilog notation in our ASM because it clearly documents what the hardware will do, which of course is the goal of a behavioral ASM.

If the instruction in the `ir` is MRI, the machine proceeds to state F4A. State F4B loads `mb` with the data that the machine may need to use to execute the memory reference instruction. For example, in the program of appendix A, when the memory reference instruction 1107 is fetched by state F3A, the machine schedules 0107 to be loaded into ma at the next rising edge of the clock. Since ir[11:9] is 1 (which is less than 6), at that next clock edge the machine proceeds to state F4A. In state F4A, ma has just become 0107, and so the contents of memory at that effective address, m[ma], can be loaded into mb. In this example, one clock cycle later (when the machine is in state F4B), mb becomes 0152.

State F4B is not necessary. It is included here as a placeholder for operations needed in the ASM to implement the complete instruction set of the PDP-8, including features not described yet.

The bottom of the ASM has a series of decisions that determines which instruction is currently in the instruction register.

This series of decisions is known as the *decoding* portion of the fetch/execute algorithm. For MRI, the decoding decisions happen in state F4B. For non-MRI, the decoding decisions happen in state F3B. Since we are implementing only four instructions in this instruction subset, there are only four decisions required to decode these instructions. The more complex the instruction set, the more difficult it is to do decoding. Most machines, including the complete PDP-8 have a long string of decisions to implement decoding. Notice, from a high-level view, decoding occurs as a series of if ... else if ... else if ... style decisions, since each instruction has a unique machine language code.

The remaining states of the machine perform certain actions required during the execution of each specific instruction.

If ir[11:9] is 1 in state F4B, the instruction is what the programmer calls a "Twos complement ADd," and so the machine proceeds to state E0TAD. In this state, the machine adds the data from memory at the effective address to the accumulator. In a complete implementation of the PDP-8, other operations are involved with a TAD, but we will ignore those details for the moment. After performing the addition in state E0TAD, the machine has completely executed the TAD instruction and is ready to fetch another instruction. Therefore, the next state after state E0TAD is state F1.

If ir[11:9] is 3 in state F4B, the instruction is what the programmer calls a "Deposit and Clear Accumulator" (DCA) and so the machine proceeds to state E0DCA. Although TAD and DCA are both MRIs, the operations involved for the DCA are more complex because the DCA instruction stores the accumulator in memory and then clears the accumulator. It takes three clock cycles to accomplish all the operations required by the DCA instruction. State E0DCA occurs during the first of these three clock cycles. The restrictions on the use of memory described at the end of section 8.3.1.4 require anything that is to be stored in memory be placed in the memory buffer register. State E0DCA schedules that the memory buffer register be assigned a copy of the value in the accumulator at the next rising edge of the clock. This is done in preparation for the next state, which is state E1ADCA.

State E1ADCA stores mb (which is now the same as the value of the accumulator) into memory at the effective address. After state E1ADCA, the next state is state E1BDCA.

State E1BDCA schedules that zero be assigned to the accumulator at the next rising edge of the clock. After scheduling the accumulator to be cleared in state E1BDCA, the machine has completely executed the DCA instruction and is ready to fetch another instruction. Therefore, the next state after state E0DCA is state F1.

If ir is 7200 in state F3B, the instruction is what the programmer calls a "CLear Accumulator"(CLA) and so the machine proceeds to state E0CLA. Note that since 7200 is not MRI, the decoding occurs earlier (in state F3B) than for the MRI examples above. Also note that for non-MRI-like 7200, all twelve bits of the instruction register must be tested, since there is no effective address. In state E0CLA, the machine schedules that zero be assigned to the accumulator at the next rising edge of the clock. After this, the machine is ready to fetch another instruction, and so the next state after state E0CLA is state F1.

If ir is 7402 in state F3B, the instruction is what the programmer calls a "HaLT," and so the machine proceeds to state E0HLT. Again all twelve bits of the instruction register must be tested in state F3B to decode this instruction. In state E0HLT, the machine schedules that one be assigned to the halt register at the next rising edge of the clock. The next state after state E0HLT is state F1, not because the machine is going to fetch another instruction, but instead because state F1 is where the test of the halt register occurs. (As mentioned above, when halt is one in state F1, the machine proceeds to state IDLE.) The final sequence of states through which the machine goes near the end of a program will be F1, F2, F3A, F3B, E0HLT, F1, IDLE, IDLE, IDLE ....

8.3.1.6 Example machine language program

Assume that the following machine language program:

```
0100/7200
0101/1106
0102/1107
0103/1110
0104/3111
0105/7402
0106/0112
0107/0152
0110/0224
0111/0510
```

is present in memory starting at address 0100 when power is turned on. (For an explanation of this program, see appendix A.) The following shows how the ASM proceeds when the external input `cont` is not asserted:

```
INIT    ma=????  mb=????  pc=????  ir=????  halt=?  ac=????
F1      ma=????  mb=????  pc=0100  ir=????  halt=1  ac=????
IDLE    ma=0100  mb=????  pc=0100  ir=????  halt=0  ac=????
```

The question marks indicate an unknown value in registers when power is first turned on.[7] State INIT initializes the `halt` flag so that the machine goes straight from F1 to IDLE. State INIT is also initialized to the starting address of our sample program. The machine stays in IDLE until the external input `cont` is asserted. When it is asserted, the following happens:

```
IDLE    ma=0100  mb=????  pc=0100  ir=????  halt=0  ac=????
F1      ma=0100  mb=????  pc=0100  ir=????  halt=0  ac=????
F2      ma=0100  mb=????  pc=0100  ir=????  halt=0  ac=????
F3A     ma=0100  mb=????  pc=0101  ir=????  halt=0  ac=????
F3B     ma=0100  mb=????  pc=0101  ir=7200  halt=0  ac=????
EOCLA   ma=0000  mb=????  pc=0101  ir=7200  halt=0  ac=????
F1      ma=0000  mb=????  pc=0101  ir=7200  halt=0  ac=0000
```

In state F1, the program counter (0100) is saved in the memory address register. In state F2, the program counter is scheduled to be incremented to become 0101 (as can be seen in state F3A) in preparation for fetching the next instruction four clock cycles later. In state F3A, the instruction register is scheduled to be loaded from memory address 0100. In state F3B, this instruction (7200) becomes available in the instruction register, and since `ir[11:9]` `>=` 6, the instruction decoding takes place. State EOCLA schedules zero to be loaded into the accumulator. Having completed the fetching and execution of the CLA instruction, the machine performs similar operations to fetch the second instruction. This time, the program counter is 0101 in state F1. The following shows how the fetching and execution of the second instruction proceeds:

---

[7] In other chapters, a similar idea is denoted with the "x" value in Verilog.

```
F2       ma=0101  mb=????  pc=0101  ir=7200  halt=0  ac=0000
F3A      ma=0101  mb=????  pc=0102  ir=7200  halt=0  ac=0000
F3B      ma=0101  mb=????  pc=0102  ir=1106  halt=0  ac=0000
F4A      ma=0106  mb=????  pc=0102  ir=1106  halt=0  ac=0000
F4B      ma=0106  mb=0112  pc=0102  ir=1106  halt=0  ac=0000
E0TAD    ma=0106  mb=0112  pc=0102  ir=1106  halt=0  ac=0000
F1       ma=0106  mb=0112  pc=0102  ir=1106  halt=0  ac=0112
```

In state F2 the saved program counter (0101) is visible in the memory address register at the same time the program counter is scheduled to be incremented to become 0102 (as can be seen in state F3A). In state F3A, the instruction register is scheduled to be loaded from memory address 0101. In state F3B, this instruction (1106) becomes available in the instruction register, but unlike the above non-MRI, instruction decoding does not take place in state F3B. Instead, state F3B schedules the memory address register to be loaded with the effective address (0106), derived from the instruction register. Since ir[11:9] < 6 in state F3B, the machine proceeds to state F4A, where the memory buffer register is scheduled to be loaded with the contents of memory (0112) at that effective address, as can be seen in state F4B. In state F4B, instruction decoding takes place. Since, ir[11:9] == 1, the machine proceeds to state E0TAD, where the 0112 in the memory buffer register is added to the zero in the accumulator. The remaining two TAD instructions execute in a similar fashion:

```
F2       ma=0102  mb=0112  pc=0102  ir=1106  halt=0  ac=0112
F3A      ma=0102  mb=0112  pc=0103  ir=1106  halt=0  ac=0112
F3B      ma=0102  mb=0112  pc=0103  ir=1107  halt=0  ac=0112
F4A      ma=0107  mb=0112  pc=0103  ir=1107  halt=0  ac=0112
F4B      ma=0107  mb=0152  pc=0103  ir=1107  halt=0  ac=0112
E0TAD    ma=0107  mb=0152  pc=0103  ir=1107  halt=0  ac=0112
F1       ma=0107  mb=0152  pc=0103  ir=1107  halt=0  ac=0264
F2       ma=0103  mb=0152  pc=0103  ir=1107  halt=0  ac=0264
F3A      ma=0103  mb=0152  pc=0104  ir=1107  halt=0  ac=0264
F3B      ma=0103  mb=0152  pc=0104  ir=1110  halt=0  ac=0264
F4A      ma=0110  mb=0152  pc=0104  ir=1110  halt=0  ac=0264
F4B      ma=0110  mb=0224  pc=0104  ir=1110  halt=0  ac=0264
E0TAD    ma=0110  mb=0224  pc=0104  ir=1110  halt=0  ac=0264
F1       ma=0110  mb=0224  pc=0104  ir=1110  halt=0  ac=0510
```

The accumulator now contains the sum of the three numbers (0510). The following shows the execution of the DCA instruction:

```
F2        ma=0104   mb=0224   pc=0104   ir=1110   halt=0   ac=0510
F3A       ma=0104   mb=0224   pc=0105   ir=1110   halt=0   ac=0510
F3B       ma=0104   mb=0224   pc=0105   ir=3111   halt=0   ac=0510
F4A       ma=0111   mb=0224   pc=0105   ir=3111   halt=0   ac=0510
F4B       ma=0111   mb=0000   pc=0105   ir=3111   halt=0   ac=0510
E0DCA     ma=0111   mb=0000   pc=0105   ir=3111   halt=0   ac=0510
E1ADCA    ma=0111   mb=0510   pc=0105   ir=3111   halt=0   ac=0510
E1BDCA    ma=0111   mb=0510   pc=0105   ir=3111   halt=0   ac=0510
F1        ma=0111   mb=0510   pc=0105   ir=3111   halt=0   ac=0000
```

In state F2 the saved program counter (0104) is visible in the memory address register at the same time the program counter is scheduled to be incremented to become 0105 (as can be seen in state F3A). In state F3A, the instruction register is scheduled to be loaded from memory address 0105. In state F3B, this instruction (3111) becomes available in the instruction register. State F3B schedules the memory address register to be loaded with the effective address (0111), derived from the instruction register. Since $ir[11:9] < 6$ in state F3B, the machine proceeds to state F4A, where the memory buffer register is scheduled to be loaded with the contents of memory (0000) at that effective address, as can be seen in state F4B. In state F4B, instruction decoding takes place. Since, $ir[11:9] == 3$, the machine proceeds to state E0DCA, where the memory buffer register is scheduled to be assigned the value from the accumulator (0510).[8] In state E1ADCA, this value is stored in memory. State E1BDCA schedules the accumulator to be cleared. Finally, the HLT instruction executes:

```
F2        ma=0105   mb=0510   pc=0105   ir=3111   halt=0   ac=0000
F3A       ma=0105   mb=0510   pc=0106   ir=3111   halt=0   ac=0000
F3B       ma=0105   mb=0510   pc=0106   ir=7402   halt=0   ac=0000
E0HLT     ma=0002   mb=0510   pc=0106   ir=7402   halt=0   ac=0000
F1        ma=0002   mb=0510   pc=0106   ir=7402   halt=1   ac=0000
IDLE      ma=0106   mb=0510   pc=0106   ir=7402   halt=1   ac=0000
IDLE      ma=0106   mb=0510   pc=0106   ir=7402   halt=0   ac=0000
```

The value in the memory address register calculated by state F3B (this value, 0002, becomes visible in state E0HLT) is irrelevant. In hardware, unnecessarily doing a harmless computation sometimes is more efficient than having a decision avoid the computation when it is unwanted.[9] It does not slow the machine, and it is simpler always to load these bits from the instruction register into the memory address register, even, as

---

[8] Having the ASM proceed through F4A and F4B was unnecessary in this case since state E0DCA does not use the value loaded into memory buffer register by state F4A in the same way E0TAD does. This is harmless but slower than required and was done to simplify the explanation of state F3B.

[9] As the last footnote indicates, whether it is efficient depends on whether extra states, like F4A, are involved or not. Here there are no extra states involved.

in this case, when they are not needed. State E0HLT schedules the `halt` flag to become zero, which causes the machine to go to F1 and then back to IDLE, where the machine will stay (unless `cont` is pressed again).

8.3.2 Including more in the instruction set

The machine described by the ASM in section 8.3.1.5 is rather useless. It was presented only to introduce the essential aspects of the fetch/execute algorithm. Rather than implement a useless subset of instructions in hardware, let's include more of the PDP-8's instructions in our instruction set. For the extended example in this section, we will implement a machine that executes the following PDP-8 instructions:

mnemonic	octal machine language	what the mnemonic stands for
AND	0xxx	AND memory with accumulator
TAD	1xxx	add memory to accumulator (Two's Complement Add)
DCA	3xxx	Deposit accumulator in memory and Clear Accumulator
JMP	5xxx	goto a new instruction (JuMP)
CLA	7200	CLear Accumulator
CLL	7100	CLear Link
CMA	7040	CoMplement Accumulator
CML	7020	CoMplement Link
IAC	7001	Increment ACcumulator
RAL	7004	Rotate Accumulator and link Left
RAR	7010	Rotate Accumulator and link Right
CLACLL	7300	CLear Accumulator and CLear Link
SZA	7440	Skip next instruction if Zero is in Accumulator
SNA	7450	Skip next instruction if Non-zero value is in Accumulator
SMA	7500	Skip next instruction if Minus (negative) value is in Accumulator
SPA	7510	Skip next instruction if Positive (non-negative) value is in Accumulator
SZL	7430	Skip next instruction if Zero is in Link
SNL	7420	Skip next instruction if one (Non-zero) is in Link
HLT	7402	HaLT
OSR	7404	Or Switch "Register" with accumulator

These instructions are explained more fully in appendix B. The first four instructions (AND, TAD, DCA and JMP) are memory reference instructions. As in section 8.3.1, we will only consider direct page zero addressing.

The next eight mnemonics (CLA, CLL, CMA, CML, IAC, RAL, RAR, CLACLL) describe instructions that manipulate the accumulator and the link registers without referencing memory. (The link register was not considered in section 8.3.1.5 because it requires some details that are discussed in section 8.3.2.2.) Although the full PDP-8 instruction set allows for 256 combinations of these operations, we will only consider the eight listed here.

The skip instructions allow conditional execution of the following instruction. If the condition is met, the following instruction does not execute. If the condition is met, the skip acts like a NOP.

The HLT instruction causes the machine to stop executing a program and instead proceed to states that allow the machine to interface with its programmer. Unlike the example in section 8.3.1.5, the ASM in this section will include interface states after the HLT instruction that allow an arbitrary program to be loaded at an arbitrary address any time the programmer wishes. The programmer communicates with the halted machine using an external 12-bit input, sr. In the original PDP-8 documentation, the sr is known as the switch "register"; however sr is not a register. sr is an external input bus, very much like the buses x and y in the division machine of chapter 2. In the physical realization of the PDP-8, the sr is simply a set of twelve switches (one for each bit).

The OSR instruction is an unusual kind of input instruction unique to the PDP-8, which is ideal for our purposes in this section. The OSR instruction ORs input coming from the external sr bus with the contents of the accumulator. This allows a discussion here of software input without the need for machine language instructions 6xxx.

Even though the PDP-8 is one of the simplest instruction sets ever designed, and we have still chosen to implement only about half of it, you may have a feeling of panic about whether you will ever be able to design such a machine. Have faith—top-down design will come to the rescue.

8.3.2.1 ASM implementing additional instructions
Here is the ASM for the improved machine that implements the instructions listed above:

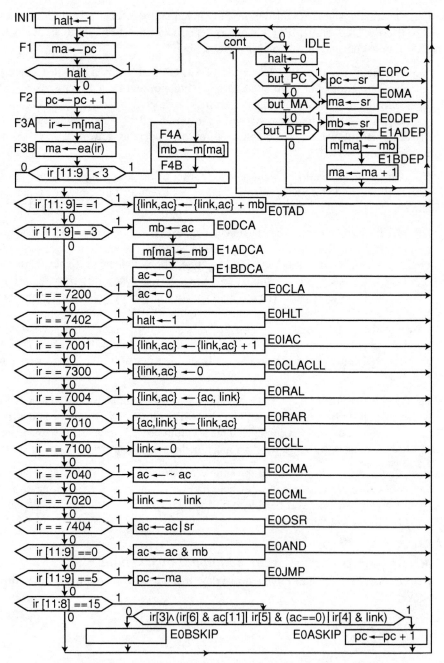

Figure 8-8. ASM implementing more instructions of the PDP-8.

Section 8.3.2.2 describes the states required to execute these additional instructions. Section 8.3.2.3 describes the additional states that allow the programmer a manual interface when the machine is halted.

8.3.2.2 States for additional instructions

Many of the execute states (E0CLA, E0HLT, E0DCA, E1ADCA and E1BDCA) of the improved ASM in section 8.3.2.1 are identical to the states with the same names of the useless ASM in section 8.3.1.5. Therefore, these states will not be discussed here.

Also, the fetch states (F1, F2, F3A, F3B, F4A and F4B) of the improved ASM in section 8.3.2.1 are identical to the states with the same names in the useless ASM of section 8.3.1.5, except the decision whether to go to state F4A is different. As the DCA instruction (3111) used in the example machine language program of section 8.3.1.6 illustrates, it is not necessary for some memory reference instructions to have the memory buffer register initialized. Although it is harmless to do so, it slows the machine down. Therefore, in this section, we will only perform state F4A if it is required. For direct addressing, only AND(0xxx), TAD (1xxx), ISZ (2xxx) and JMS (4xxx) require state F4A. Since JMS and ISZ are not part of the instruction subset implemented in this section (JMS and ISZ are left as exercises), the condition can be restated as $ir[11:9] < 3$. If the full PDP-8 instruction set with all addressing modes were implemented, this condition would be more complicated.

8.3.2.2.1 Instruction described with concatenation

One place where there is a noticeable difference between the useless ASM and the improved ASM is in state E0TAD. This difference is due to the fact that the TAD instruction treats the `link` and the `ac` together as a 13-bit entity. One way to describe this is to say that value inside the CPU is $4096*link + ac$, but the CPU never performs such a wasteful computation. Instead, inside the CPU the 12-bit bus coming out of the `ac` and the one-bit bus coming out of the `link` are joined together to form a 13-bit bus. We need a notation to describe this joining together, technically known as concatenation. Although the material in this section does not depend on any knowledge of Verilog, Verilog does indeed have such a concatenation notation: `{link,ac}` is a 13-bit value. The most significant bit (bit 12) of `{link,ac}` is `link`, the next to most significant bit (bit 11) is `ac[11]`, ... and the least significant bit (bit 0) is `ac[0]`. As a different illustration of concatenation, note that `ir[11:9]` is the same as `{ir[11],ir[10],ir[9]}`.

State E0TAD of the improved ASM properly shows that the TAD instruction affects and is affected by the link. It is perfectly legal for a concatenation to be on the left-hand side of a register transfer. The operation `{link,ac}` ← `{link,ac}` + mb means that the 12-bit mb is extended to have 13 bits (by implicitly concatenating a 0 on the left). This is added to the 13-bit {link,ac}. The respective portions of the 13-bit sum are stored back into link and ac. The following table shows four examples of before and after state E0TAD:

before			after		
link	ac	mb	link	ac	mb
0	0040	4001	0	4041	4001
0	4040	4002	1	0042	4002
1	0040	4003	1	4043	4003
1	4040	4004	0	0044	4004

For instance, the last line shows $\{1,4040_8\}+4004_8 == 1100000100000_2 + 0100000000100_2 == 10000000100100_2 == 20044_8$, which is too big to fit in 13 bits, and so the result is 0 in the link and 0044_8 in the accumulator.

There are four other instructions (IAC, CLACLL, RAL and RAR) that are most easily described using concatenation. If the instruction register contains 7001 in state F3B, the instruction is what the programmer calls "Increment ACcumulator," and so the machine proceeds to state E0IAC, which is similar to E0TAD, except one rather than mb is added to {link,ac}. If the instruction register is 7300 in state F3B, the instruction is what the programmer calls a "CLear Accumulator, CLear Link," and so the machine proceeds to state E0CLACLL, where a 13-bit zero is assigned to {link,ac}.

If the instruction register is 7010 in state F3B, the instruction is what the programmer calls a "Rotate Accumulator and link Right," and so the machine proceeds to state E0RAR, where the 13-bit {link,ac} is rotated right. Similarly, if the instruction register is 7004 in state F3B, the instruction is what the programmer calls a "Rotate Accumulator and link Left," and so the machine proceeds to state E0RAL, where the 13-bit {link,ac} is rotated left. Concatenation is the simplest way to describe rotation. Recall that:

```
{link,ac}==
{link,ac[11],ac[10],ac[9],ac[8],ac[7],ac[6],ac[5],ac[4],ac[3],ac[2],ac[1],
ac[0]}
```

and

```
{ac,link} ==
{ac[11],ac[10],ac[9],ac[8],ac[7],ac[6],ac[5],ac[4],ac[3],ac[2],ac[1],ac[0],link}
```

The single 13-bit wide register transfer:

```
{link,ac} ← {ac,link}
```

is a more succinct way to describe 13 separate register transfers, each one bit wide:

```
link     ← ac[11]
ac[11]   ← ac[10]
ac[10]   ← ac[9]
ac[9]    ← ac[8]
         . . .
ac[2]    ← ac[1]
ac[1]    ← ac[0]
ac[0]    ← link
```

Observe that, except for the link, the bits are shifted over one place to the left. The old value of the link "rotates" around to the least significant bit of the accumulator. The following table illustrates examples what is in the link and the accumulator before and after state E0RAL:

	before		after	
	link	ac	link	ac
	0	1001	0	2002
	0	2002	0	4004
	0	4004	1	0010
	1	1001	0	2003
	1	2002	0	4005
	1	4004	1	0011

Although the software uses of the previous instructions were fairly obvious, the RAL instruction may seem a bit strange. In fact, RAL has two uses: arithmetic and logical. The first three lines above illustrate the arithmetic use: if the programmer has previously cleared the link, RAL is like multiplication by two (with overflow in the link). The remaining examples above illustrate the logical use: to rearrange bits without losing any information.

RAR is the inverse of RAL, and the concatenation notation makes this clear:

```
{ac,link} ← {link,ac}
```

In the following, the second, third and last three lines use data from the previous (RAL) table to illustrate that RAR is the inverse of RAL (the rotates do not lose information, they simply rearrange it):

	before			after	
	link	ac		link	ac
	0	1001		1	0400
	0	2002		0	1001
	0	4004		0	2002
	1	1001		1	4400
	1	2002		0	5001
	1	4004		0	6002
	0	2003		1	1001
	0	4005		1	2002
	1	0011		1	4004

The first three examples above illustrate the arithmetic use of RAR: if the programmer clears the link, RAR is like unsigned division by two (with the remainder in the link).

8.3.2.2.2   Additional non-memory reference instructions

If the instruction register is 7100 in state F3B, the instruction is what the programmer calls a "CLear Link," and so the machine proceeds to state E0CLL, where zero is assigned only to the link (the accumulator is left alone). If the instruction register is 7040 in state F3B, the instruction is what the programmer calls a "CoMplement Accumulator," and so the machine proceeds to state E0CMA, where ~ac is assigned only to the accumulator (the link is left alone). If the instruction register is 7020 in state F3B, the instruction is what the programmer calls a "CoMplement Link," and so the machine proceeds to state E0CML, where ~link is assigned only to the link (the accumulator is left alone).

If the instruction register is 7404 in state F3B, the instruction is what the programmer calls "Or Switch Register," and so the machine proceeds to state E0OSR, where the external sr input is ORed with the current value of the accumulator. Here is a typical use of this instruction:

```
0002/7402
0003/7200
0004/7404
0005/3100
```

Assume the machine executed instructions, prior to 0002, which are irrelevant to this discussion. When the program wants input from the user, it halts (by executing the HLT instruction, 7402). The machine will proceed to state IDLE, but the program counter remains at 0003. While the machine is halted, the user is free to enter in whatever value is desired on the switches. When the user pushes the `cont` button, the fetch/execute algorithm proceeds with the instruction at 0003, which is a CLA instruction (7200). This is done to get rid of any extraneous value in the accumulator in preparation for the next instruction. The next instruction in fact is the OSR (7404), which ORs zero in the accumulator with the desired value from the external `sr` input. Because zero is the identity for OR (i.e., `0|sr == sr`), the input value from the switches is loaded into the accumulator. Finally, a DCA instruction stores the input value into memory. The OSR instruction, in conjunction with IDLE state and the `cont` input, provides a simple user interface that will work nicely for the software in this chapter.\

8.3.2.2.3 Additional memory reference instructions

There are six memory reference instructions in the instruction set of the PDP-8. Two of these (TAD and DCA) were described earlier. Two of these (JMS and ISZ) are left as exercises. The other two (AND and JMP) are described in this section.

If `ir[11:9]` is 0 in state F4B, the instruction is what the programmer refers to as "AND," and so the machine proceeds to state E0AND. This state is similar to E0TAD, except the AND instruction only changes the accumulator. (AND leaves the link register alone.) Recall that & is the bitwise AND operator, and so the register transfer:

```
ac ← ac & mb
```

is equivalent to:

```
ac[11] ← ac[11] & mb[11]
ac[10] ← ac[10] & mb[10]
ac[9]  ← ac[9]  & mb[9]
            . . .
ac[2]  ← ac[2]  & mb[2]
ac[1]  ← ac[1]  & mb[1]
ac[0]  ← ac[0]  & mb[0]
```

If `ir[11:9]` is 5 in state F4B, the instruction is what the programmer calls a "JuMP," and so the machine proceeds to state E0JMP. All general-purpose computers have some kind of jump (sometimes known as branch) instruction. The purpose of a jump instruction is to modify the program counter. The jump instruction allows high-level language features (such as loops and decisions) to be translated into machine language.

Although the jump instruction of the PDP-8 is categorized as a memory reference instruction, it does not actually reference memory. It simply takes the effective address (from the memory address register) and uses this as the new value of the program counter.

8.3.2.2.4 Skip instructions

High-level language programs are composed of statements like if and while. The JMP instruction by itself is not enough to translate such programs. For this reason, the PDP-8 instruction set includes several "skip" instructions. These instructions test to see whether the value in the accumulator (or link) meets certain conditions. If it does, the next instruction will be skipped. If the condition is not met, the next instruction will execute normally. The following table illustrates several skip instructions, and how they are encoded in machine language.

mnemonic	octal ir	ir[6]	ir[5]	ir[4]	ir[3]	
SMA	7500	1	0	0	0	Skip if Minus (negative) Accumulator
SZA	7440	0	1	0	0	Skip if Zero Accumulator
SNL	7420	0	0	1	0	Skip if Non-zero Link
SPA	7510	1	0	0	1	Skip if Positive Accumulator
SNA	7450	0	1	0	1	Skip if Non-zero Accumulator
SZL	7430	0	0	1	1	Skip if Zero Link

If ir[11:8] is 15 in state F3B, the instruction is one of the above skip instructions. If the condition is mct, the machine proceeds to state E0ASKIP, where the program counter is scheduled to be incremented an extra time. If the condition is not met, the machine proceeds to state E0BSKIP, where the machine leaves the program counter the way it was.

The condition is determined by ir[6:3]. ir[3] is a bit that reverses the meaning of the instruction; hence ir[3] is 0 for SMA, SZA and SNL, but ir[3] is 1 for SPA, SNA and SZL. (If you think about it, you will realize SMA, SZA and SNL, respectively, are the opposites of SPA, SNA and SZL.) `ir[6]` is 1 for SMA and SPA. `ir[5]` is 1 for SZA and SNA. `ir[4]` is 1 for SNL and SZL. Therefore, the condition that decides whether to proceed to state E0ASKIP is:

```
ir[3]^(ir[6]&ac[11]|ir[5]&(ac==0)|ir[4]&link)
```

where ^ is the exclusive OR, which reverses the meaning of the parenthesized expression when `ir[3]` is one. Note: `ac[11]` is the "sign" bit of the accumulator (the bit that indicates 12-bit negative twos complement values).

As an illustration of how a programmer uses the skip instructions in conjunction with the other instructions, consider the unsigned greater than or equal decision. Suppose `r1` (stored at 0032) and `y` (stored at 0101) are software variables that contain 12-bit unsigned numbers. Should the high-level language programmer wish to test (in either an `if` or a `while`) to see whether `r1` is greater than or equal to `y`, there are several equivalent ways to write this (given that the following is performed in 13-bit twos complement arithmetic):

```
r1                    >= y
r1                    >= {0,y}
-{0,y}      + r1 >= 0
{~0,~y}+1 + r1 >= 0
```

The last of these can be performed with the instructions described earlier. The final signed 13-bit result in {`link,ac`} can be tested with the SZL instruction, as shown below:

```
0014/7200    CLA
0015/7100    CLL       /{link,ac} = {0,0}
0016/1101    TAD y     /{link,ac} = {0,y}
0017/7040    CMA       /{link,ac} = {0,~y}
0020/7020    CML       /{link,ac} = {~0,~y}
0021/7001    IAC       /{link,ac} = {~0,~y}+1
0022/1032    TAD r1    /{link,ac} = {~0,~y}+1 + r1
0023/7430    SZL       /test whether {~0,~y}+1 + r1 >= 0
0024/5xxx    JMP xxx   /if {~0,~y}+1 + r1 < 0, goto xxx
             ...       /if {~0,~y}+1 + r1 >=0, execute here
```

8.3.2.3 Extra states for interface

The ASM shown in section 8.3.2.1 has three additional states (E0MA, E0PC and E0DEP) that allow the programmer to interface with the machine using the `sr` input when the machine is not performing the fetch/execute algorithm. These states only occur when the programmer pushes buttons (`but_MA`, `but_PC` and `but_DEP`, respectively) during the time that the machine is in state IDLE.

State E0PC allows the programmer to load the the program counter with a value previously placed on the switches. This allows a program to reside anywhere in memory, unlike the nearly useless ASM given in section 8.3.1. For this reason, state INIT no longer initializes the program counter, and instead the programmer is responsible for pushing `but_PC` appropriately prior to pushing `cont`.

The programmer uses `but_MA` and `but_DEP` together to load a program into memory at an arbitrary address. First, the programmer enters the address on the switches to indicate where in memory the programmer desires to place the program or data. Then the programmer pushes `but_MA`, which causes state E0MA to occur, where the `sr` input is assigned to the memory address register. Next, the programmer enters the first word to go into memory onto the switches, and pushes `but_DEP`, which causes state E0DEP to occur. State E0DEP assigns the `sr` input to the memory buffer register, and then the machine proceeds to state E1ADEP. In state E1ADEP the machine deposits the memory buffer (containing the programmer's desired word) into memory at the memory address. Finally, state E1BDEP increments the memory address (in case the programmer has more words to deposit). The programmer may enter as many successive words as desired with this technique. Finally, the programmer uses `but_PC` and `cont` as described above.

8.3.2.4 Memory as a separate actor

At this point, we have described the behavior of the complete general-purpose computer system. Now, we need to consider what the external inputs and outputs of this system are:

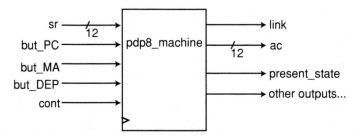

Figure 8-9. Block diagram for the PDP-8 system.

The present state and the other outputs (such as the memory buffer register, memory address register, program counter, instruction register and the halt register) are sent out from the machine primarily to allow the programmer to observe the internal operation of the machine.

8.3.2.4.1   Top-level structure of the machine

In fact, the machine hidden inside the last diagram is composed of three components: the CPU, the peripherals and the memory. Since in this chapter we are ignoring the peripherals, that leaves two separate components that must be interconnected to form the system: the CPU and the memory. Although one could consider the memory as just another component of the CPU's architecture, this is normally not done. As described in section 8.2, there are many different technologies for memory, and often the technology to implement memory is different than the technology used to implement the CPU. Therefore, we would like to physically separate the memory from the CPU and design the CPU independently from the memory. This means that memory is an independent actor, as illustrated in the following diagram:

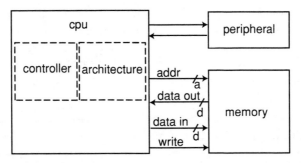

Figure 8-10. System composed of processor (controller and architecture) with memory as a separate actor.

Let's assume that we will implement this machine using an asynchronous, volatile, static, deterministic access time memory with separate data input and data output. The choice of this kind of memory device simplifies the design in several ways. First, since this memory is static, there is no need to refresh it. Second, since the access time is known, proper functioning is easily guaranteed by choosing a sufficiently slow clock. Third, since this memory has separate buses for data input and data output, there is no need to introduce the complexity of tri-state buffers.

The one design complexity that must be dealt with is the fact that this memory is asynchronous. The reason for choosing an asynchronous memory device is cost and availability. The problem with doing so is that extra care must be taken in providing the inputs to the memory device. In particular, the register transfer m[ma] ← mb must be implemented by asserting a hazard-free external (to the CPU) command output, write, as described in section 8.2.2.1. The memory address and memory buffer outputs of the CPU provide the addr and din inputs to the memory device. The dout of the memory device provides the membus input to the CPU.

8.3.2.4.2 Pure behavioral ASM with memory as a separate actor
The ASM of section 8.3.2.1 can be rewritten to reflect that memory is a separate actor. Every place (states F3A and F4A) where m[ma] is used on the right-hand side of a register transfer in section 8.3.2.1, the revised ASM will use membus instead. All other places (states E1ADEP and E1ADCA) that mention m[ma] are of the form m[ma] ← mb. These can be replaced with an assertion of the write signal, as illustrated in figure 8-11.

8.3.2.5 Our old friend: division
This book uses the childish division algorithm (first described in section 2.2) in most chapters to illustrate various ways that hardware can be designed. This algorithm is ideal as a learning example because it is simple. Although the operations used to implement this algorithm are typical of the most sophisticated algorithms, it is so elementary that any child can perform it. Unlike faster division algorithms, why it works is obvious.

There is another reason why division is the centerpiece of this book. Division has played an interesting role in the history of general-purpose computers. As mentioned previously, the very first program ever run on a general-purpose computer was the childish division algorithm. Much faster algorithms than the childish algorithm exist for division, but they are very complex and hard to understand. Many general-purpose computers throughout history (from the BINAC in 1949 up to the Pentium half a century later) have provided "divide" instructions that implement much more sophisticated division algorithms in hardware than our old friend, the childish algorithm. Despite this, many computers, including many PDP-8s,[10] have shunned division in hardware in favor of division in software. The irony of this is that a flaw in the hardware divide instruction of the Pentium general-purpose computer caused Intel great embarrassment in the mid 1990s.

---

[10] Some models of the PDP-8 had an optional hardware feature, known as EAE, that assisted in performing division.

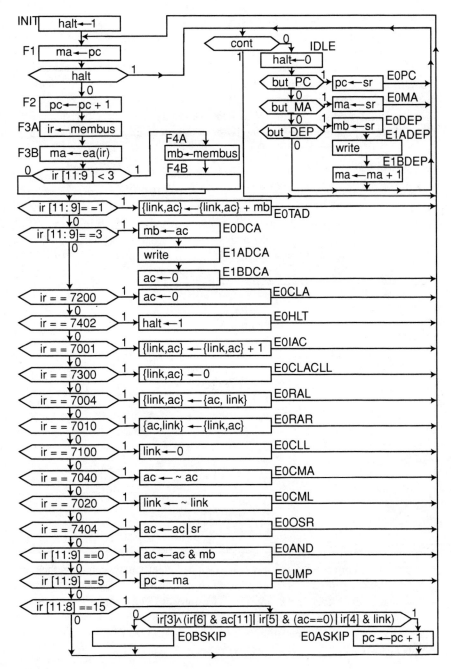

Figure 8-11. ASM with memory as a separate actor.

A major theme of this book is that speed is not the primary concern of the designer—**correctness** is. An algorithm implemented in software might be slower than the same algorithm implemented in hardware, but you should not worry about speed. The most important thing is to implement the algorithm properly, whether in hardware or software. (Sometimes speed is part of the specification of a correct design, but even then, as the Pentium incident indicates, the rest of the design must be correct before the speed matters.) And so, with an appreciation of the important role division has played in the history of computer design, let us consider how to implement our old friend, the childish division algorithm, in software with PDP-8 machine language.

8.3.2.5.1   Complete childish division program in C

A complete software program written in a high-level language typically has some input/output formatting statements, such as `scanf` and `printf` in C or READ and WRITE in Pascal (or `$readmemb` and `$display` in Verilog). For instance, a complete C program to implement the the childish division algorithm from section 2.2 might appear as follows:

```
main ()
  {
    unsigned x,y,r1,r2;

    while (1)
      {
        scanf("%04o",&x);
        scanf("%04o",&y);
        r1 = x;
        r2 = 0;
        while (r1 >= y)
          {
            r1 = r1 - y;
            r2 = r2 + 1;
          }
        printf("%04o\n",r2);
      }
  }
```

Since octal is a convenient notation that is used frequently in this chapter, the input and output are shown formatted as four-digit octal numbers by `scanf` and `printf`. Translating statements like `scanf` and `printf` into the PDP-8 instruction set requires using 6xxx machine language instructions. We have avoided implementing the 6xxx instructions of the PDP-8 because their implementation requires concepts not covered in this chapter.

8.3.2.5.2   User Interface for the software

The purely hardware implementations of the childish division algorithm in chapter 2 avoided these input/output formatting problems with:

a) two separate external data input buses (x and y), presumably connected to switches, so that the "friendly" user can toggle in the binary values desired to be input into the hardware.

b) an external status input (pb), presumably connected to a push button, so the "friendly" user can indicate the proper time for the hardware to look at the external input buses.

c) an external data output bus (r2), presumably connected to lights, so the "friendly" user can observe the computation of the quotient in binary as it progresses. (If the clock is fast enough, the user will not notice anything except the result.)

d) an external command output (READY), presumably connected to a light, so the "friendly" user can know when r2 has become the correct quotient.

This approach requires that the person who uses the hardware described in chapter 2 be "friendly," someone who is willing and able to adhere to several rules that describe how to use the machine properly. Some of these rules relating to the timing of the push button were described in chapter 2. Although chapter 2 explained the operation of this machine in decimal notation, the use of switches as a physical input medium and lights as a physical output medium additionally demands a user who is comfortable with the binary number system.

To translate the childish division algorithm into PDP-8 machine language, we are going to assume a similar "amicable" user. This "amicable" user is willing to toggle binary values into the sr to provide inputs to the algorithm (as described in section 8.3.2.2.2) and observe the binary result in the accumulator. The main difference between the "amicable" user of the algorithm implemented in PDP-8 machine language and the "friendly" user of the algorithm implemented in the hardware given in chapter 2 is how the user must operate the machine. The PDP-8 software version (which has two HLT instructions) re-uses the single sr input bus so that the "amicable" user must press the cont button twice. The hardware in chapter 2 uses two separate input buses (x and y) so that the "friendly" user only presses the pb button once.

The following summarizes the hardware of the PDP-8 that is utilized for crude input and output by the software implementation of the childish division algorithm:

a) one external data input bus (sr), connected to twelve switches, so the "amicable" user can toggle in the desired binary input values.

b) an external status input (cont), connected to a push button, so the "amicable" user can push cont after toggling each separate value in on the sr.

c) an external data output bus (`ac`), connected to twelve lights, so the "amicable" user can observe the computation of the quotient (and other things) in binary as it progresses. (If the clock is fast enough, the user will not notice anything except the result.)

d) an external command output (`present_state`), connected to lights, so the "amicable" user can know when `ac` has become the correct quotient (i.e., when `present_state` becomes IDLE).

8.3.2.5.3 *Childish division program in machine language*

On the left side of the following is the PDP-8 machine language code for the childish division algorithm. On the right side is the corresponding assembly language mnemonics and symbolic operands, in the style explained in appendix A, with comments following the slash. Only the machine language resides in memory. The commented assembly language is shown only to clarify how the program operates:

```
           / Childish division algorithm in PDP-8 machine language
           /
           / The following 2 instructions allow the user to observe
           / the previous result (r2) in the ac

0000/7300 L0, CLACLL      /{link,ac} = {0,0}
0001/1033      TAD r2      /{link,ac} = {0,r2}
           /
           / The following 4 instructions wait for the user to toggle in
           / the first value on sr (while still displaying r2 on ac)
           / When the user presses cont, this first value toggled
           / in on sr is stored into x

0002/7402      HLT         /wait for user to toggle first value
0003/7200      CLA         /           {link,ac} = {0,0}
0004/7404      OSR         /           {link,ac} = {0,0|sr}
0005/3100      DCA x       /x = sr;    {link,ac} = {0,0}
           /
           / The following 3 instructions wait for the user to toggle in
           / the second value on sr (ac is now cleared)
           / When the user presses cont, this second value toggled
           / in on sr is stored into y

0006/7402      HLT         / wait for user to toggle second value
0007/7404      OSR         /           {link,ac} = {0,0|sr}
0010/3101      DCA y       /y = sr;    {link,ac} = {0,0}
           /
           / The following 3 instructions initialize r1 and r2
           / prior to the while loop
```

Continued

```
0011/1100      TAD x      /           {link,ac} = {0,0+x}
0012/3032      DCA r1     /r1 = x;    {link,ac} = 0
0013/3033      DCA r2     /r2 = 0;
        /
        / The following 9 instructions implement
        /   while (r1>=y)

0014/7200      CLA
0015/7100      CLL        /           {link,ac} = {0,0}
0016/1101      TAD y      /           {link,ac} = {0,y}
0017/7040      CMA        /           {link,ac} = {0,~y}
0020/7020      CML        /           {link,ac} = {~0,~y}
0021/7001      IAC        /           {link,ac} = {~0,~y}+1
0022/1032      TAD r1     /           {link,ac} = {~0,~y}+1 + r1
0023/7430      SZL        / test whether {~0,~y}+1 + r1 >= 0
0024/5000      JMP L0     / if {~0,~y}+1 + r1 < 0, exit while (goto L0)
                         / if {~0,~y}+1 + r1 >=0, stay in while loop

            /The following 5 instructions implement body of the while loop

0025/3032      DCA r1     /r1 = r1 - y;  i.e. r1=~y+1+r1; ac=0
0026/1033      TAD r2     /              {link,ac} = {0,0+r2}
0027/7001      IAC        /              {link,ac} = {0,0+r2}+1
0030/3033      DCA r2     /r2 = r2+1
0031/5014      JMP L1     /continue while loop
        /
        / The following 2 words store data manipulated by
        / the childish division algorithm
        /
0032/0000 r1,  0000
0033/0000 r2,  0000
        /
        / The following 2 words store data input from the sr
        /
0100/0000 x,   0000
0101/0000 y,   0000
```

8.3.2.5.4 Analysis of childish division software

The following table summarizes how many clock cycles it takes for each part of the childish division program given in section 8.3.2.5.3 to execute:

```
High-level Operations                          Before   During   After
                                               while    while    while

    r1 = x;                                      14
    r2 = 0;                                        7
    while (r1 >= y)                                         44      44
       {
          r1 = r1 - y;                                       7
          r2 = r2 + 1                                       24
       }                                                             5
    display r2 in accumulator and halt                              18

  Total Clock Cycles                             21       75        67
```

The times are listed in three columns. The left column indicates operations that execute
just once, before the `while` loop begins. The right column indicates operations that
execute just once, upon exiting from the `while` loop. The middle column indicates
operations that occur **each** time through the loop. The `while` statement itself involves
formation of the 13-bit twos complement (32 cycles) and testing (12 cycles). The entry
44 (32+12) occurs both in the middle and right columns because this machine code
occurs each time through the loop as well as the final time when the condition `r1>=y`
becomes false. Just as in chapter 2, the number of times the loop executes is proportionate to the quotient. Neglecting how long it takes for the user to toggle in the inputs,
from the time the program actually starts computing the quotient (when the program
counter was 0011) until the machine returns to state IDLE is `21 + 67 + 75*quo-
tient` clock cycles.

8.3.2.5.5 Comparison with special-purpose implementation
This table compares different implementations of the childish division algorithm:

	section	clock cycles	12-bit registers	12-bit memory words	ctrl states
	2.2.7	3+quotient	3	0	2
special	2.2.3	2+2*quotient	2	0	4
purpose	2.2.2	3+3*quotient	2	0	5
hardware	2.2.5	2+3*quotient	3	0	5
PDP-8 software	8.3.2	88+75*quotient	5	30	31

The "section" column indicates where the ASM (and in the case of the PDP-8 software, also the machine language) is defined. The "clock cycle" column indicates how long it takes to compute the quotient, neglecting the time for the user to toggle in the binary inputs. The "12-bit registers" indicates how many **hardware** registers of this size are required (in the case of the PDP-8, this is the accumulator, instruction register, memory address register, memory buffer register and program counter). We neglect one bit registers such as halt and link as being insignificant in the total cost. We also neglect the cost of the combinational logic that interconnects the registers within the architecture (since this is the subject of section 8.4). The "12-bit memory words" indicates how many words the machine language version requires for both program and data. Special-purpose implementations of this algorithm do not need memory because the hardware registers continually hold the data, and the controller implements the algorithm. The "ctrl states" indicates how many states are required by the **hardware** ASM.

Any way you look at the above table, software appears to be a real loser. Compared to the fastest special-purpose implementation listed above (section 2.2.7), the software approaches being about seventy-five times slower for a large `quotient`:

$$\lim_{quotient \rightarrow \infty} (88+75*quotient)/(3+quotient) = 75$$

For the particular case traced above and in chapter two (quotient=14/7), the ratio is 238/5, or about 47 times slower. One reason why the hardware implementation in section 2.2.7 makes the software look so bad is because the hardware does the equivalent of three high-level operations (test `r1>=y`, `r1=r1-y` and `r2=r2+1`) in parallel during **each** clock cycle. The childish division algorithm has the potential for this parallelism, and so we ought to exploit this.

On the other hand, if we wanted to handicap the hardware to make the contest seem more sporting, the ASM of section 2.2.2 is the closest to the software implementation because it only does one high-level operation at a time. For very large `quotient`, the software approaches being about 25 times slower than section 2.2.2:

$$\lim_{quotient \rightarrow \infty} (88+75*quotient)/(3+3*quotient) = 25$$

For the particular case traced above and in chapter two (`quotient`=14/7), the ratio is 238/9, or about 26 times slower.

Even when the hardware only does one thing at a time (as in section 2.2.2), the software appears much slower. There are two reasons for this. First, it takes several PDP-8 instructions to do the equivalent of one high-level language statement (which is most noticeable in implementing the `while`). Second, the way we have implemented the ASM for the PDP-8, it takes several clock cycles (either five or seven) for each instruction to execute.

Software requires general-purpose hardware in order to run. The PDP-8 is about as simple as a general-purpose computer can be, but even so, it requires five registers. Software also requires memory for programs and data. Because of technological differences explained in section 8.1, the cost to store a bit in memory is usually several times lower than to store a bit in a register. For the sake of argument, say that the cost for a 12-bit word in memory is five times cheaper than for a 12-bit register. The storage costs are then 2*5 for section 2.2.2 hardware, 3*5 for section 2.2.7 hardware, and 5*5+30 for the PDP-8 implementation (assuming we only pay for the memory actually used to implement the childish division program). Therefore, section 2.2.2 storage cost is about one fifth that of the PDP-8 implementation, and section 2.2.7 storage cost is about one-quarter that of the PDP-8 implementation.

8.3.2.5.6   Trade-off between hardware and software

One cannot draw sweeping conclusions having examined only a single algorithm in hardware and software, and having examined the software on a single implementation of a single instruction set. The difference between hardware and software may be less pronounced when the algorithm is more complicated or when the instruction set is more capable. In particular, algorithms that require memory for storage of data structures, such as arrays, may show software performance closer to that of special-purpose hardware. However, for the childish division algorithm, we can conclude the software solution gives lower performance and costs more.

Would you pay more to buy something slower? Paradoxically, in most instances, you probably would because hardware speed and cost are often not the primary concern. Certainly in this case, speed is unimportant when you consider the problem that the childish division algorithm solves. It interactively obtains two 12-bit inputs, divides them in a very inefficient way,[11] and displays the answer. It is going to take the user several seconds to toggle in the inputs, and several more for the user to comprehend the output. Since the largest 12-bit quotient is 4095, the maximum total time for the PDP-8 implementation is 88+75*4095 = 307213 clock cycles. Although this seems awful in comparison to the 4098 clock cycles required by the hardware implementation of section 2.2.7, it is less than the blink of an eye when the clock period is 100 nanoseconds

---

[11] Regardless of the underling implementation (hardware or software), there are much better algorithms than the childish division algorithm if you really want to divide fast.

(a very moderate clock speed using current integrated circuit technology). The user will only see a brief flash before the correct answer appears. Occasionally, the specification of a problem has a real-time aspect to it. For example, if instead of our friendly user, the input came from another machine that needed to divide two thousand numbers per second, only the hardware of section 2.2.7 would be able to keep up.

In most instances however, the factor that matters more than hardware speed and cost is design speed and cost. In other words, how long does it take and how much does it cost for the designer to produce a correct design? Designers are willing to use hardware which is, in a technological[12] sense, more costly and slower than is theoretically necessary because in doing so they obtain the benefit of rapid debugging. When a designer finds an error, it is easier to change a few bits in the memory of a general-purpose computer than it is to fabricate a corrected version of a special-purpose computer. Also, many design changes occur not because of a designer's mistake but instead are required due to changing specifications. Productivity tools for general-purpose computers, such as compilers, assemblers, linkers, editors, debuggers, etc., make the software designer's task of coping with bugs and changing specifications much easier than the above machine language examples.

The situation that has existed for the last half century is designers have had the choice between using a general-purpose computer or building a special-purpose computer. If the market price (in dollars, rather than in gates) of the general-purpose computer is within budget and its speed is adequate (not the fastest, just adequate) and otherwise meets physical constraints (size, weight, power consumption, ruggedness) for the intended application, the designer typically chooses the general-purpose computer because of the advantages of rapid debugging. Although most algorithms work adequately on general-purpose computers, some demand special-purpose hardware. This has created two different economic phenomena.

First is the emergence of the general-purpose computer industry, composed of only a handful of companies worldwide that actually design CPUs. All together, these companies employ only a few hundred computer designers at best, and so few of the readers of this book will ever be employed as general-purpose computer designers. These designers face a daunting challenge: they design machines that will be used for tasks that no one has yet conceived. Programmers in the future will think of new things to do with the general-purpose machines that designers are working on today. Why does knowing how the machine will be used assist the designer? Speed is not the primary concern of a designer solving a specific problem because the designer can easily tell if the machine is fast enough. A special-purpose computer does not have to be the fastest computer in existence—it just has to be fast enough, and, of course, do its job correctly. General-purpose computer designers do not have the luxury of knowing what is fast

---

[12] Here "technological" means measuring cost in in terms of registers, gates, chip area, etc., rather than in dollars.

enough. Because of the market pressures created by this uncertainty, they have developed more efficient (but intricate) variations on the fetch/execute algorithm that allow software to approach the speed of a special-purpose computer. This has come at the cost of increased hardware by using sophisticated techniques, such as pipelining (chapters 6 and 9), and is why there is such variety among the instruction sets (chapter 10).

Second, a more recent phenomenon is the emergence of hardware description languages (chapter 3) running on general-purpose computers that allow the debugging (simulation) and synthesizing of efficient special-purpose computers to be almost as easy as the programming of software. It is the theme of this book that the worlds of hardware and software are converging. You will need to be aware of both of these to prosper in the next half century of the computer age.

8.4 Mixed fetch/execute

In order to illustrate that there is nothing mysterious about the design of a general-purpose computer once the details of the fetch/execute algorithm are specified, let's translate the pure behavioral ASM (section 8.3.2.4.2) for the PDP-8 instruction subset into the mixed stage of the top-down design process. Recall from section 2.1.5.2, the mixed stage consists of two hardware structures: a controller and an architecture.

There are many possible architectures that can implement a given pure behavioral ASM. The more complicated the ASM, the more room there is for creativity in the design of the architecture. Once the designer decides upon an architecture, the design of the controller is a relatively mechanical process.

8.4.1 A methodical approach for designing the architecture

When an ASM uses more than a handful of registers and/or states, it becomes difficult to keep track of all of the details in your head. In such an instance, it is wise to take a methodical approach to designing the architecture. To begin with, note all of the register transfers that occur in each state. Write down this information grouped together by destination. Since in section 8.3.2.4.2 there are six possible destinations (left-hand sides of ←, excluding the memory, which in section 8.3.2.4.2 is a separate actor), there are six groups to note:

a) register transfers to the accumulator and/or lind. (These are together in one group since {link,ac} often acts as a 13-bit register, and so modifications to the accumulator by itself or to the link by itself should be considered as modifications to {link,ac});

b) register transfers to the halt flag;

c) register transfers to the instruction register;

d) register transfers to the memory address register;

e) register transfers to the memory buffer register; and

f) register transfers to the program counter.

It is wise to write down the state(s) in which each transfer occurs so that you can refer back to the ASM as necessary. (When both the right-hand sides and left-hand sides of register transfers in two or more states are identical, note the names of all such states.) The following table illustrates this for the ASM in section 8.3.2.4.2:

RTN	State(s)
ac ← 0	E0CLA, E1BDCA
ac ← ac & mb	E0AND
ac ← ac | sr	E0OSR
ac ← ~ac	E0CMA
link ← 0	E0CLL
link ← ~link	E0CML
{ac,link} ← {link,ac}	E0RAR
{link,ac} ← {ac,link}	E0RAL
{link,ac} ← 0	E0CLACLL
{link,ac} ← {link,ac} + 1	E0IAC
{link,ac} ← {link,ac} + mb	E0TAD
halt ← 0	IDLE
halt ← 1	E0HLT, INIT
ir ← membus	F3A
ma ← ea(ir)	F3B
ma ← ma + 1	E0BDEP
ma ← pc	F1
ma ← sr	E0MA
mb ← ac	E0DCA
mb ← membus	F4A
mb ← sr	E0DEP

Continued

RTN	State(s)
pc ← ma	E0JMP
pc ← pc + 1	F2, E0ASKIP
pc ← sr	E0PC

Note that implicitly, the {link,ac} group should be thought of as implementing the following register transfers:

{link,ac} ← {link,0}	E0CLA, E1BDCA
{link,ac} ← {link,ac & mb}	E0AND
{link,ac} ← {link,ac | sr}	E0OSR
{link,ac} ← {link,~ac}	E0CMA
{link,ac} ← {0,ac}	E0CLL
{link,ac} ← {~link,ac}	E0CML
{ac,link} ← {link,ac}	E0RAR
{link,ac} ← {ac,link}	E0RAL
{link,ac} ← 0	E0CLACLL
{link,ac} ← {link,ac} + 1	E0IAC
{link,ac} ← {link,ac} + mb	E0TAD

These two ways of describing link and accumulator register transfers are equivalent. The former is easier for the designer to comprehend. The latter is important in the next step the designer takes.

8.4.2 Choosing register types

Here is where the creative part occurs. Whatever hardware structure the designer chooses must be capable of implementing each of the above register transfers during the state(s) indicated. The controller will take care of making sure the states happen at the proper times, so we do not have to worry about that. Our concern now is that the architecture can manipulate the data as listed above.

The first decision the designer must make is what kind of structural device will implement each register. One possibility would be to use enabled registers for every variable in the ASM (other than memory); however, this will typically cause the architecture to be more complex than if other types of registers are selected. A better approach is to look at each group (corresponding to transfers to a particular register) individually and note those register transfers where the right-hand side consists only of constants and/or the variable (or concatenated variables) on the left-hand side.

For the link and accumulator group, there are several such register transfers:

{link,ac} ← {link,0}		E0CLA, E1BDCA
{link,ac} ← {link,~ac}		E0CMA
{link,ac} ← {0,ac}		E0CLL
{link,ac} ← {~link,ac}		E0CML
{link,ac} ← 0		E0CLACLL
{ac,link} ← {link,ac}		E0RAR
{link,ac} ← {ac,link}		E0RAL
{link,ac} ← {link,ac} + 1		E0IAC

For the halt flag, both of the possible register transfers are of this kind:

halt ← 0	IDLE
halt ← 1	E0HLT, INIT

For the memory address register, only one of the register transfers meet this criteria:

ma ← ma + 1	E0BDEP

Similarly, for the program counter, there is only one register transfer that uses pc on both sides:

pc ← pc + 1	F2, E0ASKIP

For the instruction register and memory buffer register, there are no such register transfers.

The reason for identifying such register transfers is that, in theory, such transfers can be implemented internally within a register device without the need for any external data interconnection. Although such devices may be slightly more expensive, the intellectual simplification they provide to the architecture is usually worth the added cost.

For registers where no such transfers occur, it is clear that the designer should use enabled registers. Therefore, to implement an architecture for the ASM of section 8.3.2.4.2, the instruction register and the memory buffer register should be enabled registers. For these registers, whatever new data is loaded always comes from outside the enabled register.

In the case of the memory address register and the program counter, it is obvious from the above that an up counter register is the most appropriate choice. For the halt flag, a clearable enabled register (or its equivalent) is a reasonable choice because this allows the halt $\leftarrow$ 0 transfer to occur internally (leaving only the halt $\leftarrow$ 1 to be provided externally).

The choice of the register type for {link,ac} is less clear. In theory, one could imagine a device that is capable internally of doing all the operations listed for the link and accumulator. The problem is that such a contrived device is not one of the standard register building blocks discussed in appendix D. The intellectual simplification of register building blocks occurs not only because they hide details internally (hierarchical design) but also that their behavior is widely understood in industry and they can be concisely explained in a single cohesive sentence. (An up counter can hold, load, clear and increment its data. These operations are no more and no less than what is required to "count up.") It would not be wrong to build a device that does everything for the link and accumulator. (An automated synthesis tool, such as the one described in chapter 7, might take such an approach.) As a matter of good style for a manually synthesized design and out of consideration to others who attempt to understand the architecture, we will instead choose standard register building blocks of the kind described in appendix D.

Of the building blocks described in appendix D, there are two possible choices for the link and accumulator: the up counter and the shift register. If the designer chooses an up counter, the following register transfers can be implemented internally by the device:

{link,ac} $\leftarrow$ {link,0}	EOCLA, E1BDCA
{link,ac} $\leftarrow$ {0,ac}	EOCLL
{link,ac} $\leftarrow$ 0	EOCLACLL
{link,ac} $\leftarrow$ {link,ac} + 1	EOIAC

If the designer chooses a shift register, a different set of register transfers can be implemented internally by the device:

```
{ac,link} ← {link,ac}        E0RAR
{link,ac} ← {ac,link}        E0RAL
```

Of the complete group of link and accumulator register transfers, the ones that **cannot** be implemented by **either** of these building blocks include:

```
{link,ac} ← {link,~ac}       E0CMA
{link,ac} ← {~link,ac}       E0CML
{link,ac} ← {link,ac & mb}   E0AND
{link,ac} ← {link,ac | sr}   E0OSR
{link,ac} ← {link,ac} + mb   E0TAD
```

The best way to implement such computations that must occur outside the register building block is with a combinational ALU, such as the one described in section 2.3.4, that is capable of doing addition and logical operations. Since the ALU can add an arbitrary number to the accumulator (as required in state E0TAD), it can also increment the accumulator (as required in state E0IAC). The ALU can perform sixteen different logical (bitwise) operations, including AND, OR and NOT (as required in states E0AND, E0OSR, E0CMA and E0CML). The ALU can output zero, and so the clearing operations (as required in states E0CLA, E1BDCA, E0CLL and E0CLACLL) can be accomplished at no added cost. Since the ALU is suitable for either design alternative ({link,ac} as an up counter or {link,ac} as a shift register) but the ALU can do the incrementing that would otherwise require a counter, an appropriate design decision is to use a shift register for {link,ac}.

Here are the register types chosen above:

```
{link,ac}     13-bit shift register
mb            12-bit enabled register
ma            12-bit up counter register
ir            12-bit enabled register
pc            12-bit up counter register
halt           1-bit clearable enabled register
```

8.4.3 Remaining register transfers

Having decided on each register type, we can eliminate those register transfers that occur internally within the register device, which leaves the following:

```
ac ← 0                              E0CLA, E1BDCA

ac ← ac & mb                        E0AND

ac ← ac | sr                        E0OSR

ac ← ~ac                            E0CMA

link ← 0                            E0CLL

link ← ~link                        E0CML

{link,ac} ← 0                       E0CLACLL

{link,ac} ← {link,ac} + 1           E0IAC

{link,ac} ← {link,ac} + mb          E0TAD


halt ← 1                            E0HLT,E0INIT


ir ← membus                         F3A


ma ← ea(ir)                         F3B

ma ← pc                             F1

ma ← sr                             E0MA


mb ← ac                             E0DCA

mb ← membus                         F4A

mb ← sr                             E0DEP


pc ← ma                             E0JMP

pc ← sr                             E0PC
```

The remaining {link,ac} transfers are listed above in their original form to provide documentation that more closely matches the original ASM. For example, ac ← 0 is more concise than {link,ac} ← {link,0}.

8.4.4 Putting the architecture together

In choosing the shift register, we also determined that every one of the remaining {link,ac} transfers (listed in section 8.4.3) can be implemented by the ALU. One of the inputs to the ALU will be the 13-bit {link,ac}. The other will be a 13-bit mux that selects between $\{0, sr\}$, $\{0, mb\}$ and the constant one. It is important to note that although the $\{link, ac\}$ is a unified thirteen-bit shift register, the link and accumulator portions are controlled separately. Therefore, it is possible to load just the accumulator, just the link or both of them. The controls for the $\{link, ac\}$ are as follows:

link_ctrl bits	link_ctrl symbol	ac_ctrl bits	ac_ctrl symbol	action
00	`HOLD	00	`HOLD	do nothing
00	`HOLD	11	`LOAD	ac ←alubus[11:0]
11	`LOAD	00	`HOLD	link ← alubus[12]
11	`LOAD	11	`LOAD	{link,ac} ← alubus
10	`LEFT	10	`LEFT	{link,ac} ← {ac,link}
01	`RIGHT	01	`RIGHT	{ac,link} ← {link,ac}

The default (when link_ctrl and ac_ctrl are not mentioned in a state) is to hold the accumulator and link as they are.

The ea(ir) function can be implemented by trivial combinational logic. We leave this as a separate device since there are other addressing modes not implemented here that are described in appendix B and that are left as exercises.

There is only one register transfer left for the halt flag, and so its input is a constant one. Similarly, there is only one register transfer for the instruction register, and so its input is the memory bus (which provides m[ma] to the architecture from the external memory device).

The remaining register transfers can be provided for by placing muxes on the inputs of the appropriate registers. The input to the memory buffer register is a 12-bit mux that selects among sr, the accumulator and memory bus. The input to the memory address register is a 12-bit mux that selects among sr, ea(ir) and the program counter. The input to the program counter is a 12-bit mux that selects between the sr and the memory address register.

Here is the block diagram of the architecture that was just derived for the subset PDP-8:

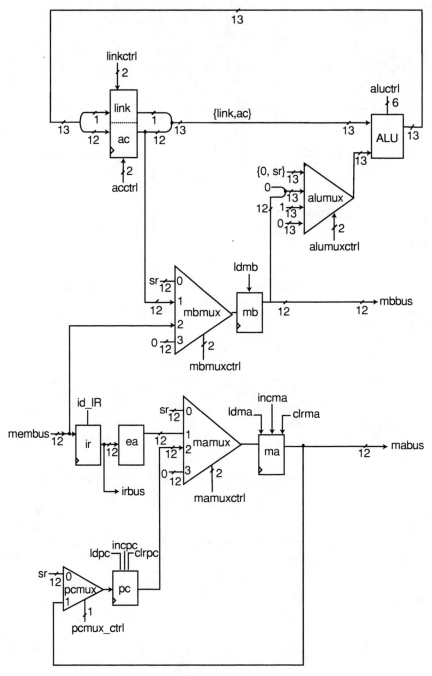

Figure 8-12. Architecture for PDP-8 subset.

Figure 8-12. Architecture for PDP-8 subset (continued).

The fourth inputs to the memory address and memory buffer muxes are not required and therefore tied to zero. It is left as an exercise to show that these fourth inputs will help to implement more of the instructions given in appendix B.

8.4.5 Implementing the decisions

The ASM of section 8.3.2.4.2 has several decisions. Some of these test external status inputs (cont, but_PC, but_MA and but_DEP) that have nothing to do with the architecture. The remaining decisions test data contained in the registers of the architecture (link, ac, ir and halt). Although it would be possible to implement each of these decisions using a comparator, another easier way to implement these decisions exists.

Recall from section 2.1.3.1.2 that a multi-bit external status signal which is only tested against constants can be rewritten as a nested series of decisions that test the individual bits of the status. Using this approach, the internal status inputs to the controller are simply link, ac, ir and halt with no need for comparators in the architecture. In particular, since the instruction register is used in so many decisions, it is prudent to make it an input to the controller.

8.4.6 Mixed ASM

Here is the mixed ASM for the architecture of section 8.4.4 that implements the register transfers of section 8.3.2.4.2:

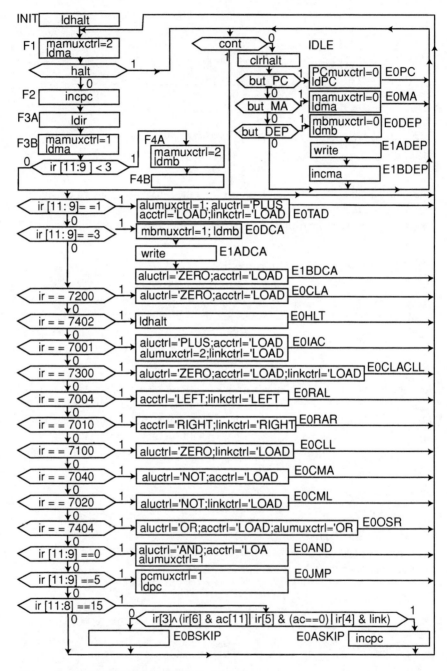

Figure 8-13. Mixed ASM for PDP-8 subset.

As discussed in the previous section, the decisions, which are implemented entirely in the controller, can (and should) be documented in the most understandable way. This means using `ir == 7200` rather than the equivalent twelve individual bit decisions that the controller actually uses. As is shown elsewhere in this book, modern synthesis tools can translate decisions like `ir == 7200` into the details required in the controller.

8.4.7 Block diagram

The following block diagram shows how to put the entire structure together:

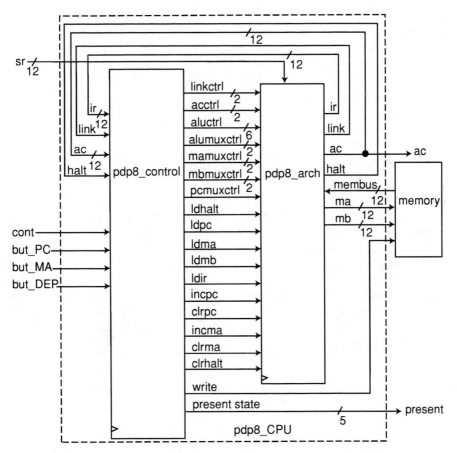

Figure 8-14. Block diagram for PDP-8 subset.

It is a simple, but tedious, matter to use hierarchical design to fill in the details of the controller from the ASM of section 8.4.6. Happily, synthesis tools can also aid the designer with this process.

8.5. Memory hierarchy

As described in section 8.2.3, the design of large and fast memories has been a challenge since the time of the earliest electronic computers. By the end of the twentieth century, these issues became of increasing concern because improving silicon technologies[13] allowed general-purpose processors to run at ever-higher clock frequencies. Large low-cost memory, such as dynamic memory chips, was unable to keep up with increasing processor speeds.

If general-purpose computers accessed memory in a completely random and haphazard fashion such that we could not make any kind of accurate prediction for which word in memory the processor would access next, this mismatch of processor and memory speed would be unsolvable. Happily, because of the nature of the fetch/execute algorithm and the nature of most machine language programs interpreted by the fetch/execute algorithm, we can predict, with reasonably good odds, what word the processor might fetch next. This solution to the mismatch between processor and memory speed has been recognized since 1962, when Kilburn and others at the University of Manchester designed the Atlas computer to take advantage of the fact that not all words in memory are accessed with the same frequency.

Kilburn's solution, which has endured with minor variations for more than a third of a century, is to design a hierarchy of memories of different speeds, sizes and costs. The hierarchy might have several different levels, each containing a different memory technology. The lowest level has a memory technology that costs the least per bit. This memory will have the largest number of words since we can afford to buy quite a lot of such cheap memory. Such inexpensive memory necessarily has a slow access time. Each higher level in the hierarchy has a kind of memory which is faster than lower levels in the hierarchy. Because the faster memories are more expensive per bit than the memories in the lower levels in the hierarchy, we can only afford smaller memory sizes in the upper levels of the heriarchy.

---

[13] Primarily smaller chip dimensions which mean lower propagation delays.

The memory hierarchy is usually effective because, statistically speaking, most memory accesses occur to words that have already been accessed before. If the system keeps the few words that are more likely to be accessed in the fast but small-sized higher levels of the hierarchy, and all the other words that are less likely to be accessed in the lower levels, we observe two benefits. First, the cost of the system is not significantly higher than if it were built entirely of slow cheap memory. Second, the speed of the system is not significantly slower than if it were built entirely of fast expensive memory. In essence, we almost get the best of both alternatives. However, this good cost and performance occurs only in a statistical sense: the "average" program will on "average" execute almost as fast as if the system used a fast memory. The program you are interested in may actually execute considerably slower, depending on the pattern in which that program accesses memory for the particular data you give it and depending on the details of the memory hierarchy you use.

There are two common kinds of memory hierarchy. The first of these is known as *cache memory*, which is discussed in the next section. The second of these, which is what Kilburn used, is known as *virtual memory*. The idea of virtual memory is to keep less frequently used parts of memory on disk. The access time for the data on disk is many orders of magnitude slower than for data in semiconductor memory. It is also very non-deterministic because of the unpredictable distance the disk has to rotate to be positioned on the proper data.[14] Although conceptually, virtual memory is very similar to cache memory, its implementation requires complicated hardware and software. Hardware implementation of virtual memory requires a disk controller, and the management of virtual memory is usually intertwined with the software details of an operating system. Since hardware disk controllers and software operating systems are beyond the scope of this book, we will not consider virtual memory.

8.5.1 Cache memory

Cache memory is the fastest part of the memory hierarchy. It is built out of several components. The cache needs its own controller, which we will ignore for the moment. Of course, the cache needs high-speed memory for the data to be stored in the cache, but the cache also needs a *tag memory* which indicates the address associated with each portion of the cache. The data and tag memories of the cache are usually composed of expensive high-speed static memory that can be accessed in significantly less than one clock cycle. When the propagation delay of the rest of the system is considered, this still allows data to be fetched from the cache in one clock cycle.

---

[14] Also, there is the chance the disk head has to move, which can take a significant fraction of a second.

The tag memory is needed because a particular part of the cache may be associated with more than one address at different times during the operation of the cache. In contrast, a particular part of an ordinary (main) memory will always be associated with one particular constant address. As explained in section 8.2.2.3.1, such a main memory can be thought of as a mux which selects one of several register values. Each cell in a main memory is always associated with its particular address because that address specifies the port of the mux to which the corresponding register is wired.

There are two common approaches to designing a cache. In the *direct mapped* approach, there is only one tag memory and one data memory. In the multi-way set associative approach, there are several parallel tag and corresponding data memories. The direct mapped approach is simpler and therefore allows a faster access time. On the other hand, the direct approach is often not as successful in keeping the appropriate words in the cache as the multi-way approach, and so even though the access time of the multi-way approach is slower, it may be faster overall for some programs than the direct approach. This section, however, will concentrate on the direct mapped approach, which is easier to comprehend.

The typical cache memory uses the low-order bits of the address bus to select information out of both the data and tag portions of the cache. In order for a memory access to be fast, the information fetched from the tag memory must match the address bus.[15] If it does not, the cache must be updated from some lower level of the memory hierarchy. Commercial computer systems often have more than one level of cache. In such systems, the first level is often on the same chip as the processor to maintain the highest (single clock) speed. The second level (referred to as L2) is contained on separate chips that allow access in a small number of cycles. The main memory is composed of dynamic memory, with an access time of many clock cycles. In this section, however, there will only be two levels in the memory hierarchy: the direct mapped cache and the main memory.

In this chapter, we will assume each element of the cache content memory is a single word. Often, in commercial systems, each element of the cache content memory is a group of several contiguous words, known as a *line*. Using a line composed of several words may improve the performance of the cache, but including such details here would obscure the idea being discussed in this section: how a cache is a cost-effective way to improve the performance of a general-purpose computer.

For example, assume a cache size of four words[16] with the following simple program that goes through a loop eight times producing nine values[17] (7760, 7762, ... 7776 and 0000) in the accumulator:

---

[15] In an actual implementation, only the high-order bits need to be stored in the tag memory and checked against the high-order bits of the address bus, but we will ignore this detail for now.

[16] This is too small for practical use but will illustrate how a cache works.

[17] These are the nine decimal values -16, -14, ... -2 and 0.

```
0000/7300              CLACLL       // ac = -16
0001/1006              TAD A
0002/1011    L,        TAD B        // ac = ac + 2
0003/7510              SPA          // if ac>=0, halt
0004/5002              JMP L        // if ac<0, stay in loop
0005/7402              HLT
0006/7760    A,        7760         // equivalent to decimal -16
                       *0011
0011/0002    B,        0002         // +2
```

Assuming the instructions of this program are loaded into memory in the same order as listed above, at the time the fetch/execute cycle begins, the cache will contain:

```
        cache                main memory
   tag       data            0000/7300
   0/0004    0/5002          0001/1006
   1/0011    1/0002          0002/1011
   2/0006    2/7760          0003/7510
   3/0003    3/7510          0004/5002
                             0005/7402
                             0006/7760
                             0011/0002
```

The words shown in bold for the main memory are the ones currently in the cache. When the processor fetches the first instruction, the memory access will be slow because address 0000 is not currently in the cache. This is known as a *cache miss*. The cache has to bring in this word (7300) from the main memory, and so the cache now looks like:

```
        cache                main memory
   tag       data            0000/7300
   0/0000    0/7300          0001/1006
   1/0011    1/0002          0002/1011
   2/0006    2/7760          0003/7510
   3/0003    3/7510          0004/5002
                             0005/7402
                             0006/7760
                             0011/0002
```

Fetching the next instruction (1006) causes another cache miss:

```
         cache                    main memory
    tag       data               0000/7300
    0/0000    0/7300             0001/1006
    1/0001    1/1006             0002/1011
    2/0006    2/7760             0003/7510
    3/0003    3/7510             0004/5002
                                 0005/7402
                                 0006/7760
                                 0011/0002
```

However, when this TAD instruction is executed, the cache already has the data 7760 required by the processor. This is known as a *cache hit*. The second memory access during this instruction is fast because it is a cache hit.

Fetching and executing the next instruction (1011) causes two cache misses:

```
         cache                    main memory
    tag       data               0000/7300
    0/0000    0/7300             0001/1006
    1/0011    1/0002             0002/1011
    2/0002    2/1011             0003/7510
    3/0003    3/7510             0004/5002
                                 0005/7402
                                 0006/7760
                                 0011/0002
```

Fetching and executing the SPA instruction (7510) causes a cache hit, and so this memory access is fast. Since the accumulator is negative, the skip does not occur, and the processor needs to fetch the next (5002) instruction. This causes another cache miss:

```
         cache                    main memory
    tag       data               0000/7300
    0/0004    0/5002             0001/1006
    1/0011    1/0002             0002/1011
    2/0002    2/1011             0003/7510
    3/0003    3/7510             0004/5002
                                 0005/7402
                                 0006/7760
                                 0011/0002
```

From this point on, as long as the program stays inside this three-instruction loop (TAD; SPA; JMP), all of the instruction and data accesses are cache hits. The final cache miss occurs when the program halts:

```
          cache              main memory
    tag        data           0000/7300
  0/0004     0/5002           0001/1006
  1/0005     1/7402           0002/1011
  2/0002     2/1011           0003/7510
  3/0003     3/7510           0004/5002
                              0005/7402
                              0006/7760
                              0011/0002
```

In total, there are six cache misses[18] and 29-cache hits in this example. With the given value of A, this is a 17% miss rate and an 83% hit rate, although the hit rate would increase for values of A that are more negative.[19]

The good performance that the above program exhibits using this little cache depends heavily on how the instructions and data are arranged. For example, if B were located at address 0007, there would be 20 cache misses and only 15 cache hits, which is a 57% miss rate and 43% hit rate. A larger cache size will often improve performance. If the program with B at address 0007 runs on a machine with a cache size of eight, the hit rate becomes 100% because this entire tiny program can reside in the cache. If the program with B at address 0011 runs on a machine with a cache size of eight, the hit rate is 94% (two misses) because the program cannot all fit in the cache at once (0001 and 0011 cannot reside in a direct mapped cache of size eight at the same time).

8.5.2 Memory handshaking

Regardless of whether a machine uses cache memory, virtual memory or both in its memory hierarchy, one thing is clear: the access time is non-deterministic. Although we expect the majority of memory accesses to occur in a single cycle, some accesses will take additional cycles. The ASM chart for fetch/execute given in section 8.3.2.4.2 assumes that every memory access can occur in one cycle, which is not the case for a memory hierarchy. A more sophisticated ASM is required that waits for the memory

---

[18] The number of misses is the same in this program regardless of the value of A and therefore of how many times the loop executes.

[19] The number of hits depends on how many times the loop executes.

hierarchy to provide the requested instruction or data. To coordinate the operation of the memory hierarchy with the CPU requires using what is called a *partial handshaking* protocol.

In the partial handshake protocol, there is an extra command signal, memreq, that the CPU sends to the memory when the CPU requests a particular word of the memory hierarchy. Unlike the simple memory described in section 8.2.2.3.1, the memory hierarchy might ignore the address bus when memreq is not asserted. Only when memreq is asserted does the memory take notice of the address bus and respond accordingly.

In the partial handshake protocol, there is also an extra status signal, the memory read acknowledge (memrack), that the memory sends back to the CPU to acknowledge that the memory hierarchy has obtained the word desired by the CPU. The CPU must continue to assert its memory request until the memory responds with its acknowledge signal. If the desired word is already in the cache, the memory hierarchy will instantly[20] assert memrack. Having the memory hierarchy assert memrack within the same cycle that the CPU first asserts memreq means that only one cycle is spent on the memory access. If the desired word is not already in the cache, the memory hierarchy will wait however long is necessary before asserting memrack. The ASM for the CPU must stay in a wait state prior to when the memory hierarchy asserts the memrack status signal. For example, consider the portion of the ASM from section 8.3.2.4.2 (consisting of states F3B, F4A and F4B) shown on the left:

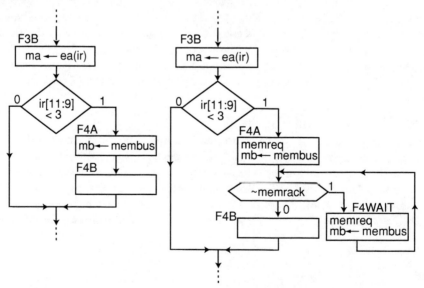

Figure 8-15. ASMs without and with memory read handshaking.

---

[20] Ignoring a trivial amount of propagation delay, as was done in earlier portions of this chapter.

As explained in section 8.3.1.5, these states fetch the operand for PDP-8 instructions, such as TAD, on the assumption that every memory access can occur in one clock cycle. In state F3B, the memory address register is scheduled to change to the effective address of the instruction. As with all ←, the change does not take effect until the next rising edge of the clock. Assuming the instruction is a TAD, the CPU will be in state F4A when that next rising edge occurs. In state F4A, the memory buffer register is scheduled to be loaded with the corresponding contents of memory that comes via the memory bus. In this ASM, there is never more than one clock cycle for the memory to give the correct data to the CPU.

In contrast, the more complicated ASM on the right uses handshaking to adapt to the speed of each particular memory access. This requires introducing an extra state, F4WAIT, and asserting memreq in both states F4A and F4WAIT. Also, there is a decision involving memrack that occurs in both states F4A and F4WAIT. If the memory access is fast, the CPU never goes to state F4WAIT, and so the state transitions of the left ASM are identical to the state transitions of the right ASM. On the other hand, if the memory access is slow, the machine goes to state F4WAIT where it will loop until the memory hierarchy asserts memrack. Note the data that the memory hierarchy provides on membus must be valid **before** the hierarchy can assert memrack.

A similar handshaking approach is required for memory writes, except the memory hierarchy responds back with a memory write acknowledge (memwack):

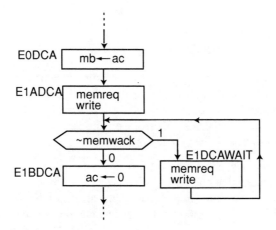

Figure 8-16. ASM with memory write handshaking.

8.5.3 Architecture for a simple memory hierarchy

The memory hierarchy is a separate actor from the CPU, and so the memory hierarchy needs its own controller and architecture. Assuming we keep the names the same as in the earlier sections of this chapter, the memory hierarchy and the CPU communicate address and data using `mabus`, `mbbus` and `membus`. More importantly, the CPU tells the memory hierarchy what it needs done to memory with `write` and `memreq`, and the memory responds back when the requested operation is complete using either `memrack` or `memwack`. Both the CPU and the memory hierarchy are fed the system clock.[21] Here is a diagram that illustrates this interconnection:

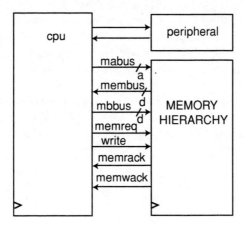

Figure 8-17. Connection of processor to memory hierarchy.

We are going to design a memory hierarchy consisting of a main memory and a cache. There are many choices available to the designer of a cache. Although the cache could be either set associative or direct mapped, we will use the direct mapped technique since it is easier to understand. Also, there is a choice about how the cache treats writes: either a *write-through cache* or a *write-back cache* .

A write-back cache waits until it is necessary to write data back into the main memory. This has the advantage that operations on values such as loop counters do not have to wait for the slower access time of the main memory. The problem with the write-back approach is that the main memory and the cache can become inconsistent with each other. Because *cache consistency* is not guaranteed at all times with the write-back approach, a request from the CPU for a memory read may also cause a write to the main memory (that restores cache consistency). This makes a write-back cache considerably harder to design than a write-through cache. It may even make the write-back

---

[21] This is a requirement of the partial handshake protocol.

cache slower than a comparable write-through cache.[22] The complexity of write-back caches are even more pronounced when multiple CPUs share the same main memory but have distinct caches.

In contrast, write-through caches are simple enough[23] that the design of an elementary one can be described using only the mixed Moore ASM notation (section 2.3) or using the equivalent Verilog (section 4.2). The essential idea of a write-through cache is that when the CPU requests the memory hierarchy to do a write operation, the memory hierarchy will store the data into both the main memory and the cache. A write-through cache is guaranteed to be consistent with the main memory.

We are going to skip over the pure behavioral stage of the design, and proceed straight to the mixed stage so that we may focus on the architecture of the memory hierarchy.[24] This architecture consists of the main memory, the cache content memory and the cache tag memory. The main memory is asynchronous with a deterministic access time bounded by a known number of clock cycles. Like the one and only memory shown previously in section 8.4.7, the main memory of the hierarchy has its data input connected to mbbus and its address input connected to mabus. The distinctions between the main memory of the hierarchy and the memory of section 8.4.7 are the main memory of the hierarchy receives its mainwrite signal from the internal cache controller (rather than from the CPU) and the data output of the main memory of the hierarchy does not connect directly to the CPU. Instead, the data output of the main memory connects to the data input of the cache content memory.

The cache content and cache tag memories are synchronous memories that can be accessed within one clock cycle. The address inputs to both the cache content and cache tag memories come from the low order j bits of the CPU's memory address bus. We will use 'CACHE_SIZE to indicate the number of words that can reside in the cache, which is the same as 2^j. The data output of the cache content register (cache_content[mabus % 'CACHE_SIZE]) is connected to the memory bus that goes back to the CPU. The data output of the cache tag register (cache_tag[mabus % 'CACHE_SIZE]) is connected to a comparator. The other input of the comparator is the memory address bus from the CPU.[25] The output of the comparator is the memory read acknowledge signal (memrack). This signal is sent to

---

[22] Whether write-through or write-back is faster depends on several factors, including the pattern in which the particular program accesses memory.

[23] A write-back cache requires the Mealy notation of chapter 5.

[24] We are also avoiding the pure behavioral stage now because, even on this simple write-through cache, the pure behavioral ASM requires the Mealy notation of chapter 5. The mixed ASM does not need the Mealy notation because the memrack signal is generated by the architecture, and not the controller.

[25] Many implementations would only use the high order $a - j$ bits, but for simplicity, we use all a bits.

both the CPU (where it controls the duration of the wait loops described in section 8.5.2) and to the cache controller. The cache controller generates the `ldcont` and `ldtag` commands for the cache content and cache tag memories. When asserted, these commands indicate that the cache content and cache tag memories will be loaded with new information at the next rising edge of the clock.

The cache controller also generates the memory write acknowledge (`memwack`) signal after the main memory, the cache tag memory and the cache content memory have been updated as a result of a `write` signal from the CPU. The following shows the architecture for the memory hierarchy (using the write-through direct mapped cache described above) and the corresponding mixed Moore ASM for the cache controller:

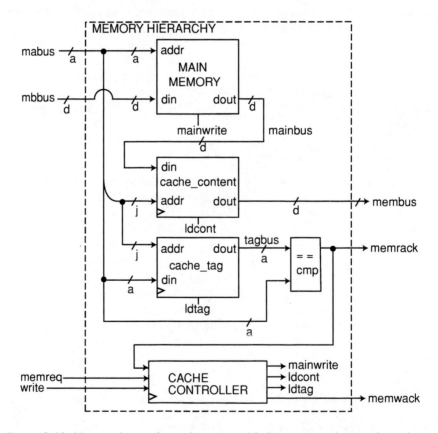

Figure 8-18. Memory hierarchy architecture with direct mapped write-through cache.

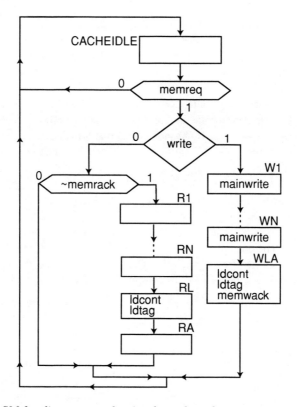

Figure 8-19. ASM for direct mapped write-through cache memory controller.

The ASM stays in state CACHEIDLE unless a memory request occurs. There are three possibilities for a memory request. Two of these possibilities are when the memory request is for a read operation (i.e., `write` is zero): either the requested data is in the cache or the requested data is not in the cache. The third possibility is a memory write request from the CPU (regardless of whether it is in the cache).

The first possibility is when the data being read by the CPU is already in the cache. In this case, `memrack` will be true during the first clock cycle that `memreq` is asserted. Because the `memrack` signal comes straight from the combinational logic comparator, both the ASM for the CPU and the ASM for the cache controller proceed without delay states. The ASM for the CPU makes a transition such as from state F4A to F4B, and the ASM for the cache controller makes a transition from CACHEIDLE back to that same state.

The second possibility is when the data being read by the CPU is not in the cache during the first clock cycle that memreq is asserted. During this clock cycle memrack will be false. It will stay false for as long as the output of the cache tag memory does not equal the memory address bus from the CPU. In a case like this, when memrack is false, the ASM for the CPU makes a transition such as from state F4A to F4WAIT, and the ASM for the cache controller makes a transition from CACHEIDLE to state R1. The ASM has an appropriate number of empty delay states (not shown) to allow for the read access time of the asynchronous main memory. Then, in state RL, the cache controller issues the ldcont command. This causes the cache content memory to be loaded at the next rising edge of the clock with the data obtained from the slow main memory. Also in state RL, the cache controller issues the ldtag command. This causes the cache tag memory to be loaded at the next rising edge of the clock with the address being provided by the CPU. Because of this change to the tag memory, when the cache controller proceeds to the empty state, RA, the architecture will for the first time assert memrack. The one empty state, RA, is all that is necessary to allow the CPU to make a transition such as from state F4WAIT to F4B. Of course, the cache controller makes a transition during that same clock cycle from state RA back to CACHEIDLE.

The third possibility is when the CPU makes a memory write request. The ASM for the cache controller proceeds from state CACHEIDLE to state W1 during the same clock cycle that the ASM for the CPU proceeds from a state such as E1ADCA to E1DCAWAIT. The ASM has an appropriate number of delay states (not shown) that each assert mainwrite. This allows for the write access time of the asynchronous main memory.[26] Finally, in state WLA, the cache controller asserts ldcont, ldtag and memwack. The assertion of ldcont and ldtag is not necessary for this write operation but is required for any future read operations to be fast. Therefore, a separate empty state for write acknowledgement is not necessary here as was the case for read acknowledgement. Because memwack is asserted in state WLA, at the same time that the ASM for the cache controller makes a transition from state WLA to state CACHEIDLE, the ASM for the CPU makes a transition such as from state E1DCAWAIT to E1BDCA.

The following example is a program that adds two numbers together and stores the sum in memory. Both state machines (CPU and memory controllers) cooperate to fetch instructions and data and to store results back in memory. This example illustrates each of the three possibilities explained above. The first two instructions, as well as the first word of data fetched, are already in the cache. In such an instance (shown in bold) the cache state remains in CACHEIDLE and the CPU does not need a wait state. This situation is signaled by memreq and memrack both being one during the same clock cycle.

---

[26] The read and write access times need not be the same.

CPU state	ma	mb	pc	ir	h	ac	CACHE state	mem req	mem rack	mem wack
F1	0002	1006	0000	xxxx	0	xxxx	CACHEIDLE	0	0	0
F2	0000	1006	0000	xxxx	0	xxxx	CACHEIDLE	0	1	0
F3A	**0000**	1006	0001	xxxx	0	xxxx	CACHEIDLE	1	1	0
F3B	0000	1006	0001	**7300**	0	xxxx	**CACHEIDLE**	0	1	0
E0CLACLL	0100	1006	0001	7300	0	xxxx	CACHEIDLE	0	0	0
F1	0100	1006	0001	7300	0	0000	CACHEIDLE	0	0	0
F2	0001	1006	0001	7300	0	0000	CACHEIDLE	0	1	0
F3A	**0001**	1006	0002	7300	0	0000	CACHEIDLE	1	1	0
F3B	0001	1006	0002	**1006**	0	0000	**CACHEIDLE**	0	1	0
F4A	**0006**	1006	0002	1006	0	0000	CACHEIDLE	1	1	0
F4B	0006	**1000**	0002	1006	0	0000	**CACHEIDLE**	0	1	0
E0TAD	0006	1000	0002	1006	0	0000	CACHEIDLE	0	1	0
F1	0006	1000	0002	1006	0	1000	CACHEIDLE	0	1	0

As execution proceeds, the next instruction causes a cache miss, which causes the memory controller to leave CACHEIDLE and causes the CPU to enter a wait state (shown in italic):

CPU state	ma	mb	pc	ir	h	ac	CACHE state	mem req	mem rack	mem wack
F2	0002	1000	0002	1006	0	1000	CACHEIDLE	0	0	0
F3A	**0002**	1000	0003	1006	0	1000	CACHEIDLE	1	0	0
F3WAIT	*0002*	*1000*	*0003*	*1000*	*0*	*1000*	*R1*	*1*	*0*	*0*
F3WAIT	*0002*	*1000*	*0003*	*1000*	*0*	*1000*	*R2*	*1*	*0*	*0*
F3WAIT	*0002*	*1000*	*0003*	*1000*	*0*	*1000*	*R3*	*1*	*0*	*0*
F3WAIT	*0002*	*1000*	*0003*	*1000*	*0*	*1000*	*RL*	*1*	*0*	*0*
F3WAIT	*0002*	*1000*	*0003*	*1000*	*0*	*1000*	*RA*	**1**	*1*	*0*
F3B	0002	1000	0003	**1007**	0	1000	**CACHEIDLE**	0	1	0

Fetching the operand of this instruction causes a cache hit, but fetching the following instruction causes a cache miss:

```
F4A        0007 1000  0003 1007 0  1000   CACHEIDLE   1  1  0
F4B        0007 2000  0003 1007 0  1000   CACHEIDLE   0  1  0
E0TAD      0007 2000  0003 1007 0  1000   CACHEIDLE   0  1  0
F1         0007 2000  0003 1007 0  3000   CACHEIDLE   0  1  0
F2         0003 2000  0003 1007 0  3000   CACHEIDLE   0  0  0
F3A        0003 2000  0004 1007 0  3000   CACHEIDLE   1  0  0
F3WAIT     0003 2000  0004 2000 0  3000   R1          1  0  0
F3WAIT     0003 2000  0004 2000 0  3000   R2          1  0  0
F3WAIT     0003 2000  0004 2000 0  3000   R3          1  0  0
F3WAIT     0003 2000  0004 2000 0  3000   RL          1  0  0
F3WAIT     0003 2000  0004 2000 0  3000   RA          1  1  0
F3B        0003 2000  0004 3005 0  3000   CACHEIDLE   0  1  0
```

Because this example uses a write-through cache, executing the DCA instruction causes the memory controller to leave state CACHEIDLE and causes the CPU to enter a wait state:

```
E0DCA      0005 2000  0004 3005 0  3000   CACHEIDLE   0  0  0
E1ADCA     0005 3000  0004 3005 0  3000   CACHEIDLE   1  0  0
E1DCAWAIT  0005 3000  0004 3005 0  3000   W1          1  0  0
E1DCAWAIT  0005 3000  0004 3005 0  3000   W2          1  0  0
E1DCAWAIT  0005 3000  0004 3005 0  3000   W3          1  0  0
E1DCAWAIT  0005 3000  0004 3005 0  3000   W4          1  0  0
E1DCAWAIT  0005 3000  0004 3005 0  3000   WLA         1  0  1
E1BDCA     0005 3000  0004 3005 0  3000   CACHEIDLE   0  1  0
F1         0005 3000  0004 3005 0  0000   CACHEIDLE   0  1  0
```

Fetching the final instruction causes another cache miss:

```
F2         0004 3000  0004 3005 0  0000   CACHEIDLE   0  0  0
F3A        0004 3000  0005 3005 0  0000   CACHEIDLE   1  0  0
F3WAIT     0004 3000  0005 7300 0  0000   R1          1  0  0
F3WAIT     0004 3000  0005 7300 0  0000   R2          1  0  0
F3WAIT     0004 3000  0005 7300 0  0000   R3          1  0  0
F3WAIT     0004 3000  0005 7300 0  0000   RL          1  0  0
F3WAIT     0004 3000  0005 7300 0  0000   RA          1  1  0
F3B        0004 3000  0005 7402 0  0000   CACHEIDLE   0  1  0
E0HLT      0002 3000  0005 7402 0  0000   CACHEIDLE   0  1  0
```

8.5.4 Effect of cache size on the childish division program

There are many alternatives that a designer must choose from when implementing a cache. It is often hard to predict manually what effect these choices will have on the speed of the system when it is running a particular program. This is a case where simulation is essential to allow the designer to estimate the effects different design decisions will have on the overall performance of the system. For example, one could simulate to observe the effect of cache size. One of the reasons HDLs such as Verilog have become popular is because designers need to conduct such simulations before building their machines.

The ASMs for the PDP-8 (section 8.3.2.4.2) and the cache controller (section 8.5.3) were translated into Verilog code (not shown), and the childish division program (section 8.3.2.5.3) was run for $x=14$ and $y=7$ with various cache sizes. In each case, there are 53 read accesses and seven write accesses. In this simulation, all write accesses and any cache misses cause the CPU to wait for five clock cycles. Here are hit and miss ratios for reads in this simulation:

cache size	clocks	#miss	miss ratio	hit ratio
8	541	51	96%	4%
16	396	22	58%	41%
32	296	2	4%	96%

8.6 Conclusion

General-purpose computers implement the fetch/execute algorithm, which in turn allows the hardware to interpret other algorithms coded in machine language that is stored in memory. Memory is the critical component for a general-purpose computer to be useful, and various technologies have been used to implement memory during the last half century. Static memories are fast, but dynamic memories are cheaper. Memory hierarchies that include a cache offer the best compromise between speed and cost. From an abstract behavioral viewpoint, all memory technologies can be thought of as arrays of binary words, but in reality, memory devices are independent actors that operate in parallel to the CPU. When the access time is non-deterministic, there must be handshaking between the memory and the CPU so that the CPU can adjust its speed to that of the memory.

The design process for a general-purpose CPU is similar to that of special-purpose hardware. The example used in this chapter of the PDP-8 was implemented at the behavioral and mixed stages of the design process. A methodical architecture was presented, and a variation using a direct mapped cache was considered. These designs were benchmarked using the childish division program to show that software running on the CPU designed in this chapter is slower and less efficient than when the childish division algorithm is implemented in special-purpose hardware. The next chapter will look at how this performance discrepancy can be diminished.

8.7 Further reading

BELL, C. GORDON and A. NEWELL, *Computer Structures: Readings and Examples*, McGraw-Hill, New York, NY, 1971. Chapter 5 is the definitive description of the PDP-8 from the man who also invented the first HDL (a language known as ISP).

BELL, C. GORDON, J. C. MUDGE and JOHN E. MCNAMARA, *Computer Engineering: A DEC View of Hardware Systems Design*, Digital Press, Bedford, MA, 1978. Chapter 8.

LAVINGTON, S., *Early British Computers: The Story of Vintage Computers and the People Who Built Them*, Digital Press/Manchester University Press, Bedford, MA, 1980. Describes the work of Kilburn, Williams, Turing, Wilkes and other British pioneers.

The Origins of Digital Computers: Selected Papers, 2nd ed., Edited by B. Randell, Springer-Verlan, Berlin, 1982. Reprints of original papers by computer pioneers.

PATTERSON, DAVID A. and JOHN L. HENNESSY, *Computer Organization and Design: The Hardware/Software Interface,* Morgan Kaufmann, San Mateo, CA, 1994. Chapter 7 explains virtual memory and multi-way set associative caches.

PROSSER, FRANKLIN P. and DAVID E. WINKEL, *The Art of Digital Design: An Introduction to Top down Design*, 2nd ed., Prentice Hall PTR, Englewood Cliffs, NJ, 1987. Chapter 7 describes an elegant central ALU architecture for the complete PDP-8 instruction set.

SLATER, ROBERT, *Portraits in Silicon*, MIT Press, Cambridge, MA, 1987. Gives biographies of several important pioneers including Babbage, Zuse, Atanasoff, Turing, Aiken, Eckert, Mauchly, von Neumann, Forrester, Bell and Noyce.

WAYNER, P., "Smart Memory," *BYTE*, June 1995, p. 190.

WOLF, WAYNE, *Modern VLSI Design: A Systems Approach*, 2nd ed., Prentice Hall PTR, Englewood Cliffs, NJ, 2nd ed., 1994, p. 356-370. Shows how to layout a VLSI chip that implements a PDP-8 architecture.

8.8 Exercises

8-1. Revise the ASM of section 8.3.2.1 to include the ISZ instruction described in appendix B.

8-2. Revise the architecture of section 8.4.7 to correspond to problem 8-1.

8-3. Revise the mixed ASM of 8.4.6 to correspond to problem 8-2.

8-4. Revise the ASM of section 8.3.2.1 to include the JMS instruction described in appendix B.

8-5. Revise the ASM of problem 8-4 to include all the addressing modes described in appendix B.

8-6. Revise the ASM of problem 8-5 to include the interrupt instructions ION and IOF and associated hardware described in appendix B.

8-7. Revise the architecture of section 8.4.7 to correspond to problem 8-6.

8-8. Revise the mixed ASM of 8.4.6 to correspond to problem 8-7.

8-9. Suppose a direct mapped write-through cache of size four contains the contents of addresses 0004, 0001, 0002 and 0003 when starting to run the following PDP-8 program:

```
                          0000/7200
                .         0001/1004
                          0002/3006
                          0003/7402
                          0004/1000
```

   a) How many cache read hits occur?
   b) How many cache read misses occur?
   c) What will be in the cache_tag and cache_contents when the program halts?

8-10. Translate the ASM of figure 8-8 into behavioral Verilog. Use test code that loads and runs the childish division program of section 8.3.2.5.3 using the `sr`.

8-11. Translate the architecture of figure 8-12 together with the mixed ASM of figure 8-13 into Verilog. Use the same test code as problem 8-10.

8-12. Modify problem 8-11 to include the direct mapped cache designed in section 8.5. Assume it takes five clock cycles to read or write to the main memory. Use similar Verilog test code.

9. PIPELINED GENERAL-PURPOSE PROCESSOR

The fetch/execute algorithm described in section 8.3.1.5 typically requires five clock cycles to execute each instruction. In the terminology of chapter 6, that ASM uses a multi-cycle approach. The clock is fast because its frequency is determined by the maximum propagation delay of a single combinational unit, most likely the ALU. On the other hand, the effective speed is approximately one-fifth of what could be achieved if pipelining were used instead.

In order to pipeline an algorithm that makes decisions (as fetch/execute must do in order to decode instructions), we need to use a Mealy ASM with ovals. (See chapter 5 for details about Mealy ASMs.) A Mealy approach is required because the pipeline will process different stages of independent instructions at the same time. Later stages depend upon the completion within one clock cycle of the earlier stages. In a Mealy ASM, a conditional computation begins instantly and is ready one clock cycle after the decision. In a Moore ASM, a conditional computation cannot begin until one clock cycle after the decision, and the result is not ready until two clock cycles after the decision. This would be too late for a pipelined fetch/execute, and so a Mealy ASM is required to describe the overall behavior of a pipelined general-purpose computer.

The existence of the NOP instruction (7000) is important to the design of the pipelined fetch/execute. By putting a NOP in the pipeline when none existed in the original program, it will be possible to cope with several special situations. The essential goal of the pipelined machine is to end up with the same answer in memory and the accumulator as would be obtained from a non-pipelined version. Since a NOP leaves both the accumulator and memory alone, NOP provides for a safe way to stall later stages of the pipeline while earlier stages of the pipeline are being filled. This is quite advantageous, since it can eliminate the need for "FILL" and "FLUSH" states of the kind described in chapter 6.

9.1 First attempt to pipeline

The following is a somewhat flawed attempt to design a pipelined ASM that is equivalent to the multi-cycle ASM of 8.3.1.5. This ASM is for a three-stage pipeline consisting of instruction fetch, operand fetch and instruction execution. Ideally, in each clock cycle independent instructions are being fetched, having an operand fetched and being executed. It is important to understand that what is being pipelined is the fetch/execute algorithm itself and not the software algorithm implemented by the machine language program (which may not even be possible to pipeline). The efficiency of a software algorithm running on a pipelined general-purpose computer will be better than the same software running on a multi-cycle general-purpose computer, but not as good as the efficiency of a special-purpose computer that implements the same algorithm in hardware rather than in software.

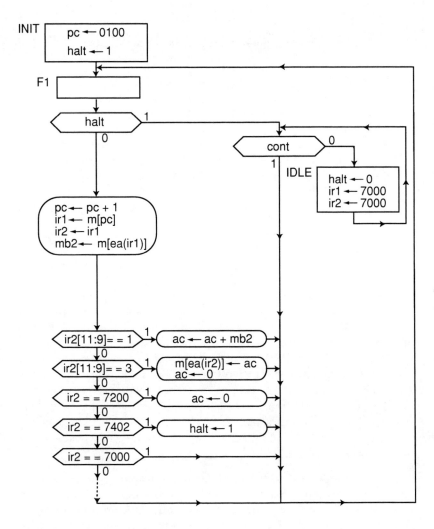

Figure 9-1. Incorrect pipelined ASM.

Here is a portion of the implicit style Verilog corresponding to this ASM:

```
...
forever
  begin
    @(posedge sysclk) enter_new_state('F1);
      if (halt)
        ...
      else
        begin
          pc <= @(posedge sysclk) pc + 1;
          ir1 <= @(posedge sysclk) m[pc];
          ir2 <= @(posedge sysclk) ir1;
          mb2 <= @(posedge sysclk) m[ea(ir1)];
          if      (ir2[11:9] == 1)
            ac <= @(posedge sysclk) ac + mb2;
          else if (ir2[11:9] == 3)
            begin
              m[ea(ir2)] <= @(negedge sysclk) ac;
              ac <= @(posedge sysclk) 0;
            end
          else if (ir2 == 12'o7200)
            ac <= @(posedge sysclk) 0;
          else if (ir2 == 12'o7402)
            halt <= @(posedge sysclk) 1;
          else if (ir2 == 12'o7000)
          ...
        end
    end
```

In this ASM, the operations of states F2, F3A, F3B, F4A, F4B, E0CLA, E0TAD, E0DCA, E1ADCA, E1BDCA and E0HLT from the ASM of 8.3.1.5 have been merged into state F1 using the Mealy notation. The most noticeable change is that the set of registers used is somewhat different than before. The memory address register has been eliminated altogether so that three separate things can be done in parallel to memory during the same clock cycle: an instruction can be fetched, an operand can be fetched and the accumulator can be stored. A single memory address register would not allow all of these to happen in parallel.

The instruction register of section 8.3.1.5 is now replaced by two registers, ir1 and ir2, so that the machine can distinguish the instructions as they travel through the pipeline. The memory buffer register is no longer used for writing to the memory, and instead is used only to hold an operand fetched from memory in the previous clock cycle. The memory buffer register has been renamed as mb2, reminding us that it is the operand for the instruction (ir2) in the final stage of the pipeline.

State INIT is identical to the one in the multi-cycle ASM. Since state INIT makes `halt` equal to one, when state F1 executes for the first time, the machine goes to state IDLE, where it waits for `cont` to become true. Note that state F1 does not do anything unconditionally, thus the program counter remains the way state INIT initialized it.

In state IDLE, `halt` becomes zero. There is an additional detail in state IDLE which was not present in section 8.3.1.5. The instruction pipeline (`ir1` and `ir2`) gets initialized to NOPs. After `cont` becomes true, it will take two clock cycles for the pipeline to fill with actual instructions from the program before the first instruction can execute. By putting NOPs in `ir1` and `ir2`, the machine can execute the imaginary NOPs harmlessly while the pipeline is filling with actual instructions. This eliminates the need for "FILL" states of the kind discussed in chapter 6.

When the machine is not halted in state F1, there are three separate things that occur in parallel:

a)   The youngest instruction is fetched into `ir1` (`pc ← pc + 1` and `ir ← m[pc]` occur in parallel),

b)   The operand for the middle aged instruction is fetched as that instruction moves down the pipeline (`ir2 ← ir1` and `mb2 ← m[ea(ir1)]` occur in parallel)

c)   The oldest instruction is decoded and executed (decisions similar to section 8.3.1.5 but involving `ir2`)

The execution of each instruction must be described in a Mealy oval. When `ir2` contains a TAD instruction, the accumulator is scheduled to be updated by adding it to the operand fetched in the previous stage. When `ir2` contains a DCA instruction, the accumulator is scheduled to be stored (`m[ea(ir2)] ← ac`) in parallel to scheduling that the accumulator be cleared.

9.2   Example of independent instructions

The ASM of section 9.1 is only able to execute certain PDP-8 programs correctly. By "correctly," we mean that the pipelined version produces (in fewer clock cycles) the same result that the multi-cycle version (section 8.3.1.5) produces in more clock cycles. Since the multi-cycle and the pipelined versions proceed differently, we have to wait until both machines are halted to check if the results are the same. The limitation on the kind of machine language program that figure 9-1 will execute properly is that each instruction is independent of the others. In other words, there are no data dependencies. (This is the only kind of pipelining discussed in chapter 6.) An example of such a program is the one given in appendix A, which is used with the multi-cycle ASM in section 8.3.1.6:

```
0100/7200
0101/1106
0102/1107
0103/1110
0104/3111
0105/7402
0106/0112
0107/0152
0110/0224
0111/0000
```

Here is what happens when the first pipelined ASM executes this program:

```
INIT   pc=xxxx irl=xxxx mb2=xxxx ir2=xxxx ac=xxxx h=x      59
F1     pc=0100 irl=xxxx mb2=xxxx ir2=xxxx ac=xxxx h=1     159
IDLE   pc=0100 irl=xxxx mb2=xxxx ir2=xxxx ac=xxxx h=1     259
F1     pc=0100 irl=7000 mb2=xxxx ir2=7000 ac=xxxx h=0     359
F1     pc=0101 irl=7200 mb2=xxxx ir2=7000 ac=xxxx h=0     459
F1     pc=0102 irl=1106 mb2=xxxx ir2=7200 ac=xxxx h=0     559
F1     pc=0103 irl=1107 mb2=0112 ir2=1106 ac=0000 h=0     659
F1     pc=0104 irl=1110 mb2=0152 ir2=1107 ac=0112 h=0     759
F1     pc=0105 irl=3111 mb2=0224 ir2=1110 ac=0264 h=0     859
F1     pc=0106 irl=7402 mb2=0000 ir2=3111 ac=0510 h=0     959
F1     pc=0107 irl=0112 mb2=xxxx ir2=7402 ac=0000 h=0    1059
F1     pc=0110 irl=0152 mb2=xxxx ir2=0112 ac=0000 h=1    1159
IDLE   pc=0110 irl=0152 mb2=xxxx ir2=0112 ac=0000 h=1    1259
IDLE   pc=0110 irl=7000 mb2=xxxx ir2=7000 ac=0000 h=0    1359
```

In the above, bold shows how the first and third TAD instructions travel through the pipeline, and italics show how the CLA, the second TAD, and the DCA instructions travel through the pipeline. In the first clock cycle after leaving IDLE ($time 359), irl and ir2 contain NOPs, so nothing happens to the accumulator. In the next clock cycle, irl contains the first instruction (7200), but ir2 still contains a NOP. Only in the third clock cycle after leaving IDLE ($time 559) does an actual instruction from the program execute—in this case the accumulator is scheduled to be cleared. This action becomes visible at $time 659. At that same time the first TAD instruction is ready to execute. In the previous clock cycle, the operand (0112) needed for this TAD instruction was scheduled to be loaded into mb2. Therefore, at $time 659 the ac ← ac + mb2 can be scheduled. The sum (0000+0112) becomes visible at $time 759. The remaining TAD instructions have filled the pipeline, so they can execute one per clock cycle. This is possible because the operands (0152 available at $time 759 and 0224 available at $time 859) have also been fetched. At $time 959, the correct sum (0510) is stored into memory at address 0111.

9.3 Data dependencies

What happens if the instructions are not independent of each other? For software to do practical things, often one instruction needs to depend on results computed by previous instructions. This is known as a *data dependency*. For example, a slight variation of the program from appendix A:

```
0100/7200
0101/1106
0102/1107
0103/3111   ← this is different from appendix A
0104/1111   ← this is also different
0105/7402
0106/0112
0107/0152
0110/0224
0111/0000
```

illustrates the problem that the above ASM has with instructions that are not independent. In this program, instead of doing a third TAD at 0103, the DCA (3111) occurs. This is followed by a TAD (1111) from this same location. The TAD instruction at 0104 is dependent on the DCA instruction at 0103. Here is the wrong result that figure 9-1 produces:

```
INIT   pc=xxxx irl=xxxx mb2=xxxx ir2=xxxx ac=xxxx h=x     59
F1     pc=0100 irl=xxxx mb2=xxxx ir2=xxxx ac=xxxx h=1    159
IDLE   pc=0100 irl=xxxx mb2=xxxx ir2=xxxx ac=xxxx h=1    259
F1     pc=0100 irl=7000 mb2=xxxx ir2=7000 ac=xxxx h=0    359
F1     pc=0101 irl=7200 mb2=xxxx ir2=7000 ac=xxxx h=0    459
F1     pc=0102 irl=1106 mb2=xxxx ir2=7200 ac=xxxx h=0    559
F1     pc=0103 irl=1107 mb2=0112 ir2=1106 ac=0000 h=0    659
F1     pc=0104 irl=3111 mb2=0152 ir2=1107 ac=0112 h=0    759
F1     pc=0105 irl=1111 mb2=0000 ir2=3111 ac=0264 h=0    859
F1     pc=0106 irl=7402 mb2=0000 ir2=1111 ac=0000 h=0    959
F1     pc=0107 irl=0112 mb2=xxxx ir2=7402 ac=0000 h=0   1059
F1     pc=0110 irl=0152 mb2=xxxx ir2=0112 ac=0000 h=1   1159
IDLE   pc=0110 irl=0152 mb2=xxxx ir2=0112 ac=0000 h=1   1259
IDLE   pc=0110 irl=7000 mb2=xxxx ir2=7000 ac=0000 h=0   1359
```

In the above, italics show how the instruction at 0104 travels through the pipeline. Everything looks fine until $time 959. The mb2 register (shown in bold italics) should contain the operand needed in the next clock cycle for the TAD (1111) instruction. Unfortunately, at $time 859 when mb2 was scheduled to be loaded, memory at address 0111 still contains the zero put there originally. The DCA (3111) instruction that is going to put the correct value (0264) into memory has not yet finished executing. By $time 1059, this error in the accumulator (also shown in bold italics) becomes obvious. The accumulator is supposed to contain 0264, but instead it contains 0000.

9.4   Data forwarding

The problem illustrated in the last section is known as data dependency. *Data dependency* means that the machine needs an operand that has not yet been stored in memory because a previously fetched instruction has not yet finished executing. To overcome this data dependency, we can introduce the idea of *data forwarding* into the following improved version of the pipelined ASM. This ASM is nearly identical to the earlier one, except how mb2 is computed depends on what is in the pipeline. Under most situations, mb2 comes from memory as it did earlier (mb2 ← m[ea(ir1)]). In one special situation, the current value of the accumulator is "forwarded" to the mb2 register. This situation occurs when the oldest instruction (ir2, the one currently executing) is a DCA and the effective address of that instruction is the same as the effective address of the instruction (ir1) that will execute in the next clock cycle. The ASM in figure 9-2 uses data forwarding.

The following shows in bold how the Verilog must be changed to implement the data forwarding given in the ASM:

```
    if (halt)
      ...
    else
      begin
        pc  <= @(posedge sysclk) pc + 1;
        ir1 <= @(posedge sysclk) m[pc];
        ir2 <= @(posedge sysclk) ir1;
        if ((ir2[11:9] == 3)&&(ea(ir1)==ea(ir2)))
          mb2 <= @(posedge sysclk) ac;
        else
          mb2 <= @(posedge sysclk) m[ea(ir1)];
        ...
```

The following shows how data forwarding solves this problem:

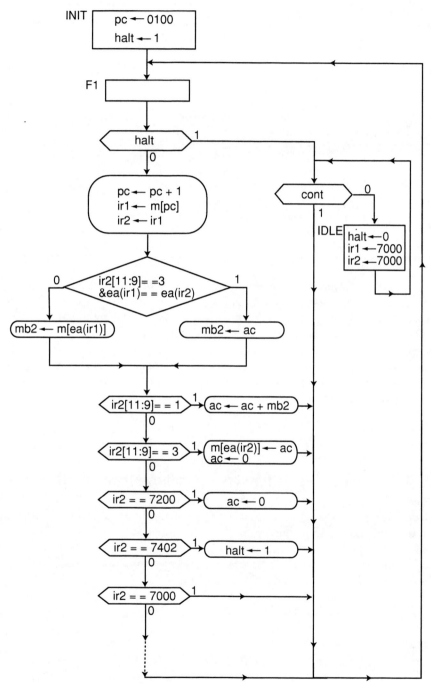

Figure 9-2. Pipelined fetch/execute with data forwarding.

```
INIT    pc=xxxx ir1=xxxx mb2=xxxx ir2=xxxx ac=xxxx h=x       59
F1      pc=0100 ir1=xxxx mb2=xxxx ir2=xxxx ac=xxxx h=1      159
IDLE    pc=0100 ir1=xxxx mb2=xxxx ir2=xxxx ac=xxxx h=1      259
F1      pc=0100 ir1=7000 mb2=xxxx ir2=7000 ac=xxxx h=0      359
F1      pc=0101 ir1=7200 mb2=xxxx ir2=7000 ac=xxxx h=0      459
F1      pc=0102 ir1=1106 mb2=xxxx ir2=7200 ac=xxxx h=0      559
F1      pc=0103 ir1=1107 mb2=0112 ir2=1106 ac=0000 h=0      659
F1      pc=0104 ir1=3111 mb2=0152 ir2=1107 ac=0112 h=0      759
F1      pc=0105 ir1=1111 mb2=0000 ir2=3111 ac=0264 h=0      859
F1      pc=0106 ir1=7402 mb2=0264 ir2=1111 ac=0000 h=0      959
F1      pc=0107 ir1=0112 mb2=xxxx ir2=7402 ac=0264 h=0     1059
F1      pc=0110 ir1=0152 mb2=xxxx ir2=0112 ac=0264 h=1     1159
IDLE    pc=0110 ir1=0152 mb2=xxxx ir2=0112 ac=0264 h=1     1259
IDLE    pc=0110 ir1=7000 mb2=xxxx ir2=7000 ac=0264 h=0     1359
```

As in the last example, italics show how the instruction at 0104 travels through the pipeline. In this case, data forwarding only occurs at $time 859, because ea(1111)==ea(3111) & ir2[11:9]==3. The underlining emphasizes the parts of ir1 and ir2 that must be identical for data forwarding to occur. During that clock cycle, the accumulator (shown in non-italic bold) contains 0264. The effect of the data forwarding becomes visible at $time 959, when mb2 (shown in italic bold) becomes 0264, which is correct. At $time 1059, we see that the accumulator (shown in italic bold) has the correct value because of this data forwarding.

9.5   Control dependencies: implementing JMP

The multi-cycle ASM of section 8.3.2.1 implemented several additional instructions. Of these, the JMP (5xxx) instruction presents a problem to implement with a pipelined ASM. If we do not do something special, two of the instructions that follow the JMP will execute prior to executing the instruction being jumped to. To avoid this error, the following ASM does not fetch these instructions after the JMP, but instead it puts two NOPs in the instruction pipeline:

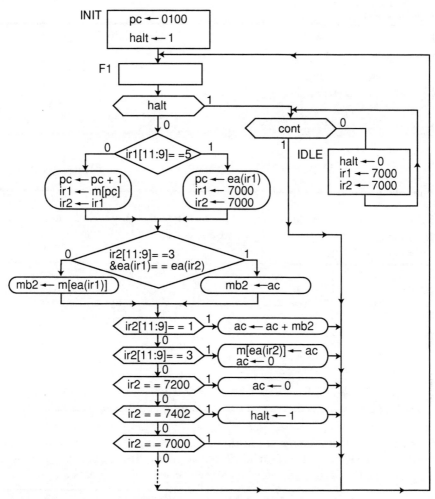

Figure 9-3. Pipelined fetch/execute with JMP.

The following shows in bold how the Verilog must be changed to implement JMP properly for the pipelined ASM:

```
if (halt)
  ...
else
  begin
    if (ir1[11:9] == 5)
      begin
        pc <= @(posedge sysclk) ea(ir1);
```

Continued

```
          ir1 <= @(posedge sysclk) 12'o7000;
          ir2 <= @(posedge sysclk) 12'o7000;
        end
    else
      begin
        pc <= @(posedge sysclk) pc + 1;
        ir1 <= @(posedge sysclk) m[pc];
        ir2 <= @(posedge sysclk) ir1;
      end
    if ((ir2[11:9] == 3)&&(ea(ir1)==ea(ir2)))
        ...
```

This occurs when the instruction in ir1 is a JMP (rather than waiting until ir2 contains the JMP). To illustrate how this works, consider the following variation of the program in appendix A:

```
0100/7200
0101/1106
0102/5105   ←   This is different from appendix A
0103/1110
0104/3111
0105/7402
0106/0112
0107/0152
0110/0224
0111/0000
```

Instead of a TAD instruction at address 0102, there is a JMP (5105) instruction that avoids executing the TAD (1110) instruction at address 0103 and the DCA instruction at 0104. The following shows how figure 9-3 executes this program correctly:

```
INIT   pc=xxxx irl=xxxx mb2=xxxx ir2=xxxx ac=xxxx h=x      59
F1     pc=0100 irl=xxxx mb2=xxxx ir2=xxxx ac=xxxx h=1     159
IDLE   pc=0100 irl=xxxx mb2=xxxx ir2=xxxx ac=xxxx h=1     259
F1     pc=0100 irl=7000 mb2=xxxx ir2=7000 ac=xxxx h=0     359
F1     pc=0101 irl=7200 mb2=xxxx ir2=7000 ac=xxxx h=0     459
F1     pc=0102 irl=1106 mb2=xxxx ir2=7200 ac=xxxx h=0     559
F1     pc=0103 irl=5105 mb2=0112 ir2=1106 ac=0000 h=0     659
F1     pc=0105 irl=7000 mb2=7402 ir2=7000 ac=0112 h=0     759
F1     pc=0106 irl=7402 mb2=xxxx ir2=7000 ac=0112 h=0     859
F1     pc=0107 irl=0112 mb2=xxxx ir2=7402 ac=0112 h=0     959
F1     pc=0110 irl=0152 mb2=xxxx ir2=0112 ac=0112 h=1    1059
IDLE   pc=0110 irl=0152 mb2=xxxx ir2=0112 ac=0112 h=1    1159
IDLE   pc=0110 irl=7000 mb2=xxxx ir2=7000 ac=0112 h=0    1259
```

At $time 559, the pipeline is filled with instructions as before. At $time 659, the first value (0000) becomes visible in the accumulator, and the operand (0112) is in mb2 for the first TAD instruction (1106). What is different at $time 659 is that ir1 contains a JMP (5105) instruction. Instead of incrementing the program counter as was done in the earlier ASMs, the program counter is loaded with the effective address (0105) of the JMP instruction. The pipeline cannot contain any fetched instruction at $time 759. Therefore, the same decision at $time 659 that changes the program counter must also schedule the instruction pipeline to be loaded with NOPs. At $time 759, the accumulator (0000+0112) contains the correct sum from the previous TAD instruction, but it will take two clock cycles before another instruction from the program can execute. At $time 959, the HLT instruction is ready to execute.

9.6   Skip instructions in a pipeline

The conditional skip instructions of the PDP-8, such as SPA (Skip on Positive Accumulator, 7510) and SMA (Skip on Minus Accumulator, 7500), were described as incrementing the program counter of the multi-cycle implementation given in chapter 8. To implement these instructions with a pipelined machine requires a different approach because the program counter changes during **every clock cycle** of pipelined execution.

One of the important ideas of this book is the meaning of the non-blocking assignment. Regardless of what kind of machine you are building, in any clock cycle there can be only **one** non-blocking assignment to a particular register. It is impossible for two values to be stored in the same register during the same clock cycle. Therefore, we cannot describe a skip instruction in a pipelined implementation as incrementing the program counter yet again. The program counter at that stage is preparing to fetch the instruction **after the one to be skipped**, which will execute regardless of the skip. It is already too late to increment the program counter by two at the time we realize that the next instruction is to be skipped because the instruction to be skipped has already been fetched into ir1. At the time it would have been appropriate to increment the program counter by two, we would not yet know whether the accumulator will be positive or negative. We need a different way to think about the skip instruction.

The overall effect of incrementing the program counter yet again in the multi-cycle implementation is to *nullify* the instruction that follows the skip. In a pipelined implementation, we can accomplish the same thing by replacing the instruction to be nullified with a NOP (7000):

```
always
 begin
  @(posedge sysclk) enter_new_state('INIT);
  pc <= @(posedge sysclk) 12'o0100;
  halt <= @(posedge sysclk) 1;
  forever
   begin
    @(posedge sysclk) enter_new_state('F1);
     if (halt)
      begin
       $stop;
       while (~cont)
        begin
         @(posedge sysclk) enter_new_state('IDLE);
         halt <= @(posedge sysclk) 0;
         ir1 <= @(posedge sysclk) 12'o7000;
         ir2 <= @(posedge sysclk) 12'o7000;
        end
      end
     else
      begin
       if ((ir2 == 12'o7510) && (~ac[11]) ||
           (ir2 == 12'o7500) && (ac[11]))
        begin
         pc <= @(posedge sysclk) pc + 1;
         ir1 <= @(posedge sysclk) m[pc];
         ir2 <= @(posedge sysclk) 12'o7000;
        end
       else
        if (ir1[11:9] == 5)
         begin
          pc <= @(posedge sysclk) ea(ir1);
          ir1 <= @(posedge sysclk) 12'o7000;
          ir2 <= @(posedge sysclk) 12'o7000;
         end
        else
         begin
          pc <= @(posedge sysclk) pc + 1;
          ir1 <= @(posedge sysclk) m[pc];
          ir2 <= @(posedge sysclk) ir1;
         end
        if ((ir2[11:9] == 3)&&(ea(ir1)==ea(ir2)))
         mb2 <= @(posedge sysclk) ac;
        else
         mb2 <= @(posedge sysclk) m[ea(ir1)];
```

```
              if       (ir2[11:9] == 1)
               ac <= @(posedge sysclk) ac + mb2;
              else if (ir2[11:9] == 3)
               begin
                 m[ea(ir2)] <= @(negedge sysclk) ac;
                 ac <= @(posedge sysclk) 0;
               end
              else if (ir2 == 12'o7200)
               ac <= @(posedge sysclk) 0;
              else if (ir2 == 12'o7402)
               halt <= @(posedge sysclk) 1;
              else if (ir2 == 12'o7041)
               ac <= @(posedge sysclk) -ac;
              else if (ir2 == 12'o7001)
               ac <= @(posedge sysclk) ac + 1;
              else if (ir2 == 12'o7000)
               ;
              else if (ir2 == 12'o7510)
               ;
              else if (ir2 == 12'o7500)
               ;
              else
               $display("other instructions...");
           end
        end
    end
```

The decision whether to nullify the instruction that follows the skip must occur at the top of the algorithm. This is because each register, such as ir2, can only have one value stored into it during each clock cycle. The normal behavior of the pipeline (transferring ir1 into ir2) cannot occur when the next instruction is to be nullified. Similarly, if that next instruction (in ir1) is a JMP (as is likely), the skip needs to take precedence over the JMP. Therefore the precedence of the decisions at the top of the algorithm is:

    a) a skip instruction in ir2 that is to be taken
    b) a JMP instruction in ir1
    c) normal pipelined behavior

Any other precedence would be incorrect. At the time the algorithm makes this decision, `ir2` already contains the skip instruction. Therefore, the bottom of the algorithm (which executes in parallel) needs to treat the 7500 or 7510 as a NOP, regardless of whether or not the following instruction will be nullified.

The above also includes the IAC (Increment ACcumulator, 7001) and CIA (Complement and Increment Accumulator, 7041) instructions. These non-memory reference instructions are similar to the CLA (7200) instruction in that the pipeline follows its normal behavior. To achieve simple pipelined behavior here with the CIA instruction, we assume that the ALU can form the twos complement negation of the accumulator in a single clock cycle.

9.7 Our old friend: division

The recurring example in this book is the childish division algorithm, introduced in section 2.2. It is used in chapter 2 to illustrate Moore ASMs, used in chapter 3 to illustrate Verilog test code, used in chapter 4 to illustrate behavioral, mixed and structural Verilog, used in chapter 5 to illustrate Mealy ASMs, used in chapter 6 to illustrate propagation delay and used in chapter 8 to benchmark the multi-cycle general-purpose PDP-8 against the special-purpose hardware of earlier chapters. The conclusion in chapter 8 is that special-purpose hardware implementations of the childish division algorithm were considerably faster and cheaper than the same algorithm running as software on the multi-cycle implementation of the general-purpose PDP-8. Yet most algorithms are implemented in software rather than hardware because software is easier to design and maintain. Pipelining allows a designer to create a more expensive general-purpose computer where the speed of its software comes closer to that of special-purpose hardware.

To illustrate what we have achieved by pipelining the PDP-8 as described in the previous sections, recall the description of the childish division algorithm in C:

```
r1 = x;
r2 = 0;
while (r1 >= y)
   {
       r1 = r1 - y;
       r2 = r2 + 1;
   }
```

For simplicity, we will assume x and y already have their values stored in memory, and that these values are less than 2048.[1]

As has been illustrated many times in earlier chapters, the while loop serves two roles: it avoids entering the loop and thus keeps r2 zero when x<y, or otherwise it stops the loop when it has repeated the proper number of times. In chapter 8, this was implemented as a skip and JMP at the top of the software loop and an unconditional JMP at the bottom. Such an approach is the easiest way to translate to machine language, but it has the cost of requiring additional instructions to execute each time through the loop. We need to find as good a machine language translation of this algorithm as we can. Such a machine language program will make the best use of the pipelined machine. The following uses an SPA instruction at the top to cause the loop to be entered the first time, and an SMA instruction at the bottom to cause the loop to exit:

```
              *0100
0100/7200     CLA
0101/1126     TAD X       // ac = 0+x
0102/3124     DCA R1      // r1 = x
0103/3125     DCA R2      // r2 = 0
0104/7200     CLA
0105/1127     TAD Y       // ac = 0+y
0106/7041     CIA         // ac = -y
0107/1124     TAD R1      // ac = r1-y
0110/7510     SPA         // if (r1-y >= 0) goto L1
0111/5123     JMP L2      //            else goto L2
0112/3124 L1, DCA R1      //    r1 = r1-y
                          //       depends on ac containing r1-y
                          //       on both paths to this inst.
0113/1125     TAD R2      //    ac = 0+r2
0114/7001     IAC         //    ac = r2+1
0115/3125     DCA R2      //    r2 = r2+1
0116/1127     TAD Y       //    ac = 0+y
0117/7041     CIA         //    ac = -y
0120/1124     TAD R1      //    ac = r1-y
0121/7500     SMA         //    if (r1-y < 0) goto L2
0122/5112     JMP L1      //            else goto L1
0123/7402 L2, HLT        // done
                          //
0124/0000 R1, 0000
0125/0000 R2, 0000
0126/0016 X,  0016        //   These must be < 2048 (3777 octal)
0127/0007 Y,  0007
```

---

[1] Since we have not implemented the link register of the PDP-8 in this pipelined version, larger values of x and y could cause the program to malfunction.

The execution of the above software on the pipelined PDP-8 illustrates how the skip instructions work:

```
INIT    pc=xxxx ir1=xxxx mb2=xxxx ir2=xxxx ac=xxxx h=x        59
F1      pc=0100 ir1=xxxx mb2=xxxx ir2=xxxx ac=xxxx h=1       159
IDLE    pc=0100 ir1=xxxx mb2=xxxx ir2=xxxx ac=xxxx h=1       259
F1      pc=0100 ir1=7000 mb2=xxxx ir2=7000 ac=xxxx h=0       359
F1      pc=0101 ir1=7200 mb2=xxxx ir2=7000 ac=xxxx h=0       459
F1      pc=0102 ir1=1126 mb2=xxxx ir2=7200 ac=xxxx h=0       559
F1      pc=0103 ir1=3124 mb2=0016 ir2=1126 ac=0000 h=0       659
F1      pc=0104 ir1=3125 mb2=0000 ir2=3124 ac=0016 h=0       759
F1      pc=0105 ir1=7200 mb2=0000 ir2=3125 ac=0000 h=0       859
F1      pc=0106 ir1=1127 mb2=xxxx ir2=7200 ac=0000 h=0       959
F1      pc=0107 ir1=7041 mb2=0007 ir2=1127 ac=0000 h=0      1059
F1      pc=0110 ir1=1124 mb2=xxxx ir2=7041 ac=0007 h=0      1159
F1      pc=0111 ir1=7510 mb2=0016 ir2=1124 ac=7771 h=0      1259
F1      pc=0112 ir1=5123 mb2=7510 ir2=7510 ac=0007 h=0      1359
F1      pc=0113 ir1=3124 mb2=7402 ir2=7000 ac=0007 h=0      1459
F1      pc=0114 ir1=1125 mb2=0016 ir2=3124 ac=0007 h=0      1559
F1      pc=0115 ir1=7001 mb2=0000 ir2=1125 ac=0000 h=0      1659
F1      pc=0116 ir1=3125 mb2=xxxx ir2=7001 ac=0000 h=0      1759
F1      pc=0117 ir1=1127 mb2=0000 ir2=3125 ac=0001 h=0      1859
F1      pc=0120 ir1=7041 mb2=0007 ir2=1127 ac=0000 h=0      1959
F1      pc=0121 ir1=1124 mb2=xxxx ir2=7041 ac=0007 h=0      2059
F1      pc=0122 ir1=7500 mb2=0007 ir2=1124 ac=7771 h=0      2159
F1      pc=0123 ir1=5112 mb2=7200 ir2=7500 ac=0000 h=0      2259
F1      pc=0112 ir1=7000 mb2=3124 ir2=7000 ac=0000 h=0      2359
F1      pc=0113 ir1=3124 mb2=xxxx ir2=7000 ac=0000 h=0      2459
F1      pc=0114 ir1=1125 mb2=0007 ir2=3124 ac=0000 h=0      2559
F1      pc=0115 ir1=7001 mb2=0001 ir2=1125 ac=0000 h=0      2659
F1      pc=0116 ir1=3125 mb2=xxxx ir2=7001 ac=0001 h=0      2759
F1      pc=0117 ir1=1127 mb2=0001 ir2=3125 ac=0002 h=0      2859
F1      pc=0120 ir1=7041 mb2=0007 ir2=1127 ac=0000 h=0      2959
F1      pc=0121 ir1=1124 mb2=xxxx ir2=7041 ac=0007 h=0      3059
F1      pc=0122 ir1=7500 mb2=0000 ir2=1124 ac=7771 h=0      3159
F1      pc=0123 ir1=5112 mb2=7200 ir2=7500 ac=7771 h=0      3259
F1      pc=0124 ir1=7402 mb2=3124 ir2=7000 ac=7771 h=0      3359
F1      pc=0125 ir1=0000 mb2=xxxx ir2=7402 ac=7771 h=0      3459
```

At $time 1259, when ir1 gets loaded with the first skip instruction (7510), we do not yet know whether the accumulator will be positive or negative, so the pipeline continues filling normally. At $time 1359, there is a decision that must be made because ir1 contains the JMP instruction (5123) and ir2 contains the SPA instruction

(7510).[2] As described above, the skip is given precedence over the JMP. Therefore, whether the next instruction (currently in ir1) will be nullified is based on ac[11]. In this case, ac[11] == 0, so the SPA will nullify the following instruction. At $time 1459, ir2 has become NOP (7000), but 3124 was fetched normally into ir1 so that the algorithm can proceed sequentially.

A different situation occurs at $time 2259. Here the SMA (7500) does not nullify the JMP instruction (5112) because the accumulator is not negative, so the behavior described in section 9.5 occurs. Both ir1 and ir2 are loaded with NOPs (7000), as is visible at $time 2359. The machine does not start executing useful instructions after the JMP until $time 2559 because of the time required to fill the pipeline.

Finally, at $time 3259, the SMA (7500) does nullify the JMP instruction (5112) because the accumulator is negative, so only ir2 has a NOP (7000) at $time 3359. This allows sequential execution of the HLT (7402) at $time 3459.

Between $time 359 and $time 3459 are 32 clock cycles. In general, if the quotient >= 1, the number of clock cycles is 12+10*quotient. The following table summarizes implementations of the childish division algorithm given in this and earlier chapters:

```
    max pipe kind hardware software
    int      of ASM section  section    clock cycles

    4095  n  Moore  2.2.7    n/a        3 +   quotient
    4095  n  Moore  2.2.3    n/a        2 + 2*quotient
    4095  n  Moore  2.2.2    n/a        3 + 3*quotient
    4095  n  Moore  2.2.5    n/a        2 + 3*quotient
    4095  n  Mealy  5.2.1    n/a        2 + 2*quotient
    4095  n  Mealy  5.2.3    n/a        3 +   quotient
    4095  n  Mealy  5.2.4    n/a        2 +   quotient
    4095  n  Moore  8.3.2.1  8.3.2.5.3  88+75*quotient
    2047  n  Moore  8.3.2.1  9.7        55+55*quotient
    2047  y  Mealy  9.6      9.7        12+10*quotient
```

The first seven lines above are for special-purpose computers whose ASMs implement the childish division algorithm. The last three lines are for general-purpose computers (whose ASMs implement fetch/execute) that need a machine language program to implement division. The "max int" column shows the maximum allowable integer input, which is 2047 for the software given in this section. The "pipe" column indicates whether the hardware is pipelined. The "kind of ASM" indicates whether the ASM uses condi-

---

[2] The 7510 in mb2 is sheer coincidence.

tional commands (Mealy) or not (Moore). The "hardware section" column indicates where the ASM is described. The "software section" applies only to general-purpose computer implementations and describes the machine language for a version of the childish division algorithm. The "clock cycle" column indicates how long it takes to compute the `quotient`, neglecting input/output time, if possible. [The "friendly user" assumptions cause the special purpose machines to do useful work (clearing `r2`) during this time that cannot be neglected. On the other hand, the software results ignore these times.]

The next to the bottom line shows how long the software given in this section takes to run on the multi-cycle hardware designed in section 8.3.2.1. This is shown to make a fair comparison of the effects of pipelining given on the bottom line. As `quotient` gets large, the speedup of the software running on the pipelined PDP-8 versus the same software running on the multi-cycle machine of section 8.3.2.1 approaches 55/10=5.5. But still, the speed of the special-purpose hardware in chapter 2 can be up to ten times faster than the speed of the pipelined PDP-8. As is discussed in the next section, the speed of the pipelined PDP-8 comes at the cost of a special kind of memory, known as multi-port memory, which is several times more expensive than the single-ported memory described in section 8.2.2.

9.8   Multi-port memory

In order to realize the pipelined fetch/execute ASM in hardware, it must be possible to do three things simultaneously with memory: fetch an instruction, fetch data and store data. The memory devices discussed in section 8.2.2 would not allow this to happen, because they are restricted to at most one read or write per clock cycle. To allow multiple operations per clock cycle in memory, we need a multi-port memory, which is shown as a letter "E" on its side.

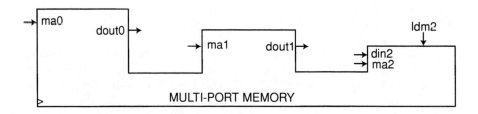

Figure 9-4. Symbol for multi-port memory.

The reason for this unconventional figure is to illustrate the fact that the multi-port memory acts like several independent devices that share a common foundation (the contents of memory). In each clock cycle, three separate operations occur in parallel:

```
always @ (m[ma0])
   dout0 = m[ma0];

always @ (m[ma1])
   dout1 = m[ma1];

always @ (posedge sysclk)
   begin
      if (ldm2)
         m[ma2] = din2;
   end
```

so that the architecture that instantiates the multi-port memory may do three things to memory in parallel.

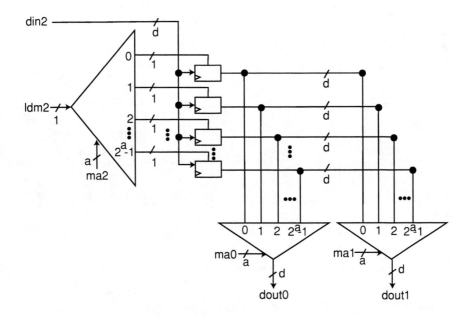

Figure 9-5. Implementation of multi-ported memory.

Figure 9-5 shows a block diagram for a synchronous multi-port memory using a demux, two muxes and enabled registers. This block diagram is a generalization of the single-port memory shown in section 8.2.2.3.1. The distinction with the multi-port memory is that there are two muxes, each of which can access the memory cells independently.

9.9 Pipelined PDP-8 architecture

Figure 9-6 shows an architecture for the pipelined PDP-8 that uses the multi-port memory. The program counter, pc, is a counter with clrpc, incpc and ldpc command signals. The other registers (ir1, ir2, mb2 and the accumulator) are enabled registers with load signals (ldir1, ldir2, ldmb2, ldac). There are two muxes that allow ir1 and ir2 to be loaded with NOPs (7000). There is another mux that allows for data forwarding from the accumulator to mb2. Also, there is a comparator that detects when ea(ir1) equals ea(ir2).

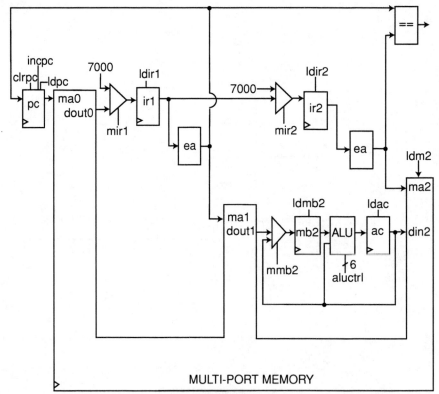

Figure 9-6. Architecture for pipelined PDP-8.

9.10 Conclusion

The pipelined PDP-8 designed in this chapter can run software in some situations about five times faster than the multi-cycle PDP-8 given in the last chapter. Because the propagation delays (which determine the clock frequency) in the pipelined and multi-cycle versions are nearly identical, there are two other factors that determine the speed. First is the number of clock cycles per instruction. (In chapter 8, most instructions take five clock cycles, but in this chapter instructions other than JMP take only one cycle.) Second is the the the mix of instructions in the program, such as how frequently JMPs occur. (The example here is the childish division algorithm, which may or may not be representative of how the algorithm you want to implement will perform.)

The major cost of the pipelined approach in this chapter is the multi-port memory, which allows simultaneous access to memory for instructions and data. The problem is that even with pipelining, this approach provides one-tenth the speed of the specialized hardware for the childish division algorithm.

When you consider both cost and speed, special-purpose hardware is much better than software running on a pipelined PDP-8, at least for this example. Although the relative performance of other algorithms might be different, this example points out that other techniques beyond pipelining of the PDP-8 are going to be required if software speed is going to approach that of special purpose hardware. The next chapter illustrates some of these techniques.

9.11 Further reading

PATTERSON, DAVID A. and JOHN L. HENNESSY, *Computer Organization and Design: The Hardware/Software Interface,* Morgan Kaufmann, San Mateo, CA, 1994. Chapter 6 gives more details about implementing data forwarding with an instruction set more complicated than the PDP-8.

STERNHEIM, ELIEZER, RAJVIR SINGH and YATIN TRIVEDI, *Digital Design with Verilog HDL,* Automata Publishing, San Jose, CA, 1990. Chapter 3 gives a different approach to modeling a pipelined general-purpose computer in Verilog.

9.12 Exercises

9-1. Modify the behavioral design in section 9.6 to include the ISZ instruction of the PDP-8 (described in appendix B). Including an ISZ instruction in a program should only increase its execution time by one clock cycle for each time the ISZ is executed. Simulate the modified design with the following programs:

```
0000/7200
0001/2007
0002/2007
0003/1007
0004/2007
0005/5003
0006/7402
0007/7774
```

9-2. Using the ISZ instruction, it is possible to implement a version of the childish division program that is about twice as fast as the one given in section 9.7. Implement such a program and use test code with the behavioral Verilog for problem 9-1 to measure how long it takes to divide x by seven, as the test code varies x from 0 through 28. Derive a mathematical formula for clock cycles comparable to those listed in section 9.7 which generalizes the data observed by the Verilog test code. Hints: The machine language program needs to precompute $-y$, and r1 should reside in the accumulator rather than in memory. The skipping action of the ISZ is irrelevant to this program.

9-3. Draw a modified architecture for problem 9-1. The only extra devices needed are two input muxes, a comparator and an incrementor.

9-4. Modify the behavioral design in section 9.6 to include the JMS instruction of the PDP-8 (described in appendix B). Simulate your modified design.

9-5. Draw a modified architecture for problem 9-4.

9-6. Modify the behavioral design in problem 9-4 to include the indirect page zero addressing mode of the PDP-8 (described in appendix B). You may assume there is an additional port to memory. How many stages will be in your pipeline?

9-7. Draw a modified architecture for problem 9-6.

9-8. Modify the behavioral design in problem 9-4 to include interrupts.

9-9. The behavioral design of section 9.4 does not execute self-modifying programs properly. Modify the design to process properly the two kinds of dependencies that are possible in such programs.

9-10. Draw a modified architecture for problem 9-9.

10. RISC PROCESSORS

All general-purpose computers implement fundamentally the same algorithm in hardware: fetch/execute. A multi-cycle implementation of fetch/execute, which is described in chapter 8 using the PDP-8, takes several clock cycles per instruction. Since the PDP-8 has a simple instruction set with a single-accumulator and a single memory address, it often takes several instructions to execute a single high-level language statement, as illustrated in section 8.3.2.5.3. It is possible to pipeline fetch/execute, as was shown in chapter 9, so that most instructions appear to execute in a single clock cycle, but a single-accumulator machine still requires several instructions to perform typical high-level language statements. The speed of a pipelined single-accumulator machine is several times slower than the special-purpose hardware of chapter 2, at least for the division example used throughout this book.

In the beginning, most general-purpose computers, such as the Manchester Mark I, adhered to this single-accumulator, single-address style of instruction set. The PDP-8 is one of the best and purest illustrations of this very simple approach. The reason for introducing it in the preceding chapters is to illustrate how the same design process used for special-purpose computers also works for general-purpose computers. The simplicity of the PDP-8 allows us to focus on using ASM charts and Verilog in the design process without having to worry about excessive complications that exist in other instruction sets.

The problem is that, after pipelining the PDP-8, we have about reached the limits of performance from a single-accumulator, single-address instruction set. To make software closer to the speed of special-purpose hardware will require specifying a different kind of machine language. The central concept of fetch/execute remains the same, but the way the machine uses the bits in the instruction register will have to be different.

10.1 History of CISC versus RISC

One attempt to increase performance of general-purpose processors that became popular in the 1970s is the idea of a Complex Instruction Set Computer (CISC). In essence, the idea is to merge a simple general-purpose machine together with special hardware (and special registers) that solve certain specific computations. The thought was that this would give the user the best of both worlds (special-purpose and general-purpose computers). To activate each special hardware unit requires including a new instruction in the instruction set. Rather than the handful of machine language instructions described in appendix B for the PDP-8, a CISC machine might have thousands of distinct instructions. Fitting all these instructions into a reasonable sized instruction register requires that some instructions occupy multiple words, which is known as a *variable length instruction set.* Such machines are aptly named CISC because the fetch/execute algorithm, although fundamentally the same, has much more complex details with a variable length instruction set. This is especially true if the machine is to be pipelined.

Two factors led to the popularity of CISC processors. First, improved fabrication technologies allowed ever-increasing amounts of hardware to fit on a chip. Second, instruction set designers had a mistaken belief that programmers and compilers would be able to utilize all this special-purpose hardware effectively.

By the early 1980s, several empirical studies had shown that CISC processors did not make effective use of all of their special-purpose hardware. As a result of these studies, several groups designed Reduced Instruction Set Computers (RISC). Like the PDP-8, RISC machines have fixed length instruction sets. This simplifies pipelined implementation. RISC instruction sets are chosen with pipelining of fetch/execute in mind, while CISC instruction sets make pipelining fetch/execute difficult. Unlike the PDP-8, RISC processors have several features that allow higher performance than is possible on a single-accumulator machine. Although CISC processors remained popular through the end of the twentieth century (the Pentium II is a CISC processor), the momentum in computer design shifted to the RISC philosophy.

10.2   The ARM

In the early 1980s, Acorn Computers, Ltd. designed an inexpensive computer for teaching computer literacy in conjunction with a BBC television program in Great Britain. The machine was originally dubbed the "Acorn RISC Microprocessor" (ARM). Several years later, Acorn entered into a consortium with more than a dozen manufacturers, including DEC[1] (the company that manufactured the PDP-8) and Apple (which uses the ARM in its Newton PDA), to promote the ARM worldwide. The ARM acronym was redefined to mean "Advanced RISC Microprocessor." The ARM is probably the most elegant RISC processor ever marketed. Its instruction set is simpler than most of the other RISC processors with which it can be compared. Although, as explained below, it does not have the performance bottlenecks of a single-accumulator machine, its superb simplicity is in some respects reminiscent of the PDP-8. This chapter will use only a small subset of ARM instructions to introduce some key ideas in choosing an instruction set for maximal performance. Appendix G explains how to access the official ARM documentation for the complete instruction set. In particular, the ARM supports several different modes of operation. We will only be concerned with what is called *user mode*.

---

[1] DEC sold its rights to the StrongARM to Intel in late 1997, but the other members of the ARM consortium were not part of that agreement and continue to produce various versions of the ARM.

10.3 Princeton versus Harvard architecture

Like all other general-purpose ("stored program") computers, commercial versions of the ARM use the same memory to store both machine language programs and data. This approach is sometimes referred to as a *Princeton architecture.* The ARM, like all other popular general-purpose computers, requires memory reference instructions, akin to those of the PDP-8, to bring data into and out of the central processing unit. (The mnemonics of these ARM instructions, given in appendix G, are LDR, STR, LDM, STM and SWP.) For the moment let us ignore these memory reference instructions, and instead assume that only programs reside in memory. A machine where programs reside in a memory exclusively for programs is sometimes called a *Harvard architecture.*[2] Although the ARM is not actually a Harvard machine, it will simplify the discussion if we assume it is.

10.4 The register file

If, at least for the moment, data is not going to reside in the same memory as programs, where is it going to be? To achieve software performance that approaches that of special-purpose hardware, there is only one plausible answer: put the data in hardware registers. In contrast to the PDP-8, with its single accumulator, a RISC processor like the ARM needs many registers for storing data.

When you design a special-purpose machine, like those in chapter 2, it is usually fairly clear how to interconnect the registers to implement the transfers required by the algorithm. The designer of a general-purpose computer does not have the luxury of knowing how registers might need to be interconnected because the register transfers will be determined by software. Therefore, the registers of a RISC processor need to be lumped together into what is called a *register file*. The register file is really a small and fast synchronous multi-port memory. The ARM has sixteen registers available in its register file in user mode.[3] We will refer to these sixteen registers using the Verilog array notation r[0] through r[15]. In assembly language, the programmer refers to these as R0 through R15. Each one of these registers contains a 32-bit value.

The program counter on the ARM is actually synonymous with r[15]. We can improve the clarity of our Verilog description of the ARM using:

---

[2] Harvard architectures usually have a separate memory for data, which we will ignore. Commercial versions of the ARM actually have Princeton architectures that share the same memory for data. It is an oversimplification to think of the ARM with a Harvard architecture.

[3] There are several other registers available for so-called supervisor modes, but we will ignore these for simplicity.

```
`define PC     r[15]
```

10.5   Three operands are faster than one

The most common operations in typical algorithms are things like addition and subtraction. For example, in the childish division algorithm implemented in section 9.7, we need to compute a difference, $d = r1 - y$. On the PDP-8, $r1$ and y are data that residesin memory. The PDP-8's accumulator contains partial results as the following four instructions execute:

```
0104/7200      CLA
0105/1127      TAD Y
0106/7041      CIA
0107/1124      TAD R1
```

As with the examples in chapters 8 and 9, the above shows the address and corresponding machine language in octal. Upon completion of the second TAD instruction, the PDP-8's accumulator contains d.

To perform a similar computation with the ARM requires that all data reside in registers. Let us assume that the ARM's $r[0]$ register takes on the role served by the PDP-8's accumulator (to contain the difference, d), that the ARM's $r[1]$ register serves the same role as the R1 location in the PDP-8's memory and that the ARM's $r[4]$ register contains the value of y.

Since there are only sixteen registers to choose from, the ARM makes it possible to specify both operands of the subtraction ($r[1]$ and $r[4]$) as well as the destination register ($r[0]$) within a single 32-bit instruction. For example, given the above assumptions, the following single ARM instruction is equivalent to the four PDP-8 instructions shown earlier:

```
0000000c/e0510004     SUBS    R0,R1,R4


              ^^   ^
              ||   |
              ||   +- specifies r[4]
              |+---- specifies r[0]
              +---- specifies r[1]
```

In contrast to the PDP-8 examples, the above ARM example shows the address and corresponding machine language in hexadecimal. This is convenient since the four-bit register specifications appear as hexadecimal digits in the machine language, as shown above.

Assuming you have pipelined versions of the ARM and PDP-8 running at the same clock speed, the ARM can do four times as many subtractions as the PDP-8. On a single-accumulator machine, like the PDP-8, the accumulator serves as one of the operands as well as the destination. The single-accumulator, single-address machine requires this approach because most of the bits of the instruction register are devoted to the memory address of the other operand.

RISC machines prohibit computation on values from memory. Instead, RISC machines insist that values to be added or subtracted already reside inside the register file. Since the register file is small compared to memory (sixteen registers in the case of the ARM), it only takes a few bits (twelve in the case of the ARM) to describe two separate operands and a separate destination. Therefore, the complete subtraction can be done in one instruction.

RISC instruction sets have three advantages. First, the access time of the register file is typically faster than that of a full-sized memory. Second, the RISC instruction set typically reduces the number of instructions in a program compared to a single-accumulator machine, which often means the software will run faster. Finally, the RISC instruction set allows the designer to exploit fetch/execute techniques that are more sophisticated than simple pipelining, such as *superscalar* design. The superscalar approach allows a general-purpose computer designer to create a machine whose software speed comes much closer to that of special-purpose hardware.

There are some disadvantages to RISC instruction sets. First, making good use of a RISC instruction set requires a sophisticated programmer or compiler. Second, because of the simplicity of operations on a RISC compared to a CISC, a RISC instruction set usually requires more instructions to accomplish the same computation than a CISC instruction set. Third, because of the large fixed-sized instruction register used in a RISC, the number of bits (as opposed to the number of instructions) to encode a program is typically larger than any variable length CISC, or even a fixed length, single-accumulator machine like the PDP-8.

With the advent of large multimedia applications at the end of the 1990s, interest in variable length CISC instruction sets reemerged because of concerns about program size. Examples of such CISC designs include Sun's Java machine and ARM's Thumb. Interestingly, such machines internally translate the variable length CISC instruction set used for compact encoding of programs into fixed length RISC instructions for execution. Also, AMD's K6 has hardware that translates its extremely CISC (Pentium like) instruction set into internal RISC instructions for execution.

10.6   ARM subset

There are eleven different categories of ARM instructions described in appendix G. It is possible to do very useful things in software using only a few of these instructions, and so we can select a handful of these instructions to illustrate the design of a RISC processor. Of the eleven categories of instructions in appendix G, we will only implement the "data processing" and "branch" categories. The data processing category is subdivided into sixteen different mnemonics, and the branch category is subdivided into two different mnemonics. We will only implement four of the eighteen possible mnemonics in these two categories.

10.6.1   Data processing instructions

There are zeros in instruction register bits 27 and 26 to indicate the data processing category. Instruction register bits 24 down to 21 determine which one of the sixteen data processing mnemonics is associated with that particular instruction. For simplicity, we will only implement the following three of the sixteen possible mnemonics:

```
           decoding                             memonic

    ir[27:26]==2'b00&&ir[24:21]==4'b0100   ADD RD,OPA,OPB
    ir[27:26]==2'b00&&ir[24:21]==4'b0010   SUB RD,OPA,OPB
    ir[27:26]==2'b00&&ir[24:21]==4'b1101   MOV RD,OPB
```

To use one of the above three instructions in a program, the assembly language programmer would replace the RD, OPA and OPB with specific registers, such as R0, R1 and R4 in the SUBS instruction of section 10.5. Instruction register bits 15 down to 12 describe the destination, RD, and instruction register bits 19 through 16 describe the first operand, OPA. When instruction register bit 25 is zero, instruction register bits 3 down to 0 indicate the register for the second operand OPB. The leftmost four bits of every ARM instruction determines whether the instruction executes. Typically, these four bits contain 4'b1110. Finally bit 20 has a special meaning. When this bit is a one, the mnemonic has an "S" on the end to indicate this special meaning. For example, the instruction in section 10.5 has the mnemonic SUBS rather than simply SUB, as illustrated by the following:

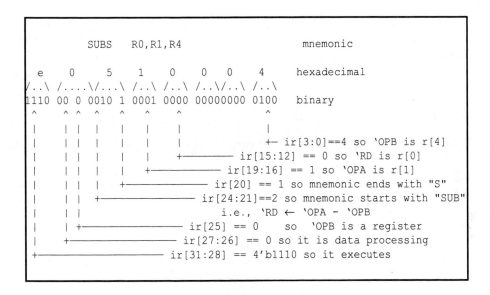

```
            SUBS    R0,R1,R4                    mnemonic

   e    0    5    1    0    0    0    4     hexadecimal
/..\ /....\/...\ /..\ /..\ /..\/..\ /..\
1110 00 0 0010 1 0001 0000 00000000 0100   binary
 ^       ^ ^  ^    ^    ^       ^
 |       | |  |    |    |       |                   |
 |       | |  |    |    |       |            +- ir[3:0]==4 so 'OPB is r[4]
 |       | |  |    |    |       +----------- ir[15:12] == 0 so 'RD is r[0]
 |       | |  |    |    +------------------- ir[19:16] == 1 so 'OPA is r[1]
 |       | |  |    +------------------------ ir[20] == 1 so mnemonic ends with "S"
 |       | |  +------------------------------ ir[24:21]==2 so mnemonic starts with "SUB"
 |       | |                   i.e., 'RD <- 'OPA - 'OPB
 |       | +----------------------- ir[25] == 0    so  'OPB is a register
 |       +----------------------- ir[27:26] == 0 so it is data processing
 +------------------------------ ir[31:28] == 4'b1110 so it executes
```

10.6.2   Branch instruction

The second major instruction category we will use for our subset of the ARM instruction set is the *branch instruction*. When instruction register bits 27 down to 25 are 3'b101, the ARM categorizes the instruction as a branch. There are two mnemonics for this category: B (branch) and BL (branch and link). Instruction register bit 24 distinguishes between the simple branch instruction and the branch and link (zero means simple branch). We will not implement the branch and link instruction here, although it is straightforward. (It utilizes R14 to save a return address, quite analogously to the way the JMS instruction on the PDP-8 uses a memory location to save a return address.)

The branch instruction on the ARM is very similar to the JMP instruction of the PDP-8, with three differences. First, the branch instruction of the ARM uses a relative addressing mode, rather than the direct addressing mode of the PDP-8's JMP instruction. In a relative addressing mode, the offset field of the branch instruction is added (as a signed twos complement value) to the program counter (rather than being moved to the program counter as occurs on the PDP-8). Second, the branch instruction of the ARM refers to the offset in terms of 32-bit words, but the program counter of the ARM refers to eight-bit bytes. (The value in the ARM's program counter is always divisible by four, and so the offset field of the branch instruction is one-quarter the value needed to add to the program counter.) Third, since the offset field is only 24 bits wide, it must be sign extended before it is added to the program counter.

The following example branch instruction forms an infinite loop by branching back to itself. Because of the relative addressing mode, this same machine language instruction will work identically regardless of the location where it occurs in a program:

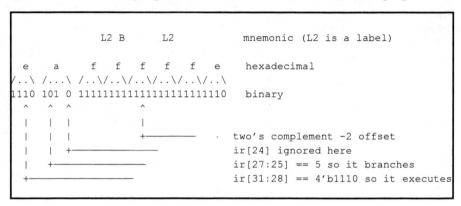

```
              L2 B       L2              mnemonic (L2 is a label)

    e    a    f   f   f   f   f   e      hexadecimal
  /..\ /...\ /..\/..\/..\/..\/..\
  1110 101 0 11111111111111111111110    binary
    ^   ^ ^                 ^
    |   | |                 |
    |   | |                 +————————  .  two's complement -2 offset
    |   | +—————————————              ir[24] ignored here
    |   +———————————————               ir[27:25] == 5 so it branches
    +—————————————                     ir[31:28] == 4'b1110 so it executes
```

It may seem a little strange, but the -2 indicates branching back to the same instruction. In other words, the new value of the program counter is the value of the program counter at the time the instruction is fetched plus $4*offset+8$, where the offset is a sign extended version of instruction register bits 23 down to 0. The reason the ARM designers chose to make -2 mean branching back to itself will become clear later in this chapter.

10.6.3 Program status register

Another detail in which the ARM is different than the PDP-8 is the way in which the ARM tests for conditions, such as testing for negative numbers. On the PDP-8, since the accumulator is the only place where a number to be tested can reside, the hardware simply uses the most significant bit of the accumulator to determine whether that number is negative or not. On the ARM, there are sixteen different registers that a programmer might choose to test, and so there are sixteen different sign bits that the hardware might need to use, which would not be economical. Instead, the ARM allows the programmer to specify a one as bit 20 of the instruction register for a data processing instruction ("S" suffix on the mnemonic). When bit 20 is a one, certain critical information about the result of the data processing instruction is saved in the *program status register*. (The "S" suffix means set the PSR.) In this chapter, we will consider bit 31 of the program status register (PSR), which is known as the "N" (negative) flag. The N flag stores the sign bit of the result of the most recent data processing instruction with an "S" suffix mnemonic.

For example, suppose `r[1] == 32'h00000007` and `r[4] == 32'h00000007` prior to the execution of the SUBS R0,R1,R4 instruction (e0510004) given in section 10.6.1. Because the result is not negative (`r[0] == 32'h00000000`) and bit 20 of the instruction register is set, the N flag becomes zero.

As a different example, suppose `r[1] == 32'h00000000` and `r[4] == 32'h00000007` prior to the second execution of the same SUBS R0,R1,R4 instruction (e0510004). Because the result is negative seven (`r[0] == 32'hfffffff9`) and bit 20 of the instruction register is set, the N flag becomes one.

If bit 20 of a data processing instruction is zero, the PSR remains unchanged. For example, suppose `r[1] == 32'h00000007` and `r[4] == 32'h00000007` prior to the execution of a SUB R0,R1,R4 instruction (e0410004) similar to the SUBS except that bit 20 of the instruction register is zero. Even though the result is not negative (`r[0] == 32'h00000000`), the N flag remains what it was (one) because bit 20 of the instruction register is zero.

There are several other bits in the PSR which we will not implement here. For example, bit 30 of the PSR is the "Z" flag, which indicates whether the result (of the most recent data processing instruction with an "S" suffix mnemonic) was equal to zero. Bit 29 of the PSR is the "C" flag, which indicates whether the result (of the most recent data processing instruction with an "S" suffix mnemonic) produced a carry (analogous to the LINK of the PDP-8). Bit 28 of the PSR is the "V" flag, which indicates whether the result (of the most recent data processing instruction with an "S" suffix mnemonic) caused a signed overflow (what should be a negative number appears positive or vice versa).

10.6.4   Conditional execution

One of the most interesting and useful features of the ARM is that every instruction can be conditional, that is, if a certain condition recorded in the PSR is not satisfied, the instruction is treated as a NOP. If that condition is satisfied, the instruction executes normally. The condition is indicated by bits 31 through 28 of the instruction register. Although there are sixteen different conditions that the actual ARM recognizes (as shown in appendix G), we will only consider the following four in the subset implemented here:

```
    decoding                cond    mnem   instruction  acts like
                            name    suffix executes if   NOP if

ir[31:28]==4'b0100 minus    MI     psr[31]==1   psr[31]==0
ir[31:28]==4'b0101 plus     PL     psr[31]==0   psr[31]==1
ir[31:28]==4'b1110 always   none       1            0
ir[31:28]==4'b1111 never    NV         0            1
```

As illustrated in earlier sections, most instructions have 4'b1110 for instruction register bits 31 down to 28 so that execution does not depend on the psr. Although the ARM documentation discourages it, for our subset, we will treat f0000000 as a NOP. (There are many other ways to form a NOP on this machine.)

Using a condition suffix like PL or MI for an instruction on the ARM is very analogous to preceding an instruction on the PDP-8 with an SMA or SPA, respectively. The only difference on the ARM is that since the condition is **part of each instruction**, only one instruction needs to be fetched, rather than two. For example, the childish division program given in section 9.7 uses an SPA prior to a JMP for the special case when the quotient is zero:

```
        0110/7510      SPA
        0111/5123      JMP L2
```

These two PDP-8 instructions are analogous to the BMI instruction (it branches when the PSR indicates minus, so it nullifies (treats like a NOP) the instruction when the PSR indicates plus):

```
        00000010/4a000003     BMI    L2
```

As another example, the PDP-8 program in section 9.7 also uses an SMA prior to a JMP for deciding whether to go through the loop another time:

```
        0121/7500      SMA
        0122/5112      JMP L1
```

The analogous ARM instruction is BPL:

```
00000020/5afffffb    BPL    L1
```

10.6.5 Immediate operands

The only practical way to put a constant value into the PDP-8's accumulator is to use a memory reference instruction. This means that two memory locations must be accessed: the one that contains the instruction and the one that contains the data.

Although the ARM does actually have memory reference instructions (which we are ignoring for now), the ARM provides a different way of working with constant values, known as immediate operands, that only requires one memory access. This is possible because the constant is part of the instruction. In a data processing instruction, when instruction register bit 25 is a one, OPB is an immediate constant, rather than the value of a register. Assuming that instruction register bits 11 down to 8 are zeros, the value of the immediate constant is given by instruction register bits 7 down to 0. (We will ignore the rotation that the full-fledged ARM does when instruction register bits 11 down to 8 are non-zero.) For example, consider the ARM instruction that adds the constant one to the R2 register without setting the PSR:

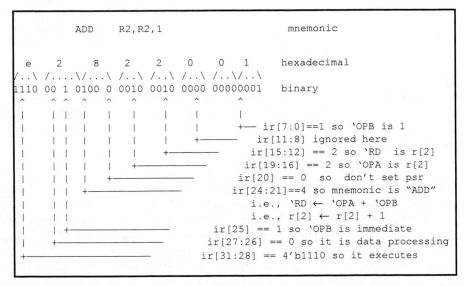

```
         ADD      R2,R2,1                    mnemonic

  e    2     8    2    2    0    0    1       hexadecimal
/..\ /....\/...\ /..\ /..\ /..\ /..\/..\
1110 00 1 0100 0 0010 0010 0000 00000001     binary
 ^    ^ ^  ^   ^  ^    ^    ^     ^
 |    | | |   | |    |    |     |
 |    | | |   | |    |    |     +--- ir[7:0]==1 so 'OPB is 1
 |    | | |   | |    |    +------- ir[11:8] ignored here
 |    | | |   | |    +---------  ir[15:12] == 2 so 'RD  is r[2]
 |    | | |   | +------------  ir[19:16] == 2 so 'OPA is r[2]
 |    | | | +--------------  ir[20] == 0  so  don't set psr
 |    | | +--------------- ir[24:21]==4 so mnemonic is "ADD"
 |    | |                  i.e., 'RD ← 'OPA + 'OPB
 |    | |                  i.e., r[2] ← r[2] + 1
 |    | +--------------- ir[25] == 1 so 'OPB is immediate
 |    +--------------- ir[27:26] == 0 so it is data processing
 +--------------- ir[31:28] == 4'b1110 so it executes
```

As another example, consider the ARM instruction that initializes the R1 register with the decimal constant fourteen:

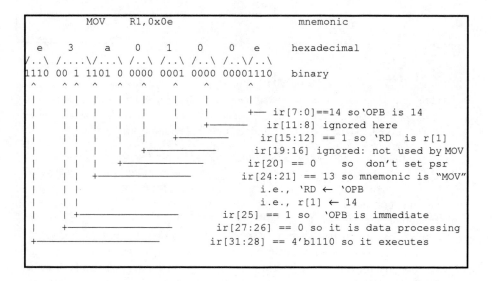

```
            MOV    R1,0x0e                        mnemonic

   e   3    a    0    1    0    0    e    hexadecimal
  /..\ /....\/...\ /..\ /..\ /..\ /..\/..\
  1110 00 1 1101 0 0000 0001 0000 00001110   binary
   ^     ^ ^  ^    ^    ^    ^    ^        ^
   |     | |  |    |    |    |    |        |
   |     | |  |    |    |    |    |        +— ir[7:0]==14 so 'OPB is 14
   |     | |  |    |    |    |    +—————     ir[11:8] ignored here
   |     | |  |    |    |    +—————          ir[15:12] == 1 so 'RD  is r[1]
   |     | |  |    |    +—————               ir[19:16] ignored: not used by MOV
   |     | |  |    +—————                    ir[20] == 0    so  don't set psr
   |     | |  +—————                         ir[24:21] == 13 so mnemonic is "MOV"
   |     | |                                 i.e., 'RD ← 'OPB
   |     | |                                 i.e., r[1] ← 14
   |     | +—————                        ir[25] == 1 so   'OPB is immediate
   |     +—————                          ir[27:26] == 0 so it is data processing
   +—————                                ir[31:28] == 4'b1110 so it executes
```

10.7   Multi-cycle implementation of the ARM subset

The multi-cycle ASM in figure 10-1, which follows the basic outline of the PDP-8's ASM given in section 8.3.1.5, implements the fetch/execute algorithm for the ARM instruction set described in section 10.6.

The state names are the same as the ones in the PDP-8's ASM, except for the execute states. In the ASM for the ARM, state E0DP occurs when a data processing instruction (such as ADD or SUB) executes, and state E0B occurs when a branch instruction (B) executes.

10.7.1   Fake SWI as a halt

The actual ARM does not have a halt instruction. Instead, it has a software interrupt (SWI) instruction (ef000000) which changes the mode from user mode to a supervisor mode. Since we are ignoring the issue of modes, and since it is helpful to keep this ASM as similar as possible to the ASM in chapter 8 for the purpose of Verilog test code, we will treat the SWI as a halt. The operation of the SWI on the actual ARM is much more complicated, as explained in appendix G.

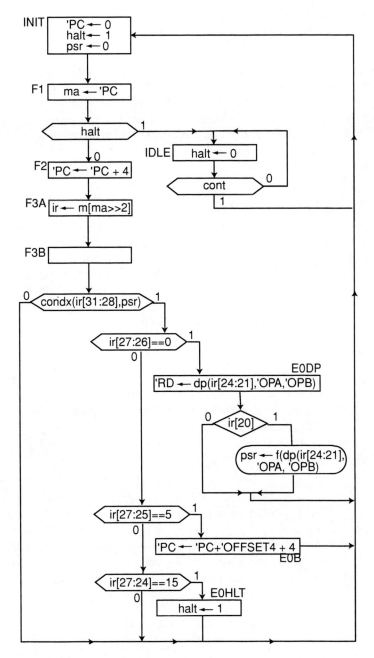

Figure 10-1 Multi-cycle ASM for ARM subset.

10.7.2 Fetch states

State INIT initializes the program counter, halt and program status registers. The machine will then proceed to state F1 and to state IDLE.

When a program executes, the normal sequence is to proceed through states F1, F2, F3A, F3B and one of the execute states. State F2 increments the program counter by four (rather than by one) because the program counter refers to an address in terms of eight-bit bytes but each 32-bit instruction is actually four bytes long. In a related way, when state F3A fetches an instruction from memory, the memory address is shifted over two bits to the right because the program counter is four times the required memory address.[4]

10.7.3 The condx function

The decoding (F3B) and executing (E0DP, E0B or E0HLT) states for the ARM are quite different than the analogous states for the PDP-8. First of all, every instruction on the ARM has the potential of being conditional, which is why instruction register bits 31 down to 28 are reserved for this purpose. The first decision that occurs in state F3B is whether the instruction should be nullified or not. On the actual ARM, this decision involves sixteen possibilities. Even though we are only going to implement four of these (4, 5, e and f), it is prudent to isolate this detail in a function which we will refer to as condx(ir[31:28],psr).

In the actual hardware, there will be some combinational logic that implements this function. The important observation is that whether an instruction is executed or is nullified depends only on two things: instruction register bits 31 down to 28 and the current information in the program status register (which, in this implementation, only contains the N flag). Because these details have been isolated inside the condx function, the other twelve conditions (0-3, 6-d) not considered here could be implemented fairly easily without having to change this ASM.

After recognizing that the condition for the instruction has been satisfied, state F3B proceeds to decode the instruction. If it is a data processing instruction (instruction register bits 27 and 26 equal zero), the ASM proceeds to state E0DP. If it is a branch instruction (instruction register bits 27 down to 25 equal 5), the ASM proceeds to E0B. If it is a SWI instruction, the machine proceeds to the PDP-8 like state E0HLT for the purpose of communicating with the Verilog top-level module that will test this machine. (As mentioned above, the actual ARM would do something more complicated for SWI.)

---

[4] The reason for this inconsistency only becomes apparent with some of the instructions we are ignoring, such as LDR and STR, that use byte-sized data in memory.

10.7.4 Data processing

State E0DP has two actions that it must perform. First, it needs to perform the requested data processing using the operands (`OPA and `OPB) and to store this result into the destination register (`RD). Second, it needs to deal with the program status register.

10.7.4.1 The *dp* function

Which kind of data processing occurs in state E0DP depends upon instruction register bits 24 down to 21. In this implementation, we are only considering three data processing operations (ADD, SUB and MOV). Once again, it is advisable to isolate such details in a function, which we will refer to here as dp, so that the implementation of the other 13-data-processing operations will be straightforward. The important observation is that the result of the data processing only depends on `OPA, `OPB and instruction register bits 24 down to 21.

10.7.4.2 Conditional assignment of *psr*

Using instruction register bit 20, the programmer can choose whether or not a data processing instruction will modify the program status register. If instruction register bit 20 is zero, state E0DP leaves the program status register the way it is. On the other hand, if instruction register bit 20 is a one (meaning the mnemonic has an "S" suffix), state E0DP has a conditional oval which assigns a new value to the program status register. This new program status register value is a function (f) of the unconditional data processing that occurs in this state (dp(ir[24:21],`OPA,`OPB)). Again, isolating details in a function will make it easier to implement the full capabilities of the ARM, should you choose to do so. For now, f simply masks off the sign bit of the result to record the N flag. This assignment to the program status register must be conditional in state `E0DP because the information to compute the new program status register properly might not exist after that clock cycle (i.e., `RD might refer to the same register as `OPA or `OPB).

10.7.4.3 Use of macros

The use of macros (`PC, `RD, `OPA and `OPB) helps conceal many of the tedious details required to implement the ARM. For example:

```
`define RD    r[ir[15:12]]
```

allows us to describe the destination register for a data processing instruction without having to mention the instruction register bits.

More importantly, macro definitions allow simple descriptions of the operands of the data processing instructions. For example, from the short explanation in section 10.6.1, one might think that 'OPA would simply be defined as r[ir[19:16]], but there are some other details to consider. For example, when 'OPA is supposed to be what the programmer refers to as R15, this is in fact the program counter. According to the specification of the ARM instruction set (appendix G), when R15 is the first operand of a data processing instruction, the value of R15 used in the computation will be eight larger than the value the program counter contained when the instruction was fetched. At the time 'OPA is evaluated, the program counter has already been incremented by four in state F2. Therefore, if the programmer wants to use R15 as 'OPA, the value used in state E0DP should be r[15]+4. On the other hand, if the programmer wants to use a different register, its value should not be incremented.

To distinguish between R15 or another register inside a macro requires using the Verilog conditional operator (? :). This feature of Verilog, which acts like the similar feature of C, works with three values: the first value occurs before the question mark, the second value occurs between the question mark and the colon, and the third value occurs after the colon. If the first value is equal to one, the result is the second value; otherwise the result is the third value. From a hardware standpoint, the Verilog conditional operator is like a mux, where the select is the first value. In this particular situation, the expression ir[19:16]!=15 chooses between two different results:

```
`define OPA  (ir[19:16]!=15?r[ir[19:16]]:r[ir[19:16]]+4)
```

As in C, parentheses are a good idea to avoid creating precedence problems when Verilog substitutes such a complicated macro.

The definition of 'OPB is even more involved because instruction register bit 25 allows the programmer to choose between an immediate value or a register value. The same problem with R15 mentioned above also must be considered:

```
`define OPB  (ir[25]?ir[7:0]:(ir[3:0]!=15?  r[ir[3:0]]:r[ir[3:0]]+4))
```

In fact, there are other issues about 'OPB that we are ignoring here. (The actual ARM allows rotation of 'OPB, which would require a more complicated expression for 'OPB.)

10.7.5 Branch

State E0B performs the relative branch by adding four plus four times the signed offset (from the low-order 24 bits of the instruction register) to the program counter.

10.7.5.1   Multiplying by four

The reason for multiplying by four is that instructions on the ARM are required to reside at byte addresses that are divisible by four. The low-order two bits are not recorded in the branch instruction in order to maximize the distance that a programmer can choose to branch with the available bits of the instruction register.

10.7.5.2   Adding four in the multi-cycle implementation

The reason for adding four in this multi-cycle implementation is to adhere to the specifications referred to in appendix G. The ARM documentation specifies that an offset of -2 in a branch instruction (eaffffe) means branching back to itself. Remember that the -2 will be multiplied by 4 to yield -8. By state E0B, the program counter has been incremented by 4. To leave the program counter the same as it was when the branch instruction (eaffffe) was fetched, an additional 4 must be added to the program counter (-8 + 4 + 4 == 0).

10.7.5.3   Sign extension macro

The offset times four needs to be treated as a signed 32-bit value. The offset in the instruction register is only 24-bits wide (bits 23 down to 0). This means the sign bit (ir[23]) must be duplicated several times on the left, in a process commonly referred to as *sign extension*. The following macro performs both the sign extension and the multiplication by four using the Verilog concatenation operator:

```
`define OFFSET4   {ir[23],ir[23],ir[23],ir[23],ir[23], ir[23],ir[23:0],2'b00}
```

10.7.6   Verilog for the multi-cycle implementation

Throughout this book, algorithms have been described using ASM charts and Verilog. For simple machines, like the childish division examples of chapter 2, the ASM and the Verilog are equivalent, and either notation gives a completely accurate description of the hardware. It is a theme of this book that both ASM charts and Verilog are important to understand, and that each notation offers the designer useful insights.

The last section illustrates the proper use of an ASM in a complex design: to provide documentation so people can see the "big picture." The last section actually uses quite a bit of Verilog notation to achieve the proper level of abstraction. This section shows that Verilog implements not only what the ASM describes but also fills in the details that should be omitted from the ASM. Verilog can do all this reasonably because it is a **textual** language.

The previous sections use many paragraphs to describe the data processing function, dp. Such a description is informal and therefore cannot be synthesized into hardware. Although such details could have been put formally into the ASM, they would have made the ASM considerably more complicated. Ultimately, such a complex ASM would lead us down the wrong path for the performance issues (such as pipelined and/or superscalar design) described later in this chapter. In Verilog, we can be precise but still set aside such details because they can occur in a different part of the source code (the function definition). Although such thinking is commonplace in modern software design, hardware designers are only beginning to realize the power of the notation available in Verilog.

In the case of the dp function, we are only implementing three of the sixteen possible operations:

```
function [31:0] dp;
   input [3:0] opcode;
   input [31:0] opa,opb;
   begin
      if       (opcode == 4'b0010)
         dp = opa - opb;
      else if (opcode == 4'b0100)
         dp = opa + opb;
      else if (opcode == 4'b1101)
         dp = opb;
      else
         begin
            dp = 0;
            $display("other DP instructions...");
         end
   end
endfunction
```

This function formally describes the SUB, ADD and MOV instructions. Except for the $display statement, this function could be synthesized into the combinational ALU required in the actual hardware. (The $display statement warns us if we attempt to execute a data processing instruction that is one of the 13 not implemented here.) This function can be reused as we improve the performance of the design. Because these details have been isolated into a function, it is easy for a designer to know where to modify the Verilog code in order to implement the remaining 13 operations.

In a similar way, we can define the Verilog function that determines whether the condition for a particular instruction to execute is true:

```
function condx;
  input [3:0] condtype;
  input [31:0] psr;
  begin
    if (condtype == 4'b1110)
      condx = 1;
    else if (condtype == 4'b0100)
      condx = psr[31];
    else if (condtype == 4'b0101)
      condx = !psr[31];
    else
      condx = 0;
  end
endfunction
```

Again, isolating this in a function makes it easy to know how to implement the remaining operations. Also, as will be shown later, defining this function will prove extremely helpful as we use more sophisticated techniques to improve performance.

For our subset of the ARM, we only implement the N flag in the program status register. The function which creates this information from the result of the ALU is trivial:

```
function [31:0] f;
  input [31:0] dpres;
  begin
    f = dpres & 32'h80000000;
  end
endfunction
```

Generating all bits of the program status register is considerably more complicated, but isolating it here helps some future designer whose job might be to do so.

A great deal of the abstraction needed for this design comes from the Verilog macros mentioned in the last section. For this multi-cycle implementation, these are:

```
`define PC        r[15]
`define RD        r[ir[15:12]]
`define OPA       (ir[19:16]!=15?r[ir[19:16]]:r[ir[19:16]]+4)
`define OPB       (ir[25]?ir[7:0]:(ir[3:0]!=15?r[ir[3:0]]:r[ir[3:0]]+4))
`define OFFSET4 {ir[23],ir[23],ir[23],ir[23],ir[23],ir[23],ir[23:0],2'b00}
```

For example, the definition of 'OPA says if the register number given in instruction register bits 19 down to 16 is not the program counter (!=15), use the value of that register; otherwise (==15) use the program counter + 4. This decision is required here because of the definition of the ARM instruction set. The Verilog conditional operator allows for compact, if somewhat cryptic, code for the decision. Recall that 'OPA will be substituted back where required in the code automatically by Verilog, and so the designer using Verilog can usually ignore these details. A designer using only an ASM would have been required to put all this detail in the ASM.

Another advantage of using Verilog here is that as we proceed to a more sophisticated technique for higher performance (such as pipelining), we can change the definition of the macros to match the more sophisticated technique, but keep the macro name the same. This is good, because the concept behind the macro is the same, even though its implementation will be different in later sections of this chapter. Given the above macros and functions, the actual translation of the ASM to implicit style Verilog is simple and is left as an exercise.

10.8   Pipelined implementation

The problem with the multi-cycle implementation is that it requires five cycles per instruction. To improve this performance, we can use a pipelined approach. There are several reasons why a pipelined implementation of our ARM instruction subset will be easier than the pipelined PDP-8 discussed in chapter 9. First, the ARM has a RISC instruction set which was designed to be pipelined. Second, we are neglecting memory reference instructions, and so the issues of operand fetch and data forwarding may be ignored here. Third, we can reuse the functions defined above without modification. Fourth, the Verilog macros given earlier can easily be redefined to match the needs of the pipelined implementation.

10.8.1   ASM for three-stage pipeline

The original versions of the ARM (referred to as the ARM1-ARM6) use a three-stage pipeline, but some of the more recent versions of the ARM use a five-stage pipeline. We will implement a three-stage pipeline because it is easier, and is similar to the three-stage examples in chapter 9. Also, since the original ARM instruction set was designed for a three-stage pipeline, implementing a pipeline of that size is most natural. The three stages are referred to as fetch, decode and execute. For our subset, decoding only considers instruction register bits 27 down to 24. The logic for decoding our subset is trivial and hardly warrants its own separate pipeline stage. The full ARM instruction set includes eleven categories, and so full decoding is rather involved, especially considering the memory reference instructions. Therefore, for our simple ARM subset, we will implement a three-stage pipeline but with the middle stage doing nothing. Later in this chapter, the middle pipeline stage will take on an important role.

As in chapter 9, it will be necessary to have multiple instruction registers (ir1 and ir2). The youngest instruction is in memory, the middle-aged instruction is in ir1 and the oldest instruction is in ir2. Also, as in chapter 9, we need to use NOPs to deal with the filling of the pipeline after a branch instruction. There are many ways to describe a NOP in ARM machine language. The simplest[5] is f0000000. Here is the pipelined ASM for our ARM subset:

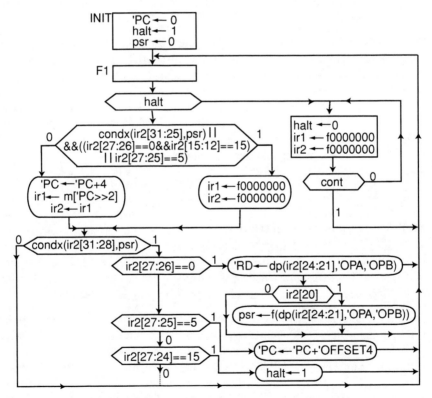

Figure 10-2. Pipelined ASM for ARM subset.

[5] Although the ARM's designers may someday redefine the meaning of this machine code to be something other than NOP, f0000000 is convenient since it is easy to recognize

Rather than translate this ASM as is into Verilog, the following sections discuss some interesting issues that it raises. These issues, such as Mealy ASMs and the non-blocking assignment, have been touched upon in earlier chapters, but with an example of this size these issues become more important.

10.8.2   Mealy ASM

There are several reasons why the pipelined ARM needs to be designed using a Mealy ASM. As in the examples in chapter 9, a decision has to be made and an assignment has to be scheduled in a single cycle because the result of the assignment must be available in the next clock cycle. With the PDP-8, this typically arises with assigning a value to the accumulator having decided already that a TAD instruction is supposed to execute. On the ARM, this same issue arises with assigning a value to the 'RD macro (which could be any of the sixteen user registers), having decided a data processing instruction is supposed to execute. The decision (that recognizes the instruction) and the resulting assignment occur in the same clock cycle.

The ARM's conditional execution feature offers an additional reason why a Mealy approach is required. Often, a data processing instruction that assigns a new value to the program status register (such as SUBS) will execute just before a subsequent instruction that depends on this result in the program status register (such as BMI). The decision whether to execute the subsequent instruction cannot occur earlier than the clock cycle when that instruction is in the final stage of the pipeline because, until that stage, the program status register does not reflect the result of the preceding instruction. If the condition is satisfied, the subsequent instruction must schedule whatever assignment(s) are associated with its execution in this same clock cycle; therefore a Mealy approach is mandatory.

10.8.3   Two parallel activities

There are two parallel activities that occur in state F1. The first determines what will be in the instruction registers and the program counter in the next clock cycle. The second, which is similar to the corresponding part of the ASM in section 10.7, deals with decoding and executing the instruction in the final stage of the pipeline.

The first parallel activity in state F1 (determination of what will be in the instruction registers and the program counter) has two cases. Let's refer to these as the "B/R15" case (for branch or modifying the program counter) and the "normal" case (increment program counter). The "B/R15" case is when the instruction in the final stage of the pipeline is an instruction that will execute (as indicated by condx(ir2[31:28],psr)) and that is either a branch instruction (bits 27 down to 25 equal to 5) or a data processing instruction that modifies r[15] (since r[15] is the same as the program counter, such an instruction is effectively like a branch instruction). The "B/R15" case is sim-ilar to the JMP instruction on the pipelined PDP-8 given in section 9.5, except that on the ARM the branch instruction cannot be executed until it reaches the last stage of the pipeline. On the PDP-8, it is possible to begin execution of the JMP when it reaches the middle stage of the pipeline because the JMP on the PDP-8 is unconditional. Since the branch on the ARM is conditional, the bits in the program status register must be valid before the decision to branch can occur. In general, the program status register will not be valid for a particular instruction until that instruction reaches the final stage of the pipeline. This is because the program status register might have been changed by the preceding instruction, which only com-pleted its execution after the preceding clock cycle. If the "B/R15" instruction ex-ecutes, the instruction registers are filled with NOPs.

The "normal" case decides what will be stored in the instruction registers and the pro-gram counter for situations other than the "B/R15" case. In the "normal" case the program counter is incremented by four, and the instructions move down the pipeline.

Besides the part of the ASM that decides what will be stored in the instruction registers and the program counter, there is the part of the ASM that decodes and executes the instruction in the final stage of the pipeline. In this part of the ASM there are five cases:

```
"nullify"     psr prevents this instruction from executing
"dp set"      execute data processing instruction that modifies psr
"dp no set"   execute data processing instruction that leaves psr alone
"ir2 is B"    modify R15
"SWI"         set halt flag
```

Except for the "ir2 is B" case, these are identical to the multi-cycle ASM given in section 10.7. In the "ir2 is B" case, four times the sign extended offset (`OFFSET4) is added to the program counter. Here is where we see that the ARM was designed to work with a three-stage pipeline. The reason that an offset of -2 means branch back to itself is that by the time the branch instruction has reached the final stage of the pipe-line, the program counter will already have been incremented twice, i.e., it is eight greater than when the branch was fetched. When the offset is -2, `OFFSET4 is -8 and so adding it to the program counter in this case puts the program counter back to where the same branch instruction will be fetched again.

10.8.4 Proper use of <=

One of the main themes of this book is the proper use of the non-blocking assignment statement. A common mistake with non-blocking assignment is to attempt to assign more than one value to a register during one clock cycle. To avoid making this mistake, a designer needs to check all possible paths through the ASM. Since there are two paths through one-half of the ASM and there are five paths through the other half that executes in parallel, there are, in theory, ten paths for the designer to check, but of these, two are contradictory. It is impossible for ir2 to contain a branch or data processing instruction that modifies R15 in the same clock cycle that it contains an SWI instruction. Also, when "ir2 is B," the "normal" case cannot occur. When these cases are eliminated, we are left with eight cases to consider. The "B/R15" case might occur in parallel with either the "nullify," "dp set," "dp no set" or "ir2 is B" case. Alternatively, the "normal" case might occur together with either the "nullify," "dp set," "dp no set" or "SWI" case.

The "B/R15" and "normal" cases are the only places where the instruction registers are scheduled to be assigned, and so there is no problem with them. The "dp set" case is the only place where the program status register is scheduled to be assigned a value, and so it is fine. Also, the "SWI" case is the only place where the halt flag is scheduled to be assigned a value; thus we do not need to be concerned with it. The danger arises with the program counter and `RD, since `RD could be r[15], which is the program counter. To avoid this danger, we must leave the program counter alone in the "B/R15" case, because the program counter is modified in parallel by the "dp no set," "dp set" or "ir is B" cases of this Mealy ASM.

10.8.5 Verilog for the pipelined implementation

Here is a partial listing of the Verilog translation for the ASM from section 10.8.1:

```
    . . .
  forever
   begin
    @(posedge sysclk) enter_new_state('F1);
     . . .
     else
      begin
       if (condx(ir2[31:28],psr) &&
        ((ir2[27:25]==3'b101)
        || (ir2[27:26]==2'b00&&ir2[15:12]==4'b1111)))
        begin // "B/R15"
         ir1 <= @(posedge sysclk) 32'hf0000000;
         ir2 <= @(posedge sysclk) 32'hf0000000;
        end
```

Continued

```
        else
         begin // "normal"
          'PC <= @(posedge sysclk) 'PC + 4;
          ir1 <= @(posedge sysclk) m['PC>>2];
          ir2 <= @(posedge sysclk) ir1;
         end
        if (condx(ir2[31:28],psr))
         begin
          if      (ir2[27:26] == 2'b00)
           begin // "dp set" or "dp no set"
            'RD <= @(negedge sysclk)
                        dp(ir2[24:21]'OPA'OPB);
            if (ir2[20]) //"dp set"
             psr <= @(posedge sysclk)
                        f(dp(ir2[24:21], 'OPA, 'OPB));
           end
          else if (ir2[27:25] == 3'b101) //"ir2 is B"
           'PC <= @(posedge sysclk) 'PC + 'OFFSET4;
          else if (ir2[27:24] == 4'b1111)//"SWI"
           halt <= @(posedge sysclk) 1;
          else
           $display("other instructions...");
         end
       end
    end
```

Some of the macros need to be redefined to take into account that ir2 is the final stage of this pipeline:

```
'define RD      r[ir2[15:12]]
'define OPA     r[ir2[19:16]]
'define OPB     (ir2[25] ? ir2[7:0] : r[ir2[3:0]])
'define OFFSET4 {ir2[23],ir2[23],ir2[23],ir2[23],ir2[23],ir2[23],ir2[23:0],2'b00}
```

Interestingly, because the original ARM was designed with a three-stage pipeline in mind, the definition of 'OPA and 'OPB are simpler than for the multi-cycle implementation. This simplification occurs since r[15] does not have to be explicitly mentioned. The value of r[15] at the time the instruction is in the final stage of the pipeline is, by definition, the correct value to use.

Execution of a data processing instruction involves non-blocking assignment to `RD, which is a macro that substitutes the subscripted Verilog array, `r[ir2[15:12]]`. This non-blocking assignment therefore uses `negedge` rather than `posedge` to be portable for the reasons explained in section 6.5.2. (Remember that, in this pipelined implementation, `ir2` changes every clock cycle.)

10.9 Superscalar implementation

The pipelined implementation given in the last section has a speed that approaches (but never quite reaches) one clock cycle per instruction. Because ARM data processing instructions have three register operands (`RD, `OPA and `OPB), one basic computation, such as incrementing `r[2]`, can be performed per clock cycle. Although this can be up to three times faster than the pipelined single-accumulator design described in chapter 9, it still is certain to be no better than the slowest special-purpose designs in chapter 2 (such as section 2.2.2). Even for a simple algorithm like childish division, it is often possible for more than one computation to occur in parallel (e.g., incrementing `r[2]` in parallel with subtracting from `r[1]`). A pipelined general-purpose processor only works because of quite a bit of parallel activity in the implementation of fetch/execute. Even so, a pipelined general-purpose computer cannot exploit the parallelism in an algorithm. Such parallelism can be exploited by special-purpose hardware (such as section 2.2.7).

Since the designer of a general-purpose computer can never be certain how fast is "fast enough," it would be desirable if the general-purpose computer could execute more than one instruction in parallel. Such an approach, known as a *superscalar* implementation, is an extension to the pipelined approach. Superscalar implementation is considerably more complex than the pipelined approach because the hardware itself must take seemingly sequential instructions and recognize when it is permissible for them to execute in parallel. In essence, some of the intelligence and skill of the hardware designer (as illustrated by the design alternatives of chapter 2) must be placed inside the hardware itself. Because the hardware of a superscalar general-purpose computer will never have as much information about the software algorithm as the designer of a special-purpose computer has about the ASM, a superscalar general-purpose machine will not be as fast as the best special-purpose hardware. Also, the complexities of superscalar design means its hardware cost may be many times the cost of the equivalent but faster special-purpose machine. However, the economies of scale for general-purpose computers have made superscalar processors viable.

10.9.1 Multiple-port register file

From a structural standpoint, a superscalar general-purpose processor can be distinguished from the slower design alternatives given earlier in this chapter (multi-cycle and pipelined) because the superscalar machine needs multiple ALUs for executing multiple instructions per clock cycle. The simplest case is to imagine that we can afford to have two ALUs, and therefore, under the best circumstances, two instructions can execute per clock cycle.

A consequence of having multiple ALUs is that the register file must be more sophisticated. If there are two ALUs, each of which might have to be fed two independent operands in each clock cycle, we need a register file with four read ports. From a behavioral standpoint, we will refer to the operands of the final stage of the pipeline the way we did in the last section ('OPA and 'OPB). However, sometimes another instruction will be executing in parallel. The operands of this parallel instruction will be referred to as 'POPA and 'POPB.

The two results of the two ALUs need two write ports into the register file. From a behavioral standpoint, we will refer to these as 'RD and 'PRD.

A register file that has four read ports and two write ports is considerably more expensive than the register file used in the pipelined implementation. There will be additional complexities with `r[15]` because it serves the role of the program counter. We will see later that the program counter in a superscalar design does some non-intuitive things.

10.9.2 Interleaved memory

In order to keep a superscalar processor going at full speed, it is necessary to provide it with as many new instructions as it is capable of executing per clock cycle. For example, if our machine is to execute two instructions per clock cycle, it will be necessary to load both `ir1` and `ir2` with instructions from memory addresses 'PC+4 and 'PC, respectively.

The single-port memory shown in figure 10-3, which can be used for the multi-cycle and pipelined implementations, does not allow more than one instruction to be fetched per clock cycle:

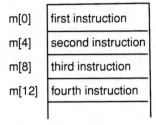

Figure 10-3. Non-interleaved memory.

Although a dual-ported memory for instructions would allow fetching of two instructions per clock cycle, such a memory is expensive. A cheaper alternative is to use an *interleaved* memory. A simple interleaved memory stores half of the instructions in one bank and the adjacent instructions in another:

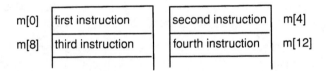

Figure 10-4. Interleaved memory.

In other words, two separate memories act as one. This approach is sufficient only because when the superscalar fetch/execute algorithm wants two instructions, they will always reside in separate banks. One of the instructions comes from an address divisible by eight; the other instruction will be plus or minus four of that address. From a behavioral standpoint, we will simply use the same kind of Verilog array notation for the two instructions that are fetched in parallel: m[(`PC+4)>>2] and m[`PC>>2].

10.9.3 Examples of dependencies

If all instructions in a program were independent of each other, such as:

```
SUB R1,R1,R4
ADD R2,R2,1
```

designing a superscalar machine would be fairly easy. For example, the above two instructions could be fetched from the interleaved memory in parallel, presented to the two separate ALUs (for subtraction and addition, respectively) in parallel and their results could be written back to the register file in parallel.

Unfortunately, in real programs, instructions are often dependent on each other. For example:

```
SUB R2,R1,R4
ADD R2,R2,1
```

It might appear that data forwarding (of R1 minus R4) could be helpful here. Such an approach would be algorithmically correct but would be slow. To make these instructions execute in parallel, the clock period would have to be slow enough allow enough time for both the ADD and the SUB:

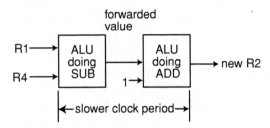

Figure 10-5. Forwarding results of dependent instructions slows clock.

Instead of data forwarding in a situation like this, it is better for the machine to execute only one instruction per clock cycle. At least this way, the clock cycle remains fast. In other words, it behaves like the simple pipeline approach of section 10.8. The hope is that after executing these two instructions sequentially, the machine will fetch some independent instructions (like the ones shown earlier) that it can execute in parallel.

Some programs have combinations of instructions that simply cannot be executed in parallel:

```
SWI
ADD R2,R2,1
```

The machine is supposed to halt (in our subset, at least) before the ADD instruction executes. In such a situation, we have to revert back to a one instruction per clock cycle (simple pipeline) approach, which allows the machine to process the SWI in exactly the order the programmer intends. On a machine that actually implements interrupts (unlike our subset), exact processing of interrupts and similar issues are significant.

A very common problem that occurs with superscalar design is that we cannot be certain, at the time when we have the hardware resources for doing so, whether we are supposed to execute an instruction (such as ADDPL):

```
        SUBS R1,R1,R4
        ADDPL R2,R2,1
```

The SUBS instruction will modify the program status register, but the ADDPL instruction needs to know that new program status information to decide whether to execute. If these instructions only executed one per clock cycle, there would be no difficulty. Also, if it were not for the "S" and "PL" suffixes on the mnemonics, there would be no problem with executing them in parallel during the same clock cycle. Although we could revert back to a simple pipeline approach (one instruction per clock cycle), the point of superscalar design is to maximize speed.

It might be tempting to try "program status forwarding" in a case like this. Such an approach would be algorithmically correct, but would have the undesirable side effect of doubling the propagation delay (the ADD cannot start until the SUB completes). This would mean the clock cycle of the machine would be twice as long, which would more than negate any advantage of our attempt at superscalar design.

10.9.4   Speculative execution

In contrast to such a flawed approach, the typical superscalar implementation uses a technique, known as *speculative execution*, to solve the problem of not knowing whether an instruction that could execute in parallel is supposed to execute. For most instructions on a RISC machine, the only irreversible consequence of executing that instruction is the storage of its result back in the register file. Speculative execution means we compute the result of an instruction at a time when it is uncertain whether or not that instruction will execute, but at that time we do not store the result back in the register file.

10.9.5   Register renaming

Instead, we put the result of an instruction that is being executed speculatively in a *rename register*. Such a register has the ability, at a later time, to take on the role of any of the user registers in the machine. In a later clock cycle, if the machine discovers that the instruction that executed speculatively in the previous clock cycle was not supposed to execute, the contents of the rename register can simply be discarded, and it will be as though the instruction never executed. If the machine discovers that the instruction was supposed to execute, the rename register takes on the role of the destination register. For our simple implementation, this will be very much like data forwarding of a single value. In a more complicated superscalar design, the renaming process could be much more sophisticated because every register in the file might have been renamed. Our simple implementation guarantees that at most one register will be renamed in any given cycle.

To implement this, there will have to be three components of the renameable register: reg_val, which indicates its 32-bit value, reg_tag, which indicates its identity if speculative execution succeeds (since it could be any of the registers) and reg_cond, which indicates the condition upon which the instruction is supposed to execute. In Verilog, these are declared as:

```
reg [31:0] reg_val;
reg [4:0] ren_tag;
reg [3:0] ren_cond;
```

It is interesting to note that ren_tag is five, rather than four, bits wide. This is required because, in addition to the sixteen user registers, we need to indicate when the renamed register is not valid. To do so, the following constant is defined:

```
`define INVALID 16
```

When an instruction cannot be executed speculatively (as in the SWI example from section 10.9.3), the machine assigns `INVALID to ren_tag. In the next clock cycle, this will cause ren_val to be ignored.

On the other hand, when an instruction can be executed speculatively, the machine assigns the destination register number to ren_tag, the condition upon which that assignment succeeds to ren_cond and the potential new value of that register to ren_val.

10.9.5.1   First special-purpose renaming example

Even though register renaming and speculative execution may appear difficult to understand at first, they are simple extensions to the idea of the non-blocking assignment which has been discussed in earlier chapters. It is still true that only one assignment can be scheduled for a particular register during a particular clock cycle and that such assignments do not take effect until the next clock cycle. In order to see how register renaming and speculative execution relate to concepts in earlier chapters, let us set aside our goal of implementing the general-purpose ARM for a moment and consider some simple **special-purpose** machines that illustrate these same concepts. In other words, we are going to describe a special-purpose machine that only executes one (nonsensical) algorithm, which we will state in terms of ARM mnemonics:

```
SUBS R2,R1,R4   ;sets psr
ADDPL R3,R3,R3  ;speculative
ADD R3,R3,1     ;R3 same as last inst
NOP             ;NOP to simplify discussion
```

Designing such a special-purpose machine is easy if all we wanted to do is to carry out the same register transfers as would occur when the above instructions execute on the pipelined implementation of the general-purpose ARM given in section 10.8.1:

```
@(posedge sysclk) #1;
  r[2] <= @(posedge sysclk) r[1] - r[4];
  psr <= f(r[1] - r[4]);
@(posedge sysclk) #1;
  if (psr[31]==0)
    r[3] <= @(posedge sysclk) r[3] + r[3];
@(posedge sysclk) #1;
  r[3] <= @(posedge sysclk) r[3] + 1;
@(posedge sysclk) #1;
```

In the first clock cycle, the sign bit of the difference is scheduled to be stored in the psr. In the next clock cycle, after this assignment has taken effect, the psr determines whether or not the doubling of r[3] will occur (which makes this a Mealy machine). In any event, incrementation of r[3] does not occur until the third clock cycle, at which $time the appropriate value (either doubled or not) will be available. The fourth clock cycle does nothing because of the NOP in the algorithm.

It is possible to cut the number of states in half by combining two actions per state; however, there is a difficulty. SUBS sets the psr, but the ADDPL that we desire to execute in parallel depends on that psr. Here is where speculative execution and register renaming come into play. The register transfers of the special-purpose machine described by the Verilog below are similar to those carried out when the equivalent instructions execute on the general-purpose superscalar ARM given in section 10.9.6; however, the following is much simpler because it only considers actions that occur related to the specific instructions:

```
@(posedge sysclk) #1;
  r[2] <= @(posedge sysclk) r[1] - r[4];
  psr <= f(r[1] - r[4]);
  ren_val <= @(posedge sysclk) r[3] + r[3];
  ren_tag <= @(posedge sysclk) 3;
  ren_cond <= @(posedge sysclk) 'PL;
@(posedge sysclk) #1;
  ren_cond <= @(posedge sysclk) 'NV;
  if ((ren_cond == 'PL)&&(psr[31]==0) ||
      (ren_cond == 'MI)&&(psr[31]==1) ||
      (ren_cond == 'AL))
    r[3] <= @(posedge sysclk) ren_val+1; //renamed
  else
    r[3] <= @(posedge sysclk) r[3]+1;     //not renamed
```

In parallel to the subtraction during the first clock cycle, the doubling of r[3] occurs before the machine can know whether the difference will be positive. Therefore, the machine saves the doubled value in ren_val, and at the same $time makes note in ren_cond of the condition ('PL) under which this speculative doubled result is to be renamed as r[ren_tag]. In the second clock cycle, after the psr resulting from the subtraction is valid, the machine makes a decision whether or not renaming occurs. If it does not, incrementation of r[3] occurs based on the value already in the register file from two or more clock cycles ago. If renaming does occur, there is a literal substitution of ren_val for r[3] in this clock cycle. Regardless of whether renaming occurs in the second clock cycle, ren_cond is set to 'NV because the NOP will not cause any renaming in the third cycle (not shown).

10.9.5.2 Second special-purpose renaming example

Let's consider a second example, similar to the last one, except the destination of the third instruction (shown in bold) is not the same as the destination of the ADDPL instruction that executes speculatively:

```
SUBS R2,R1,R4   ;//sets psr
ADDPL R3,R3,R3  ;//speculative
ADD R6,R3,1     ;//R3 same but not dest
NOP             ;//NOP to simplify discussion
```

Again, there is no problem when all we want to do is to carry out the same register transfers as would occur when the above instructions execute on the pipelined implementation of the general-purpose ARM given in section 10.8.1:

```
@(posedge sysclk) #1;
  r[2] <= @(posedge sysclk) r[1] - r[4];
  psr <= f(r[1] - r[4]);
@(posedge sysclk) #1;
  if (psr[31]==0)
    r[3] <= @(posedge sysclk) r[3] + r[3];
@(posedge sysclk) #1;
  r[6] <= @(posedge sysclk) r[3] + 1;
@(posedge sysclk) #1;
```

Of course, things get more interesting when we use speculative execution and register renaming. The register transfers of the special-purpose machine below are similar to those carried out when the equivalent instructions execute on the general-purpose superscalar ARM given in section 10.9.6:

```
@(posedge sysclk) #1;
  r[2] <= @(posedge sysclk) r[1] - r[4];
  psr <= f(r[1] - r[4]);
  ren_val <= @(posedge sysclk) r[3] + r[3];
  ren_tag <= @(posedge sysclk) 3;
  ren_cond <= @(posedge sysclk) `PL;
@(posedge sysclk) #1;
  ren_cond <= @(posedge sysclk) `NV;
  if ((ren_cond == `PL)&&(psr[31]==0) ||
      (ren_cond == `MI)&&(psr[31]==1) ||
      (ren_cond == `AL))
    begin
      r[ren_tag] <= @(posedge sysclk) ren_val; //renamed
      r[6] <= @(posedge sysclk) ren_val+1;
    end
  else
    r[6] <= @(posedge sysclk) r[3]+1; //not renamed
```

The first clock cycle is identical to the speculative example in section 10.9.5.1; thus the speculative doubling of r[3] occurs before the machine knows whether the difference of r[1] and r[4] will be positive. Again, ren_val will contain the doubled value and ren_cond will indicate the condition ('PL) when ren_val is to be renamed as r[ren_tag]. In the second clock cycle, after the psr resulting from the subtraction is valid, the machine makes a decision whether or not renaming occurs. If it does not, the assignment to r[6] occurs based on the value of r[3] already in the register file from two or more clock cycles ago. If renaming does occur, the situation is quite different than in the example of section 10.9.5.1. In this example, the destination of the third instruction (r[6]) is different than the destination of the speculative instruction (r[3]). There is still a literal substitution of ren_val for r[3], but there must also be storage

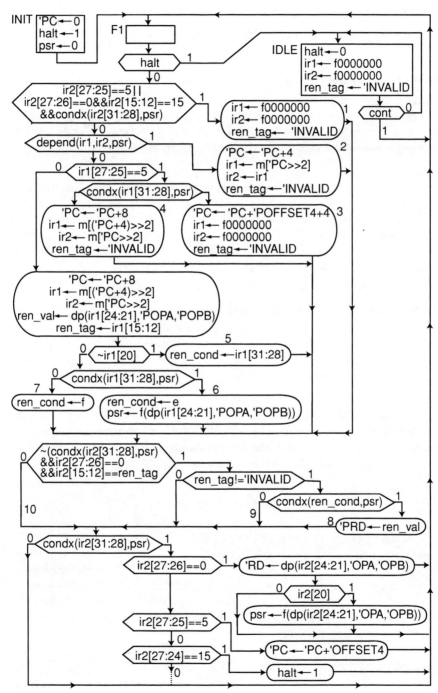

Figure 10-6. Superscalar ASM for ARM subset.

of `ren_val` in `r[ren_tag]` so that `r[3]` will contain the doubled value for future clock cycles (not shown).

10.9.6 ASM for the superscalar implementation

The preceding examples only considered speculative execution in the very limited context of a single algorithm. We want to use this in the general context of the fetch/execute algorithm for a superscalar processor. Figure 10-6 is the ASM for the superscalar implementation of the same subset of ARM instructions used earlier in this chapter.

Although the reader may not realize it at first, much of the motivation for the ARM subset used throughout this chapter was to keep this intricate ASM as simple as possible. Also, this ASM deals with the simplest case of superscalar design (two instructions per clock cycle). Common commercial superscalar processors are much more complicated than this trivial example.[6] Even so, this ASM is something to behold! The observation is that superscalar design is significantly more complicated than multi-cycle or pipelined design, even for a tiny subset of instructions.

10.9.7 Three parallel activities

The ASM in section 10.9.6 shows three parallel activities that occur in each clock cycle when the machine is in state F1. The first parallel activity (arbitrarily shown at the top of this Mealy ASM only to be consistent with earlier ASMs in this chapter) deals with the instruction registers in the pipeline and the program counter. This first parallel activity also deals with speculative execution.

The second parallel activity (arbitrarily show in the middle) deals with register renaming. The final parallel activity (arbitrarily shown at the bottom) deals with executing instructions sequentially in the final stage of the pipeline. This portion of the ASM is identical to the pipelined ASM given in section 10.8.1.

10.9.7.1 Pipeline, parallel and speculative execution

The first parallel activity deals with several different interacting components of the superscalar machine. This portion of this ASM is doing analogous work to what the first portion of the ASM of section 10.8.1 does. The essential goal is to decide what will be in the instruction register pipeline (`ir1` and `ir2`) for the next clock cycle. In the pipelined implementation, this was easy: either put NOPs into the pipeline or move the instructions down the pipe. In this superscalar implementation, there are seven distinct cases, only one of which will occur in any clock cycle. For ease of discussion, let's number these cases 1-7:

---

[6] As of 1997, despite its suitability for superscalar implementation, ARM had not yet introduced such a version of its processor, instead focusing on low-cost, low-power versions that use only pipelining.

1. `ir2` has a branch or data processing instruction that changes R15.
2. There is a dependency between `ir1` and `ir2`.
3. `ir1` has a branch that changes R15.
4. `ir1` has a branch that is nullified.
5. `ir1` has a data processing instruction that does not affect `psr`.
6. `ir1` has a data processing instruction that does affect `psr`.
7. `ir1` has a data processing instruction that is nullified.

Each one of these cases corresponds to an oval in the Mealy ASM. In all of these cases, it is guaranteed that an instruction in `ir2` will execute (as described later at the bottom the ASM). The determination of which of the above seven cases applies here is based upon whether `ir1` can be executed in parallel to `ir2`. We must know that to decide how much to increment the program counter and how to load the instruction registers for the next clock cycle. The more instructions we can execute in parallel, the more new instructions have to be fetched.

Cases 1 and 2 describe situations when it is not possible for `ir1` to execute in parallel with `ir2`. Cases 3-7 describe situations when it is possible to do something with `ir1` in parallel to `ir2`

Case 1 is similar to the analogous case of the ASM of section 10.8.1 ("B/R15"—put NOPs in the instruction register pipeline), except case 1 must also make the renamed register invalid. There is no result being computed in parallel by this case.

Case 2 is similar to the only other analogous case of the ASM of section 10.8.1 ("normal"—instructions travel down the pipeline), except case 2 must also make the renamed register invalid. There is no result being computed in parallel by this case. The decision that causes case 2 to occur is extremely intricate; thus we will defer those details until we get to the Verilog function. (The motivation here is similar but hopefully even more persuasive than earlier in this chapter: both ASMs and Verilog have their place in the toolset of the designer.) The name of this Verilog function is `depend`, and it decides whether there is a dependency (i.e., case 2) based only on the information in `ir1`, `ir2` and the program status register.

Of the remaining cases, the only one in which truly speculative execution occurs for the instruction in `ir1` is case 5. In cases 3, 4, 6 and 7, it is known whether the instruction in `ir1` will execute. The reason this is known is because of some details in the definition of the `depend` function for case 2. In other words, if you are not in case 1, 2 or 5, the machine has enough information to say with certainty whether the instruction in `ir1` will execute. (It is the responsibility of the designer to make sure this property holds, but let's ignore the details of that for a moment.)

Case 5 is interesting because it is the reason for using a renamed register. The value scheduled to be assigned to the renamed register is the result from the parallel ALU. The function computed by this ALU is based on `ir1[24:21]`. The other ALU uses `ir2[24:21]`. The condition that says whether the value in the renamed register will actually be used in the next clock cycle comes from `ir1[31:28]` in this clock cycle. The tag for the renamed register is scheduled to become the register specified as the destination in this instruction (`ir1[15:12]`). At the next rising edge of the clock after case 5 occurs, `ren_tag` will be the register number that will be modified if this speculatively executed instruction actually executes; `ren_cond` will indicate whether the register indicated by `ren_tag` should change based on the program status register in this next clock cycle and `ren_val` will be that new value.

There is a hidden detail in the `depend` function that relates to case 5. The `depend` function prevents parallel execution if both the instructions in `ir1` and `ir2` set the program status register. Because of this, distinguishing between case 5 versus cases 6 and 7 is simply a matter of looking at `ir1[20]`. If `ir1[20]` and `ir2[20]` indicate both instructions will modify the program status register, case 2 applies, and the instructions will execute sequentially. The reason for this is that both instructions cannot modify the program status register in the same clock cycle, but it is acceptable for each of them to modify the program status register in sequence.

If by the point of the decision `ir1[20]` indicates that this instruction will modify the program status register, we know that `ir2` will not. This means the program status register in the current clock cycle (rather than in the next clock cycle as was the situation for case 5) accurately reflects the information needed to decide whether `ir1` will execute. Therefore, the decision to choose between cases 6 and 7 is `condx(ir1[31:28],psr)`. Note once again the advantage of being able to reuse this Verilog function. If it is known in this clock cycle (case 6) that the instruction will execute, `ren_cond` will become `4'b1110` (always) in the next clock cycle, rather than whatever condition was present in `ir1[31:28]`. If it is known in this clock cycle (case 7) that the instruction will not execute, `ren_cond` will become `4'b1111` (never) in the next clock cycle. This way we can use the same hardware that implements speculative execution also to handle cases 6 and 7.

The reason we cannot use speculative execution here (i.e., making `ren_cond` be `ir1[31:28]`) is that case 6 changes the program status register. If a conditional instruction that changes the program status register (such as ADDPLS) executes due to the current program status information, it is possible register renaming will fail to happen in the next clock cycle because the condition is no longer true. That would prevent an instruction that is supposed to execute from actually executing. Therefore, cases 6 and 7 evaluate `condx` with the current program status register and communicate this unambiguously into the next clock cycle with the `4'b1110` or `4'b1111`.

Cases 5, 6 and 7 have quite a few things in common. In each case, two data processing instructions execute in parallel. This means the program counter needs to increment by eight rather than four. Also two instructions need to be fetched in parallel from the interleaved memory. In each case, ren_val is computed, whether or not it will actually be used later. In theory, for case 7, ren_val need not be computed, but it is easier (and harmless) to do so.

Cases 3 and 4 deal with a branch instruction in ir1. If we reach case 3 or 4, we know (because of the depend function) that the instruction in ir2 will not affect the program status register. (If it does, case 2 applies instead.) Therefore, the decision whether to take the branch can be based on condx(ir1[31:28],psr). The reason is analogous to the decision for cases 6 and 7. If the branch instruction is nullified (case 4), the program counter is incremented by eight and two instruction are fetched (as in cases 5, 6 and 7). If the branch instruction in ir1 occurs (case 3), the instruction pipeline fills with NOPs and the program counter changes (by adding 'POFFSET4 + 4, similar to the 'OFFSET+4 in multi-cycle implementation). In either case 3 or case 4, the branch instruction does not modify a user register; thus the tag for the renamed register becomes 'INVALID.

10.9.7.2   Dealing with register renaming

The second parallel activity of this ASM is to decide whether the value in the renamed register set up in the previous clock cycle (perhaps as the result of speculative execution of an instruction by case 5 of that clock cycle) should take effect permanently in the register file.

There are three conditions that cause the value in the renamed register to be discarded without being written back into the register file:

a.   ir2 has a data processing instruction that stores a newer value into that user register than what is in ren_val (refered to as case 10).
b.   ren_tag indicates 'INVALID.
c.   evaluation of ren_cond in this clock cycle fails (refered to as case 9 of this clock cycle which relates to cases 5 and 7 of the previous clock cycle).

If none of these apply, ren_val is written back into the register indicated by ren_tag (case 8). The parallel write port through which this occurs can be defined in behavioral Verilog as:

```
'define PRD        r[ren_tag]
```

It will also be necessary to define the macros for the read ports so that they forward `ren_val` during this clock cycle:

```
`define OPA  ((condx(ren_cond,psr)&&ir2[19:16]==ren_tag)?ren_val:r[ir2[19:16]])
`define OPB  (ir2[25]?ir2[7:0]:((condx(ren_cond,psr)&&ir2[3:0]==ren_tag)?ren_val:r[ir2[3:0]]))
`define POPA ((condx(ren_cond,psr)&&ir1[19:16]==ren_tag)?ren_val:r[ir1[19:16]])
`define POPB (ir1[25]?ir1[7:0]:((condx(ren_cond,psr)&&ir1[3:0]==ren_tag)?ren_val:r[ir1[3:0]]))
```

Here the only conditions that must be satisfied are that the operand comes from the register described by `ren_tag` (which is guaranteed not to be 'INVALID if `ren_tag` matches the register number) and `ren_cond` evaluates to true in this clock cycle. Even though the renamed register might be discarded after this clock cycle, if the above conditions are satisfied, the renamed register should be forwarded in this clock cycle.

10.9.8   Verilog for the superscalar ARM

This example illustrates the many advantages of using ASMs together with Verilog. It is simply not possible to put all the details of the superscalar design into a one-page ASM. Some of these details need to be placed into Verilog functions or macros. Most of the functions required, such as `condx`, were defined earlier for the multi-cycle implementation, and the fact that we can reuse them is very helpful.

10.9.8.1   The depend function

There is one function that is unique to this superscalar implementation: `depend`. This function recognizes those situations where it is not possible to execute two instructions in parallel:

a.   `ir1` operand (`'POPA`) same as `ir2` destination (`'RD`).
b.   `ir1` operand (`'POPB`) same as `ir2` destination (`'RD`).
c.   `ir1` is conditional non-dp and `ir2` sets `psr`.
d.   `ir1` and `ir2` both set `psr`.
e.   `ir1` is not branch or dp.
f.   `ir2` is not branch or dp.
g.   `ir1` has R15 as operand.

The first three of these are a form of hazard known as Read After Write (RAW). If these instructions executed sequentially, the older instruction (`ir2`) would write a value into a register (`'RD` or `psr`) which the next instruction (`ir1`) must read. To attempt to execute these instructions in parallel would mean `ir2` would read the wrong value.

The fourth condition above is a form of hazard known as Write After Write (WAW). This is a situation that has been warned against throughout this entire book: you cannot have two non-blocking assignments to the same register in the same clock cycle. As explained in section 10.9.7.1, the ASM is designed with the understanding that this situation will never occur in case 5; thus the depend function must cause the ASM to handle such situations in case 2 (i.e., ir2 and ir1 will execute sequentially).

The final three situations deal with instructions for which the ASM was not designed to execute in parallel. Here is the Verilog function that detects these seven conditions that cause the ASM to proceed to case 2:

```verilog
function depend;
 input [31:0] ir1,ir2,psr;
 begin
 depend=(ir2[15:12] == ir1[19:16]
     && ir2[27:26] == 2'b00 && ir1[27:26] == 2'b00
     && condx(ir2[31:28],psr))//POPA bad (RAW)
  || (ir2[15:12] == ir1[3:0] && ir1[25]==0
     && ir2[27:26] == 2'b00 && ir1[27:26] == 2'b00
     && condx(ir2[31:28],psr))//POPB bad (RAW)
  || (ir2[20]
     && ir2[27:26] == 2'b00
     && condx(ir2[31:28],psr)
     && ir1[31:28] != 4'b1110
     && ir1[27:26] != 2'b00)   //psr bad(RAW)non-dp
  || (ir1[20] && ir1[27:26] == 2'b00
     && ir2[20] && ir2[27:26] == 2'b00
     && condx(ir2[31:28],psr))//psr bad(WAW) dp
  || ((ir1[27:26] != 2'b00)
     &&(ir1[27:25] != 3'b101))//ir1 not dp or branch
  || ((ir2[27:26] != 2'b00)
     &&(ir2[27:25] != 3'b101))//ir2 not dp or branch
  || (ir1[27:26] == 2'b00       //ir1 has PC as ALUop
     && ((ir1[3:0] == 4'b1111 && ir1[25]==1'b0)
          ||ir1[15:12] == 4'b1111
          ||ir1[19:16] == 4'b1111));
 end
endfunction
```

Since the goal is to execute as many instructions in parallel as can be executed correctly, it is useful to ignore instructions that are known will be nullified. Since we know with certainty whether ir2 will be nullified (based on the current program status register), conditions a-d (which mention ir2) can be ANDed with `condx(ir2[31:28],psr)`. This means the `depend` function only slows the machine to one instruction per clock cycle when it is actually necessary. For example, the following two instructions

```
ADDPL   R1,R1,1
ADD     R2,R1,1
```

can be processed in parallel if the ADDPL is nullified but must execute sequentially if the ADDPL is not nullified.

10.9.8.2   Translating the ASM to Verilog

Once all the macros and functions are defined, it is easy to translate the ASM to Verilog. For example, the following is the beginning portion of the Verilog code corresponding to the first parallel activity of state F1 (parallel and speculative execution):

```
  ...
  begin
   if (condx(ir2[31:28],psr) &&
     ((ir2[27:25] == 3'b101)
     || (ir2[27:26] == 2'b00 &&
         ir2[15:12] == 4'b1111)))
    begin
    ir1 <= @(posedge sysclk) 32'hf0000000;
    ir2 <= @(posedge sysclk) 32'hf0000000;
    ren_tag <= @(posedge sysclk) `INVALID;
    `ifdef DEBUG
      $display(
       " 1. ir2 branch or R15 prevents ||",$time);
      cover(1);
    `endif
    end
   else ...
```

10.9.8.3   Code coverage

The power of Verilog is twofold. First, Verilog allows the designer to express the behavior of the hardware abstractly, without having to consider too many details. Second, Verilog allows a designer to test the design using simulation. It would be an underutilization of Verilog to synthesize a design without having ever simulated it. So, the designer has two responsibilities. The first responsibility, of course, is to design the hardware. The second responsibility, which is more important but sometimes neglected by careless designers, is the test code. This code, sometimes referred to as the *testbench*, exercises the Verilog that simulates the hardware. Ideally, we would like to try every possible case. For tiny special-purpose machines, such as the 12-bit childish division test code example in section 3.7.3, this is **barely** possible. For a **general-purpose** machine, it is **impossible to test everything**. The usefulness of simulation, however, depends on how completely the Verilog code that simulates hardware has been tested by the Verilog test code. It does not do any good to test the same correct Verilog statement a million times but ignore another statement that has a bug in it. The advantage of Verilog is that its software-like statements can be used to warn the designer that parts of the Verilog code that simulates hardware has not been tested.

The superscalar implementation given in the last section is far more complex than any of the earlier designs in this book. It is not feasible to test every possible program to see if it works correctly. We will create several programs, but then Verilog will inform us whether all of the cases we are interested in have been tested. The reason for doing this is that the designer will more than likely make a mistake in guessing what cases a moderately complex program will test. The operation of the superscalar machine is so counterintuitive (even on a small program) that it is better for Verilog to keep track of what is being tested.

10.9.8.4   Using `ifdef` for the `cover` task

The Verilog code, of which a portion was shown in section 10.9.8.2, has several statements that compile conditionally, such as:

```
`ifdef DEBUG
  $display(
    " 1. ir2 branch or R15 prevents ||",$time);
  cover(1);
`endif
```

What this means is that if the macro is defined:

```
`define DEBUG
```

the call to the system task, $display, and the cover task (described below) will be compiled with the rest of the Verilog. If this `define were omitted from the top of the file, it would be as though the $display and cover tasks were not there. The reason for using `ifdef is that once simulation is correct, the macro can be undefined, and the Verilog will no longer exhibit these test actions.

Note that `ifdef is different than an if statement (where the tasks would be compiled, but might not execute). In particular, `ifdef can be used to alter which control statements are compiled into the code. For example, cases 9 and 10 of the renaming parallel activity do nothing:

```
`ifdef DEBUG
else
 begin
  $display("10. dp overwrites renamed r%d",
    ren_tag,$time);
  cover(10);
 end
`endif
```

If 'DEBUG is not defined, there is no need for the else begin ... end to be compiled. The above shows how the scope of statements that are conditionally compiled can cross begin end boundaries. This is possible because the substitution occurs at compile time.

In addition to calling on these tasks, the cover task has to be defined. We only want to define it if the 'DEBUG macro is defined. Therefore, the task and everything associated with it will be enclosed in the `ifdef:

```
`ifdef DEBUG
  reg ['MAX_CASE_NO:0] coverage_set;
  task cover;
    input case_no;
    integer case_no;
    begin
      coverage_set = coverage_set |
          ((1 << 'MAX_CASE_NO) >> case_no);
    end
  endtask

  initial
    begin
      coverage_set = 0;
      wait(halt === 1'b0);
      wait(halt === 1'b1);
      $display("coverage=%b",
        coverage_set['MAX_CASE_NO-1:0]);
    end
`endif
```

Notice how this is completely different from an `if` statement. An `if` can only exist inside a behavioral block or task. This `'ifdef` is outside the task and the `initial` block. If 'DEBUG is not defined, the `reg`, the `cover` task and the `initial` block will not be defined.

Each case of interest in the code has a number between 1 and 'MAX_CASE_NO, which is used to identify that case in the call to `cover`. (There is no case 0.) What `cover` does is to OR the corresponding bit of the coverage set with one.

After the program has halted, the initial block prints out the coverage set. The more cases of the Verilog code that were covered by the program, the more ones there will be in the coverage set. We will run several programs in order to obtain complete coverage. (In reality, we have not considered enough test cases here to have total confidence in this design, but this task can be expanded to cover an arbitrary number of cases.)

10.9.9   Test programs

The special-purpose machines in chapters 4 and 5 are easier to test than a general-purpose machine because the test code simply has to supply test data to the machine being tested. A special-purpose machine is supposed to follow some algorithm that manipulates the data, and it is often easy to tell if the result is correct. A general-purpose machine implements a (sometimes intricate) variation of fetch/execute which in turn interprets a program that manipulates the data. It is much harder to tell if a general-purpose machine is correct.

10.9.9.1   A test of R15

One of the details of the ARM instruction set that takes special consideration in all the implementations is R15. R15 is, in fact, just another name for the program counter. The Verilog macros for all the above implementations have to take this into account. Additionally, on the pipelined implementation, R15 as a destination causes the pipeline to fill with NOPs just as though a branch had occurred. On the superscalar implementation, there are many places where special consideration is given to R15.

Therefore, we need an ARM program that exercises at least some of this Verilog code involving `r[15]`. Here is such a program:

```
'ifdef PROGRAM1 //test R15 source and destination
 arm7_machine.m[0]=32'he3b00000; //MOVS R0,0
 arm7_machine.m[1]=32'he08f1000; //ADD  R1,R15,R0
 arm7_machine.m[2]=32'he080200f; //ADD  R2,R0,R15
 arm7_machine.m[3]=32'he08f3000; //ADD  R3,R15,R0
 arm7_machine.m[4]=32'he080400f; //ADD  R4,R0,R15
```

Continued

```
arm7_machine.m[5]=32'he3b0e0ff; //MOVS R14,0xff
arm7_machine.m[6]=32'he2400008; //SUB  R0,R0,8
arm7_machine.m[7]=32'he1a0f001; //MOV  R15,R1
'endif
```

A Verilog macro, 'PROGRAM1, is defined when this is the program we want to use to test the machine with. This test code can be used with any of the implementations. For example, the pipelined implementation produces the following:

```
. . .
PC=00000024 IR1=eafffffe IR2=e1a0f001 N=0           1251
  r0=fffffff8 r1=0000000c r2=00000010 r3=00000014 r4=00000018
PC=0000000c IR1=f0000000 IR2=f0000000 N=0           1351
  r0=fffffff8 r1=0000000c r2=00000010 r3=00000014 r4=00000018
PC=00000010 IR1=e08f3000 IR2=f0000000 N=0           1451
  r0=fffffff8 r1=0000000c r2=00000010 r3=00000014 r4=00000018
PC=00000014 IR1=e080400f IR2=e08f3000 N=0           1551
  r0=fffffff8 r1=0000000c r2=00000010 r3=00000014 r4=00000018
PC=00000018 IR1=e3b0e0ff IR2=e080400f N=0           1651
  r0=fffffff8 r1=0000000c r2=00000010 r3=0000000c r4=00000018
```

Notice the contents of the registers at $time 1251 and the NOPs in the pipeline at $time 1351. Also notice the contents of the registers at $time 1651. On the other hand, the superscalar implementation produces equivalent results in an entirely different way:

```
. . .
PC=00000024 IR1=eafffffe IR2=e1a0f001 N=0            1051
r0=fffffff8 r1=0000000c r2=00000010 r3=00000014 r4=00000018
1.ir2 branch or R15 prevents ||                     1051
other DP instructions...
5.use || ALU noS ir1=f0000000 A=fffffff8 B=fffffff8 1151
PC-0000000c TR1=f0000000 IR2=f0000000 N=0           1151
r0=fffffff8 r1=0000000c r2=00000010 r3=00000014 r4=00000018
PC=00000014 IR1=e080400f IR2=e08f3000 N=0
  cond=f,ren_R 0 =00000000                          1251
r0=fffffff8 r1=0000000c r2=00000010 r3=00000014 r4=00000018
2.depend prevents ||                                1251
9.nullify renamed r 0 ||dp                          1251
6.use || ALU S ir1=e3b0e0ff A=fffffff8 B=000000ff   1351
PC=00000018 IR1=e3b0e0ff IR2=e080400f N=0           1351
r0=fffffff8 r1=0000000c r2=00000010 r3=0000000c r4=00000018
PC=00000020 IR1=e1a0f001 IR2=e2400008 N=0
```

Continued

```
   cond=e,ren_R14 =000000ff                                          1451
r0=fffffff8 r1=0000000c r2=00000010 r3=0000000c r4=00000010
2.depend prevents ||                                                1451
8.writeback renamed r14 ||dp                                        1451
```

Notice the values of the registers at $time 1051 and the NOPs in the pipeline at $time 1151. Also notice the value of the registers at $time 1451. These correspond to what was visible in the pipelined implementation at $time 1251, 1351 and 1651. This program does not execute much faster on the superscalar implementation than on the pipelined implementation because the superscalar implementation properly prevents execution of more than one instruction per clock cycle when R15 is involved. This program is important as a means of testing the Verilog code because it illustrates that case 2 happens due to the following portion of the depend function:

```
|| (ir1[27:26] == 2'b00
  && ((ir1[3:0] == 4'b1111 && ir1[25]==1'b0)
     ||ir1[15:12] == 4'b1111
     ||ir1[19:16] == 4'b1111));//ir1 has PC as ALUop
```

The coverage set for this program is 1100110101; in other words, cases 1, 2, 5, 6, 8 and 10 are covered. Notice how helpful the output from the $displays is in annotating why these cases occur.

10.9.9.2 Our old friend: division

The last program has no practical value other than that of testing the operation of certain ARM instructions. The remaining programs that we will consider will be based on the childish division algorithm used in earlier chapters. Like the PDP-8, the ARM does not have a divide instruction in the hardware; thus to do division requires some software. Although if speed were essential one would choose a faster algorithm for division than the childish algorithm, this simple algorithm illustrates many of the properties that more sophisticated algorithms also have. This is the reason why it has been implemented many times in this book, in both hardware and software. Here again is the childish division algorithm in C:

```
    r1=x;
    r2=0;
    while (r1>=y)
      {
        r1 = r1 - y;
        r2 = r2 + 1;
      }
```

Since we have not implemented the memory reference instructions of the ARM, the values of x and y must be constants. We will use immediate addressing with x and y. For example, let us assume that x is 14 and y is 7, which is a typical test case used with this algorithm in chapters 2, 4, 5, 8 and 9. It is natural for the r1 and r2 high-level variables to reside in R1 and R2, respectively. R1 and R2 are the registers that the assembly language programmer refers to, but these are r[1] and r[2] in Verilog. Also, it will be convenient for y to reside in R4. The implementations of this algorithm given in chapters 8 and 9 made use of the accumulator of the PDP-8 to contain the difference. We need to have a similar register on the ARM. In the following program, let us use R0 to serve the same role as as the accumulator. This illustrates an important property of all RISC machines (not just the ARM): there is nothing special about R0—we could have chosen any other available register to hold the difference. The following is an ARM program that implements this algorithm in the most straightforward way possible:

```
00000000/e3a0100e     MOV    R1,0x0e
00000004/e3a04007     MOV    R4,0x07
00000008/e3a02000     MOV    R2,0x00
0000000c/e0510004 L1  SUBS   R0,R1,R4
00000010/4a000002     BMI    L2
00000014/e1a01000     MOV    R1,R0
00000018/e2822001     ADD    R2,R2,0x01
0000001c/eafffffa     B      L1
00000020/ef000000 L2  SWI
```

The above is analogous to the PDP-8 program given in section 8.3.2.5.3. The first three MOV instructions set up R1, R4 and R2 to their initial values of x (14), y (7) and zero, respectively. The SUBS is the only instruction that sets the program status register. The purpose of the SUBS instruction is twofold: to compute the difference and to see if R1>=R4. The BMI makes use of this program status information. As long as R1 >= R4, the BMI is nullified and the loop continues. The difference would then be moved from R0 to R1, and R2 is incremented. The unconditional branch to the label L1 causes the test at the top of the loop to happen again. This loop repeats while the difference (in R0) is greater than zero. The BMI branches to the label L2 when the difference in R0 is negative, at which point the SWI causes the Verilog test code to finish (using the halt flag that would not exist on a an actual ARM).

By defining 'PROGRAM5 in the test code, appropriate Verilog assignment statements (not shown) would place the above machine language instructions into memory. This program can be used with any of the implementations of the ARM. For example, the pipelined implementation produces the following:

```
...
PC=00000014 IR1=4a000002 IR2=e0510004 N=0          2251
 r0=00000000 r1=00000000 r2=00000002 r3=xxxxxxxx r4=00000007
PC=00000018 IR1=e1a01000 IR2=4a000002 N=1          2351
 r0=fffffff9 r1=00000000 r2=00000002 r3=xxxxxxxx r4=00000007
PC=00000020 IR1=f0000000 IR2=f0000000 N=1          2451
 r0=fffffff9 r1=00000000 r2=00000002 r3=xxxxxxxx r4=00000007
PC=00000024 IR1=ef000000 IR2=f0000000 N=1          2551
 r0=fffffff9 r1=00000000 r2=00000002 r3=xxxxxxxx r4=00000007
PC=00000028 IR1=xxxxxxxx IR2=ef000000 N=1          2651
 r0=fffffff9 r1=00000000 r2=00000002 r3=xxxxxxxx r4=00000007
```

The loop executes two times. The third execution of the SUBS ($time 2251) produces a negative number (fffffff9), which sets the N flag. This in turn causes the BMI to branch. Not show earlier, the BMI had been nullified. On the other hand, running 'PROGRAM5 on the superscalar implementation produces equivalent results in an entirely different way:

```
...
2.depend prevents ||                                     1351
10.dp overwrites renamed r 0                             1351
PC=00000014 IR1=4a000002 IR2=e0510004 N=0
  cond=f,ren_R 0 =00000000                               1351
r0=00000000 r1=00000000 r2=00000002 r3=xxxxxxxx r4=00000007
PC=00000018 IR1=e1a01000 IR2=4a000002 N=1                1451
r0=fffffff9 r1=00000000 r2=00000002 r3=xxxxxxxx r4=00000007
1.ir2 branch or R15 prevents ||                          1451
other DP instructions...
5.use || ALU noS ir1=f0000000 A=fffffff9 B=fffffff9 1551
PC=00000020 IR1=f0000000 IR2=f0000000 N=1                1551
r0=fffffff9 r1=00000000 r2=00000002 r3=xxxxxxxx r4=00000007
PC=00000028 IR1=xxxxxxxx IR2=ef000000 N=1
  cond=f,ren_R 0 =00000000                               1651
r0=fffffff9 r1=00000000 r2=00000002 r3=xxxxxxxx r4=00000007
2.depend prevents ||                                     1651
9.nullify renamed r 0 ||dp                               1651
```

The contents of the registers in the superscalar implementation at $time 1651 are the same as the registers in the pipelined implementation at $time 2651. The output from the $display statements explain the intricate way in which the superscalar machine was able to produce the correct answer in less time. The depend function only slows the machine down (case 2) three times. For example, at $time 1351, depend detects the conditional branch (BMI in this example) following the data processing instruction that sets the program status register (SUBS in this case). At $time 1641, depend detects SWI. The Verilog coverage set for the superscalar run of 'PROGRAM5 is 1110110111. This means all but cases 4 and 7 were exercised.

10.9.9.3   Faster childish division

In section 9.7, a variation of the childish division algorithm was given that illustrates a different way of implementing a while loop in software (testing at both the top and bottom of the loop). The effect of this is to reduce the number of times the branch penalty is incurred and to reduce the number of nullified branch instructions. The following ARM program implements this approach by using both the BPL and BMI instructions:

```
00000000/e3a0100e    MOV    R1,0x0e
00000004/e3a02000    MOV    R2,0x00
00000008/e3a04007    MOV    R4,0x07
0000000c/e0510004    SUBS   R0,R1,R4
00000010/4a000003    BMI    L2
00000014/e1a01000 L1 MOV    R1,R0
00000018/e2822001    ADD    R2,R2,0x01
0000001c/e0510004    SUBS   R0,R1,R4
00000020/5afffffb    BPL    L1
00000024/ef000000 L2 SWI
```

Running on the pipelined implementation, this program (let's refer to it as 'PROGRAM4) produces at $time 2051 the same results that 'PROGRAM5 produces (also running on the pipelined implementation) at $time 2651. This illustrates that to make good use of a pipelined machine, a good compiler is essential. Manually created assembly language programs are often not as effective as the automatically created machine language from compilers. Running on this superscalar implementation, this program produces at $time 1451 the same results that 'PROGRAM5 (also running on the superscalar implementation) produces at $time 1651.

The Verilog coverage set for the superscalar run of 'PROGRAM4 is 1100110110. 'PROGRAM4 does not add to the coverage of the Verilog code provided by 'PROGRAM5 (1110110111); thus we need an additional test program to cover cases 4 and 7.

10.9.9.4 *Childish division with conditional instructions*

One of the big advantages of the ARM instruction set is that any instruction can be made to execute conditionally. In the previous program, only the branch instructions executed conditionally, but with the ARM, the programmer is free to specify that any instruction, such as a data processing instruction, should execute only when a certain condition is true. The importance of this is that conditional execution does not incur any branch penalty.

One technique that compilers use to improve the performance of high-level software running on a pipelined and/or superscalar processor is *loop unrolling*. In C programs using `for` loops, such as:

```
for(i=0;i<3;i++)
  {
    r1 = r1 - y;
    r2 = r2 + 1;
  }
```

the compiler knows a priori how many times the loop will execute; thus the unrolled code looks like:

```
r1 = r1 - y;
r2 = r2 + 1;
r1 = r1 - y;
r2 = r2 + 1;
r1 = r1 - y;
r2 = r2 + 1;
```

which can be implemented without branch penalty:

```
SUB    R1,R1,R4
ADD    R2,R2,0x01
SUB    R1,R1,R4
ADD    R2,R2,0x01
SUB    R1,R1,R4
ADD    R2,R2,0x01
```

If the program has a `for` loop that repeats too many times for it to be practical to unroll the loop completely, it can be partially unrolled. For example:

```
for (i=0;i<3000;i++)
    {
        r1 = r1 - y;
        r2 = r2 + 1;
    }
```

is the same as:

```
for (i=0;i<1000;i++)
    {
        r1 = r1 - y;
        r2 = r2 + 1;
        r1 = r1 - y;
        r2 = r2 + 1;
        r1 = r1 - y;
        r2 = r2 + 1;
    }
```

which would incur the branch penalty one-third as often.

The difficulty with the childish division algorithm (and with many practical programs) is that we do not know before we run the program how many times the loop will execute. (In the case of childish division, the number of times the loop will execute is the answer we are trying to compute.)

Here is where conditional data processing instructions come in handy. Assuming we do not care about the result in r1, the childish division algorithm can be partially unrolled as the following C code:

```
r1=x;
r2=0;
do
    {
        r1 = r1 - y;
        if (r1>=0) r2 = r2 + 1;
        if (r1>=0) r1 = r1 - y;
        if (r1>=0) r2 = r2 +1;
    } while (r1>=0)
```

Each time through this loop is equivalent to two times through the loop in the original version of the childish division program. This program (refered to as 'PROGRAM3) incurs the branch penalty half as often provided that the three if statements are translated into conditional data processing instructions:

```
00000000/e3a0100e       MOV    R1,0x0e
00000004/e3a04007       MOV    R4,0x07
00000008/e3a02000       MOV    R2,0x00
0000000c/e0511004 L1 SUBS    R1,R1,R4
00000010/52822001       ADDPL  R2,R2,0x01
00000014/50511004       SUBPLS R1,R1,R4
00000018/52822001       ADDPL  R2,R2,0x01
0000001c/5afffffa       BPL    L1
00000020/ef000000 L2 SWI
```

The reason that the ARM provides the ability to either set the program status register (bit 20 equal to one) or leave the program status register alone (bit 20 equal to zero) is so that several instructions can be made conditional on the same condition. In this program, the SUBS (and possibly the SUBPLS) determine whether $R1 >= 0$. The ADDPL and SUBPLS instructions use this program status information to decide whether to execute. Since the pipelined and superscalar implementations allow execution of conditional data processing instructions without branch penalties, such techniques can often speed up a program.

The Verilog coverage set for the superscalar run of 'PROGRAM3 is 0110110110. 'PROGRAM3 does not add to the coverage of the Verilog code provided by 'PROGRAM5 (1110110111); thus we need to do a different test to cover cases 4 and 7. One such test is identical to 'PROGRAM3, except R1 is loaded with 6 rather than 14 as the value of x.

```
00000000/e3a01006       MOV    R1,0x06
```

The coverage set for the superscalar run of this modified program is 0101111110, which covers cases 4 and 7. Therefore, all of the ten cases identified in the source code have been tested at least once. This is not to say that the overall design is correct, but at least we know we have checked all the Verilog statements translated from the original ASM. Verilog has helped us make sure that all the code has been covered.

10.10   Comparison of childish division implementations

Determining how long it takes for each of the above division programs ('PROGRAM3, 'PROGRAM4 and 'PROGRAM5) to execute is tedious, especially on the pipelined and superscalar versions of the ARM. A better approach is to let Verilog measure the time for a range of input values:

```
`ifdef DIVTEST
  cont = 0;
  t = 0;
  for (x=0; x<=42; x = x + 1)
    begin
      arm7_machine.m[0] =
        (arm7_machine.m[0] & 32'hffffff00) + x;
      arm7_machine.r[15] = 0;
      #200 cont = 1;
      #100 cont = 0;
      #400 wait(arm7_machine.halt);
      if (arm7_machine.r[2] != x/7)
        $display("error");
      $display("x=%d cl=%d r2=%d %d",x,
        ($time-t)/100,arm7_machine.r[2], $time);
      t = $time;
    end
  $finish;
`endif
```

The above Verilog modifies the MOV immediate at address 0 to initialize different values of x, that range from 0 to 42 and causes the `arm7_machine` to run the modified program. If the quotient (in r[2]) is not erroneous, the Verilog code simply prints the number of clock cycles (of period 100) elapsed since the machine language program started running for the given value of x. To use the above code, 'DIVTEST, as well as the macro for the desired machine language program ('PROGRAM3, 'PROGRAM4 or 'PROGRAM5), must be defined. When each of these three programs is run on each of the three implementations (multi-cycle, pipelined and superscalar), we obtain the following data:

run	specific clock cycles, for given quotient						upper bound for quotient > 0
	0	1	2	3	4	5 ...	
M 5	33	57	81	105	129	153 ...	24*quotient + 33
M 4	33	51	71	91	111	131 ...	20*quotient + 31
M 3	44	46	69	71	94	96 ...	12.5*quotient + 44
P 5	13	20	27	34	41	48 ...	7*quotient + 13
P 4	13	15	21	27	33	39 ...	6*quotient + 9
P 3	14	14	21	21	35	35 ...	3.5*quotient + 14
S 5	9	13	17	21	25	29 ...	4*quotient + 9
S 4	10	11	15	19	23	27 ...	4*quotient + 7
S 3	9	9	13	13	17	17 ...	2*quotient + 9

The column on the left ("run") indicates which program (3, 4 or 5) was run on which machine ("M" for multi-cycle, "P" for pipelined, or "S" for superscalar). For example, "S 3" indicates `PROGRAM3 was run on the superscalar implementation. The column on the right is an equation of an upper bound on this data for quotient > 0. Let us look back at the interesting journey we have traveled with our old friend, the childish division algorithm. The following table summarizes implementations of the childish division algorithm given in this and earlier chapters:

register f/e	bit data	P	S	kind	hard sect	soft sect	upper bound on clock cycles	
0	2	12	n	n	Mealy	5.2.4	n/a	2 +	quotient
0	2	12	n	n	Mealy	5.2.3	n/a	3 +	quotient
0	3	12	n	n	Moore	2.2.7	n/a	3 +	quotient
0	2	12	n	n	Moore	2.2.3	n/a	2 +	2*quotient
0	2	12	n	n	Mealy	5.2.1	n/a	2 +	2*quotient
4	15	31	y	y	Mealy	10.9.6	10.9.9.4	9 +	2*quotient
0	3	12	n	n	Moore	2.2.5	n/a	2 +	3*quotient
0	2	12	n	n	Moore	2.2.2	n/a	3 +	3*quotient
3	15	31	y	n	Mealy	10.8.1	10.9.9.4	14+	3.5*quotient
4	15	31	y	y	Mealy	10.9.6	10.9.9.2	9 +	4*quotient
4	15	31	y	y	Mealy	10.9.6	10.9.9.3	7 +	4*quotient
3	15	31	y	n	Mealy	10.8.1	10.9.9.3	9 +	6*quotient
3	15	31	y	n	Mealy	10.8.1	10.9.9.2	13+	7*quotient
4	1	11	y	n	Mealy	9.6	9.7	12+	10*quotient

Continued

3	15	31	n	n	Mealy 10.7	10.9.9.4	44+	12.5*quotient
3	15	31	n	n	Mealy 10.7	10.9.9.3	31+	20*quotient
3	15	31	n	n	Mealy 10.7	10.9.9.2	33+	24*quotient
4	1	11	n	n	Moore 8.3.2.1	9.7	55+	55*quotient
4	1	12	n	n	Moore 8.3.2.1	8.3.2.5.3	88+	75*quotient

The "register" columns indicate how many "data" registers and how many "f/e" (fetch/execute) registers are used. (The number of "f/e" registers is 0 for special-purpose hardware; the number of "data" registers is 1 for a single-accumulator machine like the PDP-8 and much larger (e.g., 15) for a RISC machine like the ARM.) The "bit" column shows maximum number of bits the implementation allows for x or y. (For software implementations that use the sign bit (the PDP-8's SPA/SMA or the ARM's N flag), this is one less than the register size.) The "S" column indicates whether the hardware is superscalar. The "P" column indicates whether the hardware is pipelined. (Remember a superscalar implementation also uses pipelining.) The "kind" column indicates whether the ASM uses conditional commands (Mealy) or not (Moore). The "hard sect" column indicates where the ASM is described. (The ASM implements the childish division algorithm for special-purpose hardware but the ASM implements fetch/execute for general-purpose hardware.) The "soft sect" applies only to general-purpose computer implementations and describes the machine language for the childish division algorithm. The "upper bound on clock cycles" column indicates how long it takes to compute the quotient. This table is sorted by the linear coefficient of the upper bound; thus for large quotient, the order in this table indicates the relative speed of the machines, assuming that the clock frequency of each machine is the same. (The clock frequencies may be different due to different propagation delays in each architecture, as described in chapter 6, but we will assume the clock frequency is the same here.)

There are several interesting things to note in the above table. First, special-purpose hardware is cheaper (number of registers) than general-purpose hardware, especially for the faster kinds of general-purpose hardware (pipelined and superscalar). Second, special-purpose hardware is faster than software running on general-purpose machines except that the superscalar ARM running 'PROGRAM3 is faster than the special-purpose hardware described in sections 2.2.2 and 2.2.5. Third, the expensive superscalar implementation is competitive with cheap special-purpose hardware only for 'PROGRAM3 (with its loop unrolling). This illustrates that to capitalize on sophisticated general-purpose hardware requires a good compiler. Fourth, all things being equal, Mealy machines tend to take fewer clock cycles than Moore machines. Fifth, a single-accumulator multi-cycle general-purpose machine (PDP-8) is slower than a RISC multi-cycle general-purpose machine (ARM) because the latter needs fewer instructions to carry out the algorithm. Sixth, pipelining improves the speed of a general-purpose machine. Seventh, pipelining the single-accumulator PDP-8 makes it faster than the multi-cycle ARM but slower than the pipelined ARM. The equivalent of 'PROGRAM4' on the multi-cycle PDP-8 takes `55+55*quotient` clock cycles while the same program on the pipelined PDP-8 takes `12+10*quotient` clock cycles.

10.11   Conclusions

This chapter has compared three different implementations for a RISC instruction set, using a small subset of the ARM as the example hardware and the childish division algorithm as the example software. A multi-cycle implementation requires several cycles to execute an instruction. A pipelined implementation requires one cycle to execute an instruction, except for an instruction such as branch. A superscalar implementation attempts to execute more than one instruction per clock cycle whenever possible.

A RISC machine provides a large set of registers available to the programmer, and an instruction set that allows three register operands to be specified in a single instruction. In comparison to a single-accumulator machine (like the PDP-8), this tends to reduce the number of instructions required to implement an algorithm and to enhance the chance that adjacent instructions will be independent of each other. This latter property makes the design of superscalar general-purpose machines feasible.

Superscalar implementations often use speculative execution, where the result of an instruction is computed before it is known whether that instruction will actually execute. Rather than storing the result in the actual register specified by the instruction, the physical register where this speculative result resides will be renamed to act as the destination if and only if the corresponding instruction actually executes. The superscalar example in this chapter is highly simplified. Commercial superscalar machines are often much more aggressive with speculative execution, using techniques such as *branch prediction*, where the machine executes instructions before it is known whether the branch to those instructions will actually occur, and *out of order execution*, where more instructions are issued (fetched into the pipeline) than can be retired (executed) per clock cycle. The beauty of the ARM's conditional instructions is that they allow us to illustrate the same principles of speculative execution with much simpler hardware.

The design of a superscalar processor is considerably more complex than the design of a pipelined or multi-cycle processor. Because of this, use of a hardware description language such as Verilog is helpful. Through the use of macros and functions, Verilog source code allows the designer to hide unnecessary details early in the design process, yet have those details fully specified in the final source code. Through the use of tasks, the Verilog designer can make sure that the test code covers all cases the designer considers important.

10.12 Further reading

VLSI TECHNOLOGY, INC., *Acorn RISC Machine (ARM) Family Data Manual*, Prentice Hall, Englewood Chiffs, NJ, 1990. Provides documentation on an early version known as the ARM2, which had a three-stage pipeline. How to access documentation about more current versions is given in appendix G.

10.13 Exercises

10-1. The following Verilog code for a special-purpose machine describes the register transfers carried out by four ARM instructions followed by two NOPs run on the superscalar general-purpose ARM. What are these four instructions?

```
@(posedge sysclk) #1;
 r[2] <= @(posedge sysclk) r[1] + r[4];
 psr <= f(r[1] + r[4]);
 ren_val <= @(posedge sysclk) r[3] - r[4];
 ren_tag <= @(posedge sysclk) 3;
 ren_cond <= @(posedge sysclk) `PL;
@(posedge sysclk) #1;
 ren_val <= @(posedge sysclk) r[5] + 5;
 ren_tag <= @(posedge sysclk) 5;
 if ((ren_cond == `PL)&&(psr[31]==0) ||
     (ren_cond == `MI)&&(psr[31]==1) ||
     (ren_cond == `AL))
  begin
   r[6] <= @(posedge sysclk) ren_val;
   r[ren_tag] <= @(posedge sysclk) ren_val;
  end
 else
   r[6] <= @(posedge sysclk) r[3];
@(posedge sysclk) #1;
 if ((ren_cond == `PL)&&(psr[31]==0) ||
     (ren_cond == `MI)&&(psr[31]==1) ||
     (ren_cond == `AL))
   r[5] <= @(posedge sysclk) ren_val;
```

10-2. In problem 10-1, which registers are involved with renaming?

10-3. In problem 10-1, which of the seven cases described in section 10.9.7.1 applies to each state of the special-purpose machine?

10-4. In problem 10-1, which of the instructions is executed speculatively?

10-5. The register file for the multi-cycle and pipelined ARM is a multi-port memory with two read ports and one write port (similar to that described in section 9.8), except the program counter (r[15]) must be able to be incremented (by 4) or loaded (with r[15] plus an externally supplied 26-bit signed value) independently of the operations that occur on the other ports. Also, there must be a port supplying r[15] as the address to memory. Assume that the register file has command inputs ldPC and incPC to deal with these special operations:

ldPC	incPC	action
0	0	r[15] depends on write port
0	1	r[15]←r[15]+4
1	0	r[15]←r[15]+external
1	1	impossible

The external input is 'OFFSET4 in the pipelined version and 'OFFSET4+4 in the multi-cycle version. Draw a block diagram that implements this synchronous register file.

10-6. Using a register file of the kind given in problem 10-5, design an architecture for the multi-cycle ARM subset given in section 10.7, and give the corresponding mixed ASM.

10-7. Using a register file of the kind given in problem 10-5, design an architecture for the pipelined ARM subset given in section 10.8.1, and give the corresponding mixed ASM.

10-8. The register file for the superscalar ARM is a multi-port memory with four read ports and two write ports. The program counter (r[15]) must be able to be incremented (by 4 or 8) or loaded (with r[15] plus an externally supplied 26-bit signed value plus either 0 or 4) independently of the operations that occur on the other ports. Assume that the register file has command inputs ldPC, incPC and plus4PC to deal with these special operations:

ldPC	incPC	plus4PC	action
0	0	0	r[15] depends on write ports
0	1	0	r[15]←r[15]+4
0	1	1	r[15]←r[15]+4+4
1	0	0	r[15]←r[15]+'OFFSET4
1	0	1	r[15]←r[15]+'OFFSET4+4
1	1	–	impossible

Draw a block diagram that implements this synchronous register file.

10-9. The interleaved memory described in section 10.9.2 has two conventional memories. For this problem, since the program does not change during execution, we will replace these memories with ROMs (odd_m and even_m). One of the ROMs is for words whose `addr/4` is odd. The other ROM is for words whose `addr/4` is even. The problem is we cannot predict whether the CPU will need the odd and even instructions fetched into `ir1` and `ir2` or vice versa. Give a block diagram for the interleaved memory that overcomes this problem using three muxes and an incrementor in addition to the ROMs.

10-10. Using a register file of the kind given in problem 10-8 and an interleaved memory of the kind described in problem 10-9, design an architecture for the superscalar ARM subset given in section 10.9.6, and give the corresponding mixed ASM.

10-11. As explained in appendix G, the ARM is actually a Princeton machine, which stores its program and data in the same memory. Like many other RISC machines, the ARM does not allow computation on values in memory. Rather, it only allows load and store instructions. The two most important instructions of this kind are LDR (`ir[27:26]==1&ir[20]==1`) and STR (`ir[27:26]==1&ir[20]==0`). There are several addressing modes available, but for this problem only consider the simple indexed addressing mode (`ir[24:21]` can be ignored in this problem) that accesses `m['OPA+'OPB]`. Assuming a single-port memory of the kind described in section 8.2.2.3.2, give multi-cycle behavioral Verilog to implement such LDR and STR instructions along with the other instructions described in 10.7. Create appropriate test code.

10-12. Assuming a multi-port memory of the kind described in section 9.6, modify the pipelined behavioral Verilog of section 10.8.7 to implement the LDR and STR instructions described in problem 10-11. Create appropriate test code. Unlike chapter 9, operand fetch does not occur until the execution stage of the pipeline, because the ARM has a *load* instruction (LDR), rather than the addition instruction (TAD) of the PDP-8 which required an extra stage to complete. This is important because 'OPA or 'OPB may not be available until that final clock cycle. For the same reason, a STR followed by a LDR from the same address will not require forwarding .

10-13. Rework problem 10-7 to support the instructions of problem 10-12. Note that for the STR instruction there will need to be a mux that provides `ir2[15:12]` to one of the read ports of the register file.

10-14. Using a multi-port memory like that in section 9.6 for problem 10-12 may be too expensive. Design a memory hierarchy, consisting of two direct mapped caches (section 8.5) and a main memory (that takes five cycles per access). One cache is for data manipulated by LDR and STR instructions, and uses the `read`, `write`, `memreq`, `memrack` and `memwack` signals described in section 8.5.3. The other cache is only for instructions being fetched, and uses `ireq` (which combines the roles of `read` and `memreq` for this cache) and `imemack`. You may assume that no machine language instruction will be modified during the execution of the program so that there is no need for write-through with the instruction cache.

10-15. Draw a pure behavioral ASM chart which combines problems 10-12 and 10-14. In the event that an instruction is not in the instruction cache, let NOP(s) enter the pipeline to stall until `imemack` is asserted. In the event that an LDR executes when the operand is not in the data cache, use a wait loop similar to those in section 8.5.2. For STR instructions, use a *write buffer*, which consists of registers that hold the memory address and contents while it is being written. The second of two successive STR instructions will go to a wait state only if the first is still being processed by the memory hierarchy. Because of the write buffer an STR followed by a LDR from the same address creates a dependency that will require forwarding.

10-16. Assuming a powerful multi-port memory of some kind exists, modify the superscalar behavioral Verilog of section 10.9.8.2 to implement the LDR and STR instructions described in problem 10-11. Create appropriate test code.

10-17. Modify the multi-cycle behavioral Verilog of section 10.7.6 to implement the remaining data-processing instructions described by appendix G. Give test code.

10-18. Modify the pipelined behavioral Verilog of section 10.8.7 to implement the remaining data-processing instructions described by appendix G. Give test code.

10-19. Modify the superscalar behavioral Verilog of section 10.9.8.2 to implement the remaining data processing instructions described by appendix G. Give test code.

10-20. The ARM has two multiplication instructions which are identified by `ir[27:22]==0 && ir[7:4]==9`, MUL (`ir[21]==0`) and MLA(`ir[21]==1`):

```
MUL r[ir[19:16]]<-r[ir[3:0]]*r[ir[11:8]]

MLA r[ir[19:16]]<-r[ir[3:0]]*r[ir[11:8]]+r[ir[15:12]
```

Assume that the ALU does not include a combinational multiply operation. Modify the multi-cycle behavioral Verilog of section 10.7.6 to implement the MUL and MLA instructions using a shift and add algorithm such as the one explained in problem 2-7. Test with code that computes a quadratic polynomial, `a*x*x+b*x+c`.

10-21. Modify the pipelined behavioral Verilog of section 10.8.7 to implement the MUL and MLA instructions assuming a combinational multiplier can produce one product per clock cycle. Use the same test code as problem 10-20.

10-22. Modify the superscalar behavioral Verilog of section 10.9.8.2 to implement the MUL and MLA instructions. Use the same test code as problem 10-20.

10-23. Modify `condx` and `f` to allow for all sixteen conditions. Hint: `f` will need a 33-bit input to detect overflow. Give a written justification why your test code is adequate.

10-24. Modify `OPB to include shift and rotate.

11.   SYNTHESIS

There are two common uses of Verilog: simulation and synthesis. Chapters 3, 5 and 6 describe various features of Verilog that are useful for simulation, which is the interpretation of Verilog source code on a general-purpose computer. This chapter gives examples of synthesis, which is the automated process of transforming a subset of Verilog statements into a netlist of gates whose interconnnections perform the algorithm specified by the Verilog source code.

11.1   Overview of synthesis

There are two main vehicles for the implementation of synthesized designs: custom integrated circuits (sometimes called Application-Specific Integrated Circuits or ASICs) and programmable logic. Custom integrated circuits are created by transforming the synthesized netlist for a particular design (i.e., the equivalent of gate-level structural Verilog) into a specific geometric arrangement of metal, semiconductor and insulator materials on an integrated circuit. To manufacture such circuits, an automated tool draws the physical layout of the circuit on what is called a *mask*. The mask is then used with photolithography or similar processes to mass-produce the circuit on chips.

Programmable logic is fabricated in a similar way, but the masks used by the manufacturer do not represent some specific design. Instead, a programmable logic chip consists of many building block devices together with a programmable interconnection network. After the programmmable logic chip is manufactured and sold to the designer, bits are transferred into the chip which customize the programmable logic for a specific design. Thus, the same physical hardware might be used by two different designers to implement two completely different designs. Because by itself programmable logic lacks the ability for self-modification, it is not quite general purpose in the same sense as a stored program machine (such as the Manchester Mark I, the PDP-8 or the Pentium II). Historically, the concept of rewiring a fixed set of building block units to solve different problems can be traced back to the ENIAC in the early 1940s. Modern technology now allows interconnections inside a programmable logic chip to be reconfigured by simply changing the bits, rather than having to pull out and plug in wires as was the situation with the ENIAC's plugboards. As a further convenience, Verilog synthesis tools allow the modern designer (who may be ignorant of the wiring bit patterns) to reconfigure the programmmable logic by simply changing source code. Thus, in many instances, using programmable logic with a synthesis tool provides a viable alternative to software.

Regardless of whether the designer is targeting custom integrated circuits or programmable logic, the Verilog used for synthesis is similar, except the cost of synthesizing to programmable logic is considerably cheaper than synthesizing to custom integrated circuits.[1] This chapter concentrates on programmable logic because the necessary tools should be accessible to interested readers.

11.1.1   Design flow

Figure 11-1 gives the *design flow*, which shows how various automated tools interact with one another. The design flow shows how to synthesize and test a design. The design starts as Verilog source code. It gets translated and downloaded into a kind of programmable logic, known as a Complex Programmable Logic Device (CPLD). Figure 11-1 illustrates the files (shown as rectangles) that are produced as output of various tools (shown as circles), and how, in turn, these files are used as input to other tools. The tools shown in figure 11-1 include an optional synthesis preprocessor, the main synthesis tool, a post-synthesis place and route tool targeting programmable logic and a download tool that transfers the design into the programmable logic chip. Also, a standard Verilog simulator plays an important part in the design flow.

The designer only creates two or three files: one file to be synthesized, one file for the test code and one optional file for the physical (pin number) information. The remaining files of figure 11-1 are created automatically. The file containing test code can be used with the Verilog simulator to verify the operation of the file to be synthesized and/ or the results of synthesis. The file to be synthesized, which is input to the synthesis tool (or possibly the preprocessor), contains one or more Verilog module(s) using behavioral and/or structural features of Verilog. For pure behavioral synthesis, there is only one module, which, of course, is the highest level module of the file. For structural synthesis, there could be multiple modules defined that are instantiated hierarchically inside the highest level module of the file.

Only the portlist of this highest level module determines how the physical pins of the synthesized chip will be used. The portlists of lower level modules, if any, only deal with the internal connections within the chip and therefore have no influence on the physical pins of the chip.[2]. The `input` and `output` definitions for the highest level module should include the width of each port, since some synthesis tools require this, but all synthesis tools and simulators accept this syntactic variation. Each bit of each port of the highest level module corresponds to a distinct physical pin of the synthe-

---

[1] Assuming the design will be manufactured only in small quanities.

[2] Some tools, such as PLDesigner-XL, allow a design to be partitioned onto multiple chips, but even in such a case, the partitioning is automatic and not influenced by the Verilog ports of lower level modules.

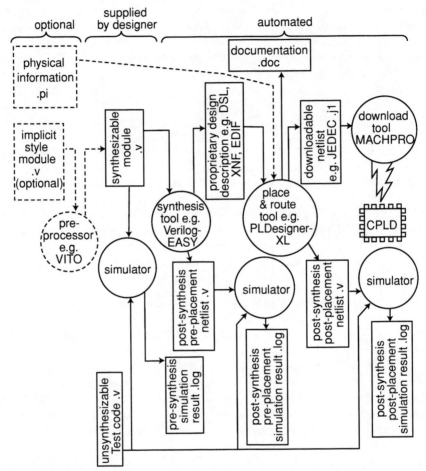

Figure 11-1 Design flow for CPLD synthesis. VITO, VerilogEASY, MACHPRO and PLDesigner-XL are specific tools discussed in section 11.1.3.

sized chip. The easiest and often most efficient approach is to let the synthesis tool choose how to connect these bits to the physical pins. In some cases, such as using a circuit board where the programmable logic chip already has its `sysclk` and similar signals soldered to specific pins, it is necessary for the synthesis tool to use specific pins. There is no standard syntax in Verilog to indicate pin numbers in the file (`.v`) that contains the highest level module, but many synthesis tools allow the designer to force the tool to connect specific bits to specific pins with the physical information file (section 11.3.6) or a similar approach.

11.1.2 Testing approaches

The file to be synthesized does not contain test code, since test code usually contains unsynthesizable statements such as `$display`. On the other hand, test code is essential for using a synthesis tool properly. Since the Verilog file used for synthesis contains a module with a portlist, running that file by itself on a simulator would produce erroneous results since the ports would be disconnected (`'bz`). When a designer does synthesis of a Verilog file, there needs to be another Verilog file with test code that does an `'include` of the synthesizable Verilog.

There are several reasonable strategies for using a simulator to test synthesizable Verilog:

        a)   before synthesis
        b)   after synthesis but before place and route
        c)   after place and route
        d)   all of the above.

What is not reasonable is to neglect testing altogether. Designers who think they are saving time by not writing test code for use in simulation are fooling themselves. It is far easier to detect bugs in simulation than after the synthesized design has been implemented in a physical chip.

The most thorough strategy is to simulate at each step in the design flow. Most vendors support some kind of backannotated output from the synthesis and/or place and route tools. This output is typically some kind of structural Verilog that gives the gate-level netlist produced by the synthesis tool. By using this output together with the test code, the designer may verify that the synthesized result behaves similarly to the original unsynthesized Verilog. The *backannotation* provides timing information to the simulator that allows the designer to predict the speed of the design without having to test the physical chip.

Although somewhat less desirable, the strategy of simulating only after synthesis makes sense when the designer needs to explore many different algorithms and design alternatives to find an acceptable solution to a problem. In such a case, the first issue is to discover which algorithms fit within the available hardware resources. If the designer makes a small mistake with an algorithm that fits within the desired chip, it is likely that the corrected version of the algorithm will also fit. After having discovered an algorithm that fits within the constraints of the hardware, it is important that the designer use simulation to determine whether there are any such bugs in the algorithm. It is much harder to debug the algorithm after it is embodied in physical hardware than when it is given only as Verilog code for simulation, and so simulation should be part of the design process.

When the physical chip will not be operated near its maximum frequency, there often is no need to simulate the back annotated Verilog resulting from synthesis.[3] In such a case, simulating only before synthesis may be reasonable. For example, the clock used in this chapter is slow enough that propagation delay is not a concern with the designs discussed below. We will do post-placement simulation only to illustrate the logical correctness of the process, and not out of concern for speed. In commercial design, speed is often an important issue, but correctness is always the first concern.

11.1.3   Tools used in this chapter

There are five specific software packages that are used as example tools in the design flow of this chapter. The details would differ slightly if other vendors' tools were used instead, but the basic principles of the design flow would be similar. The first software package, which is used only with the later examples in this chapter, is the Verilog Implicit To One hot (VITO) synthesis preprocessor described in chapter 7 and appendix F. The second software package is a synthesis tool, known as PLSynthesizer, which is available from a company called MINC Incorporated.[4] The output of PLSynthesizer is in a proprietary HDL known as DSL. PLSynthesizer also outputs structural Verilog which is logically equivalent to the DSL for post-synthesis simulation. The third software package is a place and route tool, known as PLDesigner, also from MINC, that converts the output of PLSynthesizer into what is called *JEDEC format*, which is a standard file format (`.jl`) used by several different vendors for downloading to programmable devices. PLDesigner may optionally use a physical information (`.pi`) file to indicate pin numbers. PLDesigner outputs the equivalent of the JEDEC file as a structural Verilog netlist for post-placement simulation. In addition, PLDesigner creates a documentation file (`.doc`) which indicates pin numbers and logic equations (in DSL syntax). The fourth software package is a download tool known as MACHPRO, from a CPLD manufacturer known as Vantis,[5] that reads the JEDEC file and sends the desired configuration to the programmable logic chip. The fifth software package is a simulator, such as VeriWell.

11.1.4   The M4-128/64 CPLD

The M4-128/64 is a CPLD chip manufactured by Vantis that comes in a 100-pin package, of which 64 pins are available for the designer to use.[6] Each of these 64 I/O pins may be used as one bit of an `input`, `output` or `inout` port. Internally, the M4-128/

---

[3] Provided that the subset of Verilog used means the same thing in both synthesis and simulation.

[4] MINC has made a restricted version of this technology known as VerilogEASY available to readers of this book. See appendix F for details.

[5] Vantis is a spinoff from Advanced Micro Devices (AMD), and the M4-128/64 used to be known as the AMD Mach445.

[6] The restricted version of VerilogEASY only allows 40 of these pins to be used.

64 contains 128 units, known as *macrocells*. Each macrocell contains a single OR gate, and a single optional flip flop. The output of the OR gate either feeds the flip flop or is directly available to other macrocells or I/O pins as combinational logic. The macrocell's OR gate receives its inputs from a series of AND gates. Put another way, a particular macrocell, m, can either implement sequential logic:

```
reg m;
always @ (posedge sysclk)
m =  t1_1 & t1_2 & t1_3 & ...
   | t2_1 & t2_2 & t2_3 & ...
          ...
```

or alternatively, that macrocell can implement combinational logic:

```
wire m =  t1_1 & t1_2 & t1_3 & ...
        | t2_1 & t2_2 & t2_3 & ...
               . . .
```

where the (optionally complemented) terms t1_1, t1_2, etc. are either input wires from the I/O pins or are the outputs of macrocells. In a particular macrocell, the designer does not have to use all the terms possible, but there are fairly complex internal constraints on how many and which terms may be used in particular macrocells. Because of the internal complexity of the CPLD, it is necessary for the designer to use a tool. This is true even if the designer were to create a netlist manually because the *place and route* tool must transform the original netlist into one that fits within the complex constraints of the CPLD.

Each I/O pin of the M4-128/64 has an optional flip flop, which the synthesis tool may choose to disconnect (for a combinational logic function of the input). Considering the macrocells (64 bonded to I/O pins and 64 hidden) and the I/O pins, the total number of flip flops that the M4-128/64 contains is 192. When all its macrocells are fully in use, the M4-128/64 is the equivalent of about 5000 gates.

Vantis makes a printed circuit board, known as a *demo board,* that has one M4-128/64 mounted on it together with additional hardware, such as a 1.8432MHz oscillator[7] that produces the sysclk signal. Although many similar types of devices exist,[8] the reason

---

[7] When the inputs to every macrocell only come from the internal flip flops, the M4-128/64 may be clocked up to 125 MHz. When macrocells are cascaded together to form complex combinational logic, the maximum frequency is lower. The 1.8432 MHz is slow enough to be safe for most designs.

[8] Of which the Field Programmable Gate Array (FPGA) is perhaps the most common. The FPGA uses a table rather than the AND/OR structure of a CPLD, but such details are seldom important to a designer using a synthesis tool. In the 1990s, companies such as Xilinx and Altera were leading suppliers of FPGAs.

for describing the M4-128/64 demo board here is that it is well suited for small synthesis experiments. The M4-128/64 demo board connects to a personal computer via the parallel (printer) port of that computer. This personal computer runs the synthesis tools (such as PLSynthesizer and PLDesigner) and also the MACHPRO software, which downloads the configuration of hardware determined by the synthesis tools into the M4-128/64. The downloading process changes which terms are connected to which macrocells. If a designer makes a mistake, it is a simple matter to download a corrected version of the design because the internal technology of the CPLD is similar to an EEPROM.

11.2   Verilog synthesis styles

Regardless of whether the designer wants programmable logic or custom integrated circuits, and regardless of which vendors' tools are involved, there are five basic styles of Verilog code used in synthesis: behavioral registers, behavioral combinational logic, behavioral implicit style state machines, behavioral explicit style state machines and structural instantiation. Often a particular design contains a combination of these styles.

11.2.1   Behavioral synthesis of registers

As described in sections 3.7.2.2 and 4.4.4, the synthesizable model for a register is an instantaneous assignment statement inside a block with a single time control syntax, such as @ (posedge sysclk), @ (posedge sysclk or posedge reset) or @ (posedge sysclk or negedge reset) time control. All synthesis vendors support this Verilog construct. Registers synthesize to a group of flip flops, typically D-type flip flops. Often there is combinational logic associated with a register. An example of synthesizing a register is given in section 11.3.

11.2.2   Behavioral synthesis of combinational logic

There are two ways to describe combinational logic using behavioral Verilog that all synthesis tools accept: the continuous assign statement (section 7.2.1) and an always block with a sensitivity list composed of all the variables in the block that are **not** on the left of any of the =s inside the block (section 3.7.2.1).[9] All synthesis vendors support both of these constructs. Combinational logic synthesizes to the primitive combinational units of the target hardware which are AND/OR gates for CPLDs, lookup tables for FPGAs and ROMs and arbitrary combinational gates for custom logic. An example of synthesizing combinational logic is given in section 11.4.

---

[9] With the additional requirement that none of the variables on the left of the =s occur on the right of the =s .

11.2.3 Behavioral synthesis of implicit style state machines

One of the main themes of this book is the advantage of solving problems at the highest level possible. Implicit style state machines provide this high-level approach for designing hardware. Most of the examples in this book only consider the "pure behavioral ASM" (section 2.1.5.1) or its equivalent coding with implicit style Verilog (section 3.8.2.3). Although, as noted in chapter 1, this high level of design has a history that goes back for decades before the introduction of Verilog, the implicit style has only recently begun to capture the attention of many Verilog designers. The reason that designers have become interested in this style is that it allows them to produce correct designs in less time. Unfortunately, not all synthesis vendors support this style, and there are some restrictions on the support provided by those that do. The preprocessor described in appendix F and chapter 7 allows a designer to use a reasonable subset of the implicit style even when the synthesis tool does not support it. Synthesizable implicit style consists of multiple `@(posedge sysclk)` inside an `always`, with a few additional syntax restrictions that were not considered in earlier chapters. Examples are given in sections 11.5, 11.6 and 11.9.

11.2.4 Behavioral synthesis of explicit style state machines

In contrast to the implicit style, the explicit style requires the designer to specify the present state register and next state combinational logic, as explained in section 4.3. All synthesis vendors support this style. An example is given in section 11.7.

11.2.5 Structural synthesis

The most primitive form of synthesis, which all tools accept, involves structural instances. If all modules being synthesized use only structural Verilog, the result of synthesis is simply to flatten the netlists (section 2.5) and fit the design within the constraints of the chip. More typically, some of the modules being instantiated have some of the kind(s) of behavioral code described above. In this case, the behavioral code is synthesized appropriately before the netlist is flattened. An example is given in section 11.8.

11.3 Synthesizing `enabled_register`

As described in section D.6, the enabled register is one of the most important sequential building blocks used in computer design. Suppose we wish to synthesize a two-bit-wide enabled register. The behavioral Verilog for this is identical to section 4.2.1.1, except that we substitute the literal `[1:0]` to indicate the two-bit width since many synthesis tools do not work properly with parameters and do not work properly unless the size is mentioned in the `input` and `output` declarations:

```
module enabled_register(di,do,enable,clk);
 input [1:0] di;
 input enable,clk;
 output [1:0] do;
 reg [1:0] do;
 wire [1:0] di;
 wire enable, clk;
 always @(posedge clk)
  begin
   if (enable)
     do = di;
  end
endmodule
```

The above Verilog should work with any synthesis tool. If we synthesize the above with PLSynthesizer targeting the Vantis M4-128/64 chip mentioned above, we get the following preplacement structural Verilog netlist:

```
module LPM_DFF_2_x(Clk,ClkEn,D,Q);
 input Clk,ClkEn,D; output Q;
 wire net0, net1, net2, net3, net4;
 NAN2 I_2_NAN2(.I0(net0),.I1(net1),.O(net2));
 NAN2 I_3_NAN2(.I0(Q),.I1(net3),.O(net1));
 NAN2 I_4_NAN2(.I0(ClkEn),.I1(D),.O(net0));
 INV I_1_INV(.I0(ClkEn),.O(net3));
 DFF I0(.CLK(Clk),.D(net2),.Q(Q),.QBAR(net4));
endmodule
module LPM_DFF_1_x(Clk,ClkEn,D,Q);
 input Clk,ClkEn; input[1:0]D; output[1:0]Q;
 LPM_DFF_2_x I0(.Clk(Clk),.ClkEn(ClkEn),
                .D(D[1]),.Q(Q[1]));
 LPM_DFF_2_x I1(.Clk(Clk),.ClkEn(ClkEn),
                .D(D[0]),.Q(Q[0]));
endmodule
module enabled_register(di,do,enable,clk);
 input [1:0] di;   output [1:0] do;
 input enable,clk;
 LPM_DFF_1_x dox_x(.Clk(clk),.ClkEn(enable),
                   .D(di), .Q(do));
endmodule
```

11.3.1 Instantiation by name

There are two kinds of structural instantiation syntax that are legal in Verilog. The first kind is instantiation by position, as described in section 3.10. Had instantiation by position been used above, the Verilog shown in bold would have been written as:

```
LPM_DFF_1_x dox_x(clk,enable,di,do);
```

There is no other way to write this with the positional syntax of section 3.10. The other kind of syntax that is legal in Verilog is instantiation by name, which is illustrated above in bold. Like many synthesis tools, PLSynthesizer uses this alternative syntax because the modules generated by the tool may have lengthy portlists. The advantage of instantiation by name is that the ports may be rearranged in any order and the meaning is the same. For example, the following:

```
LPM_DFF_1_x dox_x(.D(di), .Q(do),
                  .Clk(clk),.ClkEn(enable));

LPM_DFF_1_x dox_x(.Q(do), .D(di),
                  .ClkEn(enable),.Clk(clk));
```

are among the twenty-four permutations that mean the same thing.

11.3.2 Modules supplied by PLSynthesizer

The modules in section 11.3.1 (such as NAN2, INV and DFF) could contain detailed gate-level timing information, but this netlist has not yet been placed. After placement inside the M4-128/64 CPLD, the netlist is likely to be considerably different than the one in section 11.3.1. Rather, the netlist in section 11.3.1 is primarily of use to show the logical correctness of the transformation carried out by the synthesis tool. To illustrate this transformation, we will define idealized versions of the modules it instantiates:

```
module NAN2(I0,I1,O);
  input I0,I1;output O;nand g1(O,I0,I1);
endmodule
module INV(I0,O);
  input I0;output O;not g2(O,I0);
endmodule
```

Continued

```
module DFF(CLK,D,Q,QBAR);
  input CLK,D;output Q,QBAR;
  assign QBAR = ~Q;
  always @(posedge CLK)Q = D;
endmodule
```

11.3.3   Technology specific mapping with PLDesigner

In addition to structural Verilog, PLSynthesizer produces the same netlist in a proprietary form, known as DSL. The place and route tool, PLDesigner, uses the DSL to generate a netlist that is fitted within the constraints of the M4-128/64 CPLD. The output of PLDesigner includes the JEDEC netlist and an equivalent post-placement Verilog netlist. Such post-placement structural Verilog more accurately reflects the result of place and route than the netlist produced by PLSynthesizer. For this example, the resulting structural Verilog[10] of the enabled register is:

```
//Model automatically generated by Modgen Version 3.8
`timescale 1ns/100ps
enabledo00(dol0r,dol1r,dil1r,enable,dil0r,clk);
 output dol0r, dol1r;
 input dil1r, enable, dil0r, clk;  supply0 GND;
 wire pin_8,pin_11,pin_12,pin_13,pin_93,pin_94,tmp12,
tmp14,tmp15,tmp16,tmp17,tmp18,tmp19,tmp20,tmp21,tmp22;
portin PI1(pin_8,dil1r); portin PI2(pin_11,enable);
portin PI3(pin_12,dil0r); portin PI4(pin_13,clk);
portout PO1(dol0r,pin_93); portout PO2(dol1r,pin_94);
mbuf B1(tmp12,pin_13); and A1(tmp15,pin_12,pin_11);
not I1(tmp17,pin_11); and A2(tmp16,pin_93,tmp17);
or O1(tmp14,tmp15,tmp16);
dffarap DFF1(pin_93, tmp12, tmp14, GND, GND);
mbuf B2(tmp18,pin_13); and A3(tmp20,pin_8,pin_11);
not I2(tmp22,pin_11); and A4(tmp21,pin_94,tmp22);
or O2(tmp19,tmp20,tmp21);
dffarap DFF2(pin_94, tmp18, tmp19, GND, GND);
endmodule
```

---

[10] This Verilog was edited slightly for brevity.

Continued

```
module enabledo(do, di, enable, clk);
    output [1:0] do;
    input [1:0] di;
    input enable, clk;
.enabledo00 U1(.do10r(do[0]), .do11r(do[1]),
              .di11r(di[1]), .enable(enable),
              .di10r(di[0]), .clk(clk));
endmodule
```

Although much of the syntax used in the above code should be familiar from chapter 3, there are a few features of Verilog used in the above code (shown in bold) that were not mentioned previously. First is `timescale which allows the simulator to attach the proper meaning to $time that corresponds to the actual physical hardware. Second is the supply0 declaration, which is a shorthand for a continuous assignment (section 7.2.1) of the one-bit wire GND to 0, which models the connection to electrical ground inside the M4-128/64. Third are some user-defined modules (portin, portout, mbuf and dffarap) supplied by PLDesigner and explained in section 11.3.4 that model hardware resources of the M4-128/64. The name of the top level module generated by PLDesigner derives from the file name (enabledo.v) rather than the behavioral module name (enabled_register), and the order of the ports of this module may differ from that of the original behavioral Verilog which is why instantiation by name is done.

In addition to the post-placement structural Verilog, PLDesigner produces a documentation file (.doc) which summarizes the logic equations implemented by the netlist:

```
do[1].D=do[1]*/enable+di[1]*enable;
do[1].CLK=clk;
do[0].D=do[0]*/enable+di[0]*enable;
do[0].CLK=clk;
```

This is a much more primitive language than Verilog. The .D and .CLK notations indicate the macrocells are being used as D-type flip flops. To put the above in the more understandable Verilog form, the notation for Boolean operations ('*', '+', '/') must be rewritten into the corresponding Verilog notation ('&', '|', '~'). The following manual translation is the equivalent behavioral Verilog:

```
always @(posedge clk)
  begin
  do[1]<= #0((di[1]&enable)|(~enable&do[1]));
  do[0]<= #0((di[0]&enable)|(~enable&do[0]));
  end
```

The assignment statements must be non-blocking (with time control of #0) and must be listed inside an `always` block with a single `@ (posedge clk)` as the time control. This non-blocking assignment is somewhat different than the one used in earlier chapters. It is used above so that the order in which the Verilog statements occur will not effect the result. Since the non-blocking assignments use #0, the effect is almost the same as plain =, except all of the right-hand values will be evaluated before any of the left-hand values are changed. Because of the single `@ (posedge clk)` at the beginning of the `always` block, `do` will only change at the rising edge of the clock. The only reason to manually rewrite these logic equations back into Verilog is to describe the meaning of the `.doc` file. This file explains the transformation that PLSynthesizer has performed on the original behavioral Verilog more succinctly than the netlist.

11.3.4   Modules supplied by PLDesigner

Like most other place and route tools, PLDesigner allows for backannotation of timing information in the netlist after the place and route phase. Such post-placement information is more accurate than post-synthesis preplacement information because the place and route tool knows how signals will be routed through the actual chip. The details of how such information gets inserted into the structural Verilog output from a place and route tool varies among different vendors. In the case of PLDesigner, the modules used in section 11.3.3, such as `portin`, `portout` and `mbuf`, can contain detailed gate-level timing information, such as `specify` blocks, to model the interconnect delays that occur between macrocells in the M4-128/64. PLDesigner also generates an `.sdf` file which includes actual min/typ/max timing information for the routed circuit. For our purposes, we are not concerned about such detailed timing information but are rather only interested in the logical correctness of the transformation carried out by the place and route tool. To illustrate these transformations, we will define idealized versions of these modules. The first three of these are simply buffers that pass the input (`i`) through unchanged as the output (`o`):

```
module portout(o,i);
  input i; output o; buf b(o,i);
endmodule
module portin(o,i);
  input i; output o; buf b(o,i);
endmodule
module mbuf(o,i);
  input i; output o; buf b(o,i);
endmodule
```

The reason PLDesigner uses all three of these is that, in the actual backannotated Verilog, these might have different delays associated with them because they correspond to different hardware units within the M4-128/64. `portin` corresponds to routing a signal from an I/O pin used as an `input` on a `wire` that connects to an internal macrocell. `portout` corresponds to taking the output of an internal macrocell and routing it to an I/O pin to be used an `output`. `mbuf` corresponds to an internal connection between macrocells. Logically, all three of the above are equivalent. The only difference would be timing, which we are ignoring in this example.

Although Verilog provides built-in gates for combinational logic, such as `and`, `or` and `not`, Verilog does not provide built-in gates for sequential logic.[11] Therefore a place and route tool must supply modules for the sequential logic resources of the target technology, in this case the M4-128/64. Recall that each macrocell contains a D-type flip flop, which can be modeled as:

```
module dffarap(Q,CLK,D,AR,AP);
   output Q;
   input  CLK,D,AR,AP;
   reg Q;
     always @(posedge CLK or
         posedge AP or posedge AR)
       begin
         if (AP)
           Q = 1;
         else if (AR)
           Q = 0;
         else
           Q = D;
       end
endmodule
```

The above models a flip flop with an asynchronous reset (`AR`) and an asynchronous preset (`AP`). Such asynchronous signals are typically only used to initialize a controller when it is first powered up (see sections 4.4.4 and 7.1.6). In this example, these asynchronous signals are not used, and so they are instantiated with a connection to `GND`.

11.3.5   The synthesized design

As explained in figure D-17, an enabled register can be described with a block diagram consisting of a mux and a simple D-type register. Since figure D-17 is going to be synthesized into a physical component, either the designer or the synthesis tool must

---

[11] For a CPLD, there is no delay attributed to a particular AND or OR gate. Rather the delay is associated with the macrocell. For this reason, PLDesigner-XL uses built-in delayless `and`, `or` and `not`. Place and route tools for FPGAs or custom logic may take a different approach.

choose which signals will go in and out of the pins of the chip. In this instance, there are four bits of information being input and two bits of information being output, as illustrated in figure 11-2.

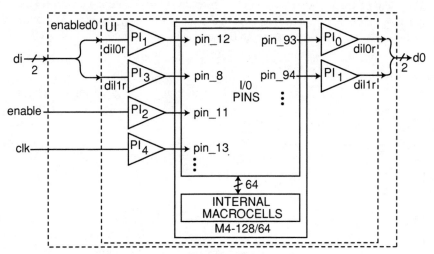

Figure 11-2. Physical pins of M4-128/64 used for two-bit enabled register.

The structural Verilog refers to the internal wires that connect to the I/O pins with the prefix `pin_`. This could be a little confusing since, for example, `pin_13` is not actually the physical pin 13 but rather is the signal from that pin after it has been buffered internally by the M4-128/64. In this design, `pin_13` is logically the same as the `clk` port, which presumably would connect to the global `sysclk` signal. The instance U1 of `enabledo00` separates the individual one-bit nets from the multi-bit ports. There are buses (which are oversimplified in this diagram) that connect the I/O pins to the macrocells. The actual implementation of the synthesized design occurs in the macrocells.

The synthesis tool has *bit blasted* the design into individual one-bit-wide *bit slices*, each one of which fits into a single macrocell, as shown in figure 11-3.

The circuit in figure 11-3 is a literal transcription of the Verilog produced by the synthesis tool. Notice how each bit slice of the mux has turned into an AND/OR gate arrangement. When `enable` (`pin_11`) is asserted, the outputs of the A2 and A4 AND gates will be zero. Thus, `di[0]` (`pin_12`) and `di[1]` (`pin_8`) will pass through their respective OR gates (O1 and O2) to become the new values of their respective flip flops (DFF1 and DFF2) at the next rising edge of `clk` (`pin_13`). When `enable` is not asserted, the old values (`pin_93` and `pin_94`) will be reloaded into their respective flip flops (DFF1 and DFF2) at the next rising edge of `clk`.

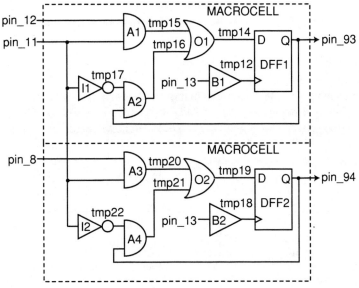

Figure 11-3. Macrocells in M4-128/64 implementing enabled register bit slices.

11.3.6 Mapping to specific pins

The physical pin numbers shown in figure 11-2 were chosen by PLDesigner. Often a designer wishes to override the choices automatically made by the place and route tool. For example, on the M4-128/64 demoboard, certain pins are attached to other hardware soldered on the board:

```
a1:93    a2: 5    a3:19    a4:31    sysclk:13
b1:94    b2: 6    b3:20    b4:32    reset:  4
c1:95    c2: 7    c3:21    c4:33     sw3:18
d1:96    d2: 8    d3:22    d4:34     sw2:54
e1:97    e2: 9    e3:23    e4:35     sw1:63
f1:98    f2:10    f3:24    f4:36     sw0:68
g1:99    g2:11    g3:25    g4:37
```

The `wires` whose names above begin with "a" through "g" are for Light Emitting Diodes (LEDs) in seven-segment displays. For example, the active low `a1 ... g1` signals control the leftmost digit. These seven segments are labeled clockwise, with `a1` at the top and `g1` at the center; thus `b1`, `c1`, `f1` and `g1` should be 0 and `a1`, `d1` and `e1` should be 1 to display the digit "4." The 1.8432MHz clock is available as `sysclk`, and a debounced push button provides the active low `reset`, which is also activated when the demoboard is powered up. There are four input DIP switches (sw0 Ö sw3) available on the demoboard.

These input pins can be named anything the designer wishes. For example, in the enabled register of section 11.3, it might be reasonable to take the `enable` from the switch on pin 54, and the `di` bus from the switches on pins 63 and 68. The two `do` bits might directly drive the `a1` and `b1` LED segments[12] on pins 93 and 94. Note that because of the active low nature of the LEDs, the light will not illuminate when the bit is a one, but it will light up when the bit is a zero. The following file, whose name must be similar to the name of the file that contains the module to be synthesized but with the extension `.pi`, is required to indicate the pin numbers to PLDesigner:

```
{MAX_SYMBOLS 0,MAX_PTERMS 0,POLARITY_CONTROL TRUE,
 MAX_XOR_PTERMS 0,XOR_POLARITY_CONTROL FALSE};
device target 'part_number amd MACH445-12YC';
 OUTPUT do[1]:93;OUTPUT do[0]:94;INPUT clk:13;
 INPUT enable:54;INPUT di[1]:63;INPUT di[0]:68;
end device;
```

11.4 Synthesizing a combinational adder

As described in section C.3, the adder is one of the most important combinational building blocks used in computer design. There are many ways (sections 3.10.5 through 3.10.7) to code an adder in Verilog, both behaviorally and structurally. Of these, the behavioral description is the easiest for the designer:

```
module addpar(s,a,b);
   output [3:0] s;
   input [3:0] a,b;
   reg [3:0] s;
   wire [3:0] a,b;
   always @(a or b)
     s = a + b;
endmodule
```

As in the last example, the `input` and `output` definitions need a size (four bits in this case). When the above is synthesized similarly to the last example, PLDesigner produces a `.doc` file that describes a series of logic equations for each bit of `s`. The following is a manual translation of this back into Verilog:

---

[12] Which just happen to be the same as the last example.

```
module addpar(s,a,b);
output [3:0] s; input [3:0] a,b;
wire [3:0] s,a,b; wire [3:3] c;
assign s[0] = ((a[0]&~b[0])|(~a[0]&b[0]));
assign s[1] = ((a[1]&~b[1]&~b[0])|(a[1]&~a[0]&~b[1])
|(a[1]&a[0]&b[1]&b[0])|(~a[1]&a[0]&~b[1]&b[0])
|(~a[1]&~a[0]&b[1])|(~a[1]&b[1]&~b[0]));
assign c[3]=((a[1]&a[0]&b[2]&b[0])|(a[2]&b[2])|(a[0]
&b[2]&b[1]&b[0])|(a[2]&a[1]&b[1])|(a[1]&b[2]&b[1]))
|(a[2]&a[1]&a[0]&b[0])|(a[2]&a[0]&b[1]&b[0]));
assign s[2] = ((a[2]&~b[2]&~b[1]&~b[0])
|(a[2]&~a[1]&~b[2]&~b[1])|(a[2]&~a[1]&~a[0]&~b[2])
|(a[2]&~a[1]&~b[2]&~b[0])|(~a[2]&~a[1]&b[2]&~b[1])
|(a[2]&a[1]&a[0]&b[2]&b[0])|(~a[2]&a[1]&~b[2]&b[1])
|(a[2]&a[0]&b[2]&b[1]&b[0])|(~a[2]&b[2]&~b[1]&~b[0])
|(~a[2]&a[1]&a[0]&~b[2]&b[0])|(a[2]&a[1]&b[2]&b[1])
|(~a[2]&~a[1]&~a[0]&b[2])|(~a[2]&~a[1]&b[2]&~b[0])
|(a[2]&~a[0]&~b[2]&~b[1])|(~a[2]&~a[0]&b[2]&~b[1])
|(~a[2]&a[0]&~b[2]&b[1]&b[0]));
assign s[3] = ((a[3]&~b[3]&~c[3])|(a[3]&b[3]&c[3])
|(~a[3]&~b[3]&c[3])|(~a[3]&b[3]&~c[3]));
endmodule
```

Of course, the designer would probably use the backannotated output from PLDesigner. This lengthy output, which is equivalent to the above assign statements, has been omitted for brevity. In this output, the internal name for the one-bit carry wire varies depending on how the module is synthesized. The name might be something like LPM_ADD_SUB_1_x_1_n0002. It might also be just c[3] as shown above.

This result from synthesis is quite a bit more complicated than one might expect when solving the same problem manually using full-adders. The above is complex because the place and route tool utilizes the wide AND/OR gates that exist in each macrocell of the M4-128/64. In the classical ripple carry adder (section 2.5), there needs to be a distinct carry signal input to each full-adder. Here the tool has eliminated the carry for all but the most significant bit by merging the logic equations for several full-adders together in a process known as *node collapsing*. This has the effect of lowering the propagation delay.

11.4.1 Test code

In any event, the designer needs to test the adder. Here is test code (sections 3.10.5 and 6.3.2) that does an exhaustive test:

a[1]a[0] b[1]b[0]

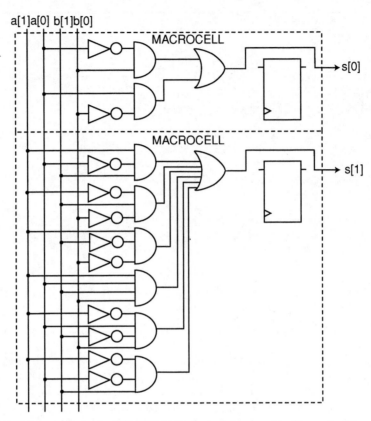

Figure 11-4. Macrocells in the M4-128/64 for low-order two-bit slices of adder

```
module test;
  integer ia,ib,numerr;
  reg [3:0] a,b;  wire [3:0] sum;
  addpar al(sum,a,b);
  initial
   begin
    numerr = 0;
    for (ia=0; ia<=15; ia=ia+1)
     for (ib=0; ib<=15; ib=ib+1)
      begin
       a=ia; b=ib;
       #1 $display("%b %b %b",a,b,sum);
       if ((ia+ib)%16 !== sum)
        begin
         $display("error");numerr=numerr+1;
```

Continued.

```
            end
          end
        $display("numerr=",numerr);
      end
   endmodule
```

The original behavioral adder, the preplacement netlist and the post-placement netlist all produce the correct results for the 256 combinations of inputs. If the width of the inputs were much larger, such an exhaustive test would be impossible.

As in the last example, we are ignoring the back annotated delay by supplying delayless modules for `portin`, `portout` and `mbuf`. If the backannotation capability were used, the #1 would have to be changed to an appropriate delay longer than the longest propagation delay of the synthesized design.

11.4.2 Alternate coding with `case`

An alternate way to describe this adder is to use behavioral statements that express the mathematics behind the ripple carry approach:

```
module addpar(s,a,b);
  output [3:0] s;
  input  [3:0] a,b;

  reg [3:0] s;
  wire [3:0] a,b;
  reg [3:0] c;

  function car;
    input a,b,c;
    begin
      case ({a,b,c})
        3'b000: car = 0;
        3'b001: car = 0;
        3'b010: car = 0;
        3'b011: car = 1;
        3'b100: car = 0;
        3'b101: car = 1;
        3'b110: car = 1;
        3'b111: car = 1;
      endcase
```

Continued

```
    end
  endfunction

  function sum;
    input a,b,c;
    begin
      case ({a,b,c})
        3'b000: sum = 0;
        3'b001: sum = 1;
        3'b010: sum = 1;
        3'b011: sum = 0;
        3'b100: sum = 1;
        3'b101: sum = 0;
        3'b110: sum = 0;
        3'b111: sum = 1;
      endcase
    end
  endfunction

  always @(a or b)
    begin
      c[0] = 0;
      s[0] = sum(a[0],b[0],c[0]);
      c[1] = car(a[0],b[0],c[0]);
      s[1] = sum(a[1],b[1],c[1]);
      c[2] = car(a[1],b[1],c[1]);
      s[2] = sum(a[2],b[2],c[2]);
      c[3] = car(a[2],b[2],c[2]);
      s[3] = sum(a[3],b[3],c[3]);
    end
endmodule
```

Here `car` is a function that models the carry required for the next higher bit position when adding three bits, and `sum` is the corresponding result in the current bit position. These functions may be coded several ways.

The `case` statement approach used above is a direct expression of the truth table for a full-adder. For synthesis, we do not consider `'bx` and `'bz` values in the cases the way that might be necessary for simulation. This is because the synthesis tool implements the `case` statement using == rather than ===, which is all that is physically possible in hardware:

```
if       ({a,b,c}==3'b000)  car = 0;
else if  ({a,b,c}==3'b001)  car = 0;
else if  ({a,b,c}==3'b010)  car = 0;
else if  ({a,b,c}==3'b011)  car = 1;
else if  ({a,b,c}==3'b100)  car = 0;
else if  ({a,b,c}==3'b101)  car = 1;
else if  ({a,b,c}==3'b110)  car = 1;
else if  ({a,b,c}==3'b111)  car = 1;
```

Either the `case` or the `if` statement approach is acceptable because for the three bits of input, all 2^3 possible cases are listed. Such a situation is known as a *full case.* The problem is that without a `default` clause in the `case` or an equivalent `else` at the end of the nested `if`s, the synthesis tool will not synthesize combinational logic properly except for a full case. For example, the following `case`, which only lists the ones of the function, is not full:

```
case  ({a,b,c})
   3'b001: sum = 1;
   3'b010: sum = 1;
   3'b100: sum = 1;
   3'b111: sum = 1;
endcase
```

A `case` like this that is not full will synthesize to what is known as a *latch,* which is an asynchronous sequential circuit, rather than the desired combinational logic. To make the `case` full requires using `default:sum=0;` in the above or using a `full case` synthesis directive. A *synthesis directive* is a comment which would be ignored by a simulator, but which causes the synthesis tool to alter its operation. Use of synthesis directives such as `full case` is common, but is dangerous because it may cause synthesis to disagree with simulation. It is better to make the `case` statement be full by supplying the appropriate `default` since that acts the same in both synthesis and simulation. Another common but dangerous directive is `parallel case`, which changes how synthesis interprets the `case` to be like `if`s without `else`s.

An alternative approach to the `case` statement would have been to use logic equations inside the functions, such as `sum=a^b^c`.

In any event, the combinational logic is defined using an `always` block having the same sensitivity list as the example in the last section that invokes the functions. Another way this could have been defined is with eight separate continuous assignment statements rather than the one `always` block:

```
        assign c[0] = 0;
        assign s[0] = sum(a[0],b[0],c[0]);
        assign c[1] = car(a[0],b[0],c[0]);
        assign s[1] = sum(a[1],b[1],c[1]);
        assign c[2] = car(a[1],b[1],c[1]);
        assign s[2] = sum(a[2],b[2],c[2]);
        assign c[3] = car(a[2],b[2],c[2]);
        assign s[3] = sum(a[3],b[3],c[3]);
```

Regardless of which of these variations we choose, the result is isomorphic to the original. This is because the wire [3:0] c is internal, and the synthesis tool can optimize it away, just as it did when synthesizing directly from a+b. The only distinction is the name chosen for c[3], but otherwise the result is an identical netlist. In this example, all the extra coding of the sum and car functions did not change the actual structure of the synthesized circuit. The details of the synthesized logic equations were dictated more by the capabilities of the CPLD exploited by the place and route tool than by anything that the designer codes. The main responsibility of the designer is to write correct Verilog. Usually, the designer should choose the modeling style which is easiest to understand (a+b in this example) and trust the synthesis tool to choose the logic equations that fit into the target device.

11.5   Synthesizing an implicit style bit serial adder

Rather than worrying about gate-level details, the designer should consider algorithmic alternatives. Although addition of two binary numbers usually is implemented as combinational logic, there are other approaches. The dependent sequence of calls to the sum and car functions inside the module addpar of section 11.4.2 makes it clear that the conventional ripple carry adder is the combinational logic required to implement one of the algorithmic variations explained in chapter 6: the single-cycle approach. Assuming we have a single register to load the sum at the next rising edge of the clock, the ripple carry adder computes in a single clock cycle all the information needed to form the next sum. In earlier chapters, we have assumed such a building block whenever we need to add. This approach for the module addpar is sometimes known as a *bit parallel* adder because all of the bits of the sum are available in parallel by the end of a single clock cycle.

Chapter 6 also describes other algorithmic alternatives besides the single-cycle approach. One of these is the multi-cycle approach. In the multi-cycle approach, each step in the dependent sequence is scheduled to occur in a different clock cycle. It is often possible to clock a multi-cycle machine faster than is possible with the single-cycle approach because less computation occurs per clock cycle. The multi-cycle approach takes several of these faster clock cycles to achieve the same result that the single-cycle approach achieves in one slower clock cycle.

11.5.1 First attempt at a bit serial adder

In order for the designer to have the freedom to explore such algorithmic variations easily, the synthesis tool should support the implicit style of Verilog. (The preprocessor described in chapter 7 and appendix F is available for those synthesis tools that do not support the implicit style.) With the implicit style, whether something occurs in parallel during a single cycle or in series during multiple cycles is simply a question of how many @(posedge sysclk)s the designer uses. In the case of the bit parallel addition algorithm given in section 11.4.2, it is a trivial matter with implicit style Verilog to change it to what is called a *bit serial* addition algorithm, which produces only one bit of the sum at a time. To do this, the assigns become non-blocking assignments, and the designer inserts @(posedge sysclk) at appropriate places inside the implicit always block:

```
`define CLK @(posedge sysclk)
`define ENS #1

  always
   begin
      . . .
    @(posedge sysclk) `ENS;
      c[0] <= `CLK 0;
    @(posedge sysclk) `ENS;
      s[0] <= `CLK sum(a[0],b[0],c[0]);
      c[1] <= `CLK car(a[0],b[0],c[0]);
    @(posedge sysclk) `ENS;
      s[1] <= `CLK sum(a[1],b[1],c[1]);
      c[2] <= `CLK car(a[1],b[1],c[1]);
    @(posedge sysclk) `ENS;
      s[2] <= `CLK sum(a[2],b[2],c[2]);
      c[3] <= `CLK car(a[2],b[2],c[2]);
    @(posedge sysclk) `ENS;
      s[3] <= `CLK sum(a[3],b[3],c[3]);
      . . .
   end
```

11.5.2 Macros needed for implicit style synthesis

In order for the implicit style to be practical, the result of simulation of implicit style Verilog before synthesis must agree with the result of simulation after synthesis (and, of course, the behavior of the physical hardware). Some synthesis tools are restricted as to the use of time control, but as discussed in section 3.8.2.1, simulators need # time control to simulate non-blocking assignment properly inside implicit style blocks. Therefore, in order that simulation agree with synthesis, it is recommended that all of the time control required for simulation be coded as macros. Only the time control needed by the synthesis tool (the @ (posedge sysclk) that denotes a state boundary outside a non-blocking assignment) is written without a macro. The other two forms of time control [the #1 and the @ (posedge sysclk) inside the non-blocking assignment] are written using macros (`ENS and `CLK). This way, they can simulate properly when the macros are defined as shown above, but they can be synthesized properly when the macros are defined as empty.

11.5.3 Using a shift register approach

A disadvantage of the code in section 11.5.1 is that it performs similar computations on different bits of the data. The synthesis tool will either have to duplicate the hardware to implement the sum and car functions multiple times, or use muxes to allow *resource sharing*, in a way analogous to the central ALU approach. To avoid this problem, we can use a shift register approach:

```
reg c;
reg [3:0] r1,r2;
  ...
@(posedge sysclk) `ENS;
  r2 <= `CLK y; r1 <= `CLK x;
  c <= `CLK 0;
@(posedge sysclk) `ENS;
  r1 <= `CLK {sum(r1[0],r2[0],c),r1[3:1]};
  c  <= `CLK car(r1[0],r2[0],c);
  r2 <= `CLK r2 >> 1;
@(posedge sysclk) `ENS;
  r1 <= `CLK {sum(r1[0],r2[0],c),r1[3:1]};
  c  <= `CLK car(r1[0],r2[0],c);
  r2 <= `CLK r2 >> 1;
@(posedge sysclk) `ENS;
  r1 <= `CLK {sum(r1[0],r2[0],c),r1[3:1]};
  c  <= `CLK car(r1[0],r2[0],c);
  r2 <= `CLK r2 >> 1;
@(posedge sysclk) `ENS;
  r1 <= `CLK {sum(r1[0],r2[0],c),r1[3:1]};
  c  <= `CLK car(r1[0],r2[0],c);
  r2 <= `CLK r2 >> 1;
```

A synthesis tool can produce a more efficient netlist from the above than from the earlier example in section 11.5.1 because the same computations, $sum(r1[0]$, $r2[0]$, c) and $car(r1[0]$, $r2[0]$, c), occur in each state. Resource sharing can now occur at no added cost. Also, in the above code there is no need to use a four-bit wire for c since a single-bit c can be reused in each state. A single-bit c variable suffices here because we are going to discard the carries in the end, anyway.

With this shift register approach, r2 starts out with the value of y. During each clock cycle, r2 is shifted over one position to the right. Therefore, during the first clock cycle, $r2[0]$ is $y[0]$. During the second clock cycle, $r2[0]$ is $y[1]$. During the third clock cycle, $r2[0]$ is $y[2]$, etc. In other words, the least significant bit is processed first, and greater significant bits are processed later. This order is essential to the bit serial technique.

The role of r1 is somewhat more complicated: r1 starts out with the value of x, but r1 is reused not just for holding the original bits of x but also for holding bits of the result. As a low-order bit of x shifts out of r2, the high-order bit of r2 is scheduled to become the corresponding $sum(r1[0],r2[0],c)$ bit. Although in the beginning, such bits are to the left of where they need to be, by the completion of the process, the result bits will have been shifted over to the proper position.

11.5.4   Using a loop

There is still room to improve the code given in section 11.5.3 because it takes many states (and therefore many flip flops in a one hot controller) to produce the answer. Although the bit serial approach necessarily takes a number of clock cycles proportionate to the number of bits in the word, the size of the controller should not also have to be proportional to the word size.

Here is where the flexibility of the implicit style is useful. With the implicit style, the designer can roll up the related computations that occur in separate states into one state inside a while loop. (Rolling up identical computations into a loop is the opposite of the loop unrolling explained in section 10.9.9.4 used by some optimizing compilers for RISC machines.) Although rolling all of these states into a single-loop state does not increase the speed of the machine, it usually will reduce the number of gates required to implement the machine. There is however, an added complication. There needs to be a loop counter that determines how many times the machine should repeat the loop state.

In previous chapters, it would have been natural to use a binary counter. Instead, here we will use a shift register (r3) to count in a unary code because with this it will be easier to understand the resulting logic equations after synthesis:

```
 1:  always
 2:   begin
 3:     ready <= `CLK 1;
 4:    @(posedge sysclk) `ENS; //ff_4
 5:     r2 <= `CLK y;
 6:     r3 <= `CLK 1;
 7:     c <= `CLK 0;
 8:     if (pb)
 9:      begin
10:        ready <= @(posedge sysclk) 0;
11:       @(posedge sysclk)`ENS; //ff_11
12:        r1 <= `CLK x;
13:        while (~r3[3])
14:         begin
15:          @(posedge sysclk) `ENS; //ff_15
16:            r1 <= `CLK{sum(r1[0],r2[0],c),r1[3:1]};
17:            c  <= `CLK car(r1[0],r2[0],c);
18:            r2 <= `CLK r2 >> 1;
19:            r3 <= `CLK r3 << 1;
20:         end
21:      end
22:   end
```

When the most significant bit of r3 becomes one, the loop stops. In other words, r3 contains the unary values 0001, 0010, 0100 and 1000 in successive clock cycles. The effect is similar to what would happen by counting 0, 1, 2 and 3. Since the computation only depends on the number of **times** the loop repeats, and not on the **value** of r3, the above unary code is just as reasonable as a binary code. A binary code might produce a somewhat smaller synthesized netlist, but the unary code will produce a synthesized circuit that typically runs faster and is easier to understand.

The above Verilog includes the friendly user interface described in sections 2.2.1 and 7.4.2. The signal ready is asserted when the machine is able to accept inputs. The user pulses pb for exactly one clock cycle to cause the machine to compute the sum, which will be available in r1 when the machine exits from the while loop.

11.5.5 Test code
The implicit style block of section 11.5.4 together with the function definitions from section 11.4.2 can be placed inside the module to be synthesized:

```
module vsyadd1(pb,ready,x,y,r1,r2,reset,sysclk);
  input [3:0] x,y; input pb,reset,sysclk;
  output ready; output [3:0] r1,r2;
  wire reset,sysclk,pb;    reg ready,c;
  wire [3:0] x,y; reg [3:0] r1,r2,r3;
  ...
endmodule
```

Prior to synthesis, it is prudent to test whether the algorithm is correct. To do such a test, we use the following test code in a different file. The test code has an instance of the module to be synthesized. Unlike the test code for the combinational logic of section 11.4.1, there needs to be a `wait` statement (section 3.7.3) so that the test code can adapt to the speed of the bit serial addition.

```
module top;
reg [3:0] x,y;   wire [3:0] sum,r2;
wire ready,sysclk; reg reset,pb;
integer numerr; time t1,t2;
cl #52000 clock(sysclk);
vsyadd1 slow_add_machine(pb,ready,
        x,y,sum,r2,reset,sysclk);
initial
 begin
  numerr = 0; pb= 0; x = 0; y = 0; reset = 1;
  #30 reset = 0;  #10 reset = 1;
  #210; @(posedge sysclk);
  for (x=0; x<=7; x = x+1)
   for (y=0; y<=7; y = y+1)
    begin
     @(posedge sysclk) pb = 1;
     @(posedge sysclk) pb = 0; t1 = $time;
     @(posedge sysclk) wait(ready); t2 = $time;
     @(posedge sysclk);
     if (x + y === sum)
      $display("ok %d",t2-t1);
     else
      begin
       $display("error x=%d y=%d x+y=%d sum=%b",
        x,y,x+y,sum);  numerr = numerr + 1;
      end
    end
  $display("number of errors=",numerr);
  $finish;
 end
```

Continued

```
always @(posedge sysclk) #20
  $display("%d r1=%d r2=%d pb=%b ready=%b",
    $time, sum,r2, pb, ready);
endmodule
```

The active low `reset` signal is necessary for the VITO preprocessor described in chapter 7 and appendix F. The test code detects no errors, so it is reasonable to synthesize `vsyadd1`.

11.5.6 Synthesizing

Since PLSynthesizer does not support the implicit style, the first step in synthesizing `vsyadd1` is to use the VITO preprocessor.[13] VITO passes through the module definitions and functions unchanged, which allows use of these names in the code generated by VITO. VITO generates a one hot controller using continuous assignment and one bit `regs` according to the principles described in chapter 7. VITO uses the line number in the names of the `wires` and `regs` generated. In this particular machine, the states correspond to `ff_4`, `ff_11` and `ff_15`. When the code generated by VITO is run through PLSynthesizer, logic equations are formed that describe the inputs to these macrocell flip flops. PLSynthesizer and PLDesigner will eliminate most of the redundant `wire` names created by VITO. The following is the manual translation of the `.doc` file into Verilog for the logic equations of the one hot controller:

```
always @(posedge sysclk or negedge reset)
 begin
  if (~reset)
   {ff_999,ff_4,ff_11,ff_15} = 0;
  else
   begin
    ff_999 <= #0 1;
    ff_4 <= #0(~ff_999|(~pb&ff_4)
      |(r3[3]&ff_15)|(r3[3]&ff_11));
    ff_11 <= #0 (pb&ff_4);
    ff_15 <= #0((~r3[3]&ff_11)|(~r3[3]&ff_15));
   end
 end
```

---

[13] The preprocessor is not necessary with synthesis tools, such as Synopsys, that support the implicit style.

If `ff_999`, `ff_4`, `ff_11` and `ff_15` are not listed in the portlist of VITO's output, PLSynthesizer will choose cryptic names for them. The above logic equations describe the conditions under which state transitions occur to the particular states. For example, a transition to state `ff_11` only occurs when the machine is in state `ff_4` and `pb` is true. There are several ways in which a transition to state `ff_4` occurs: when the machine is powered up, when the machine loops back to state `ff_4` because `pb` is false, or when the most significant bit of `r3` is one and the machine is in either state `ff_15` or state `ff_11`. The machine makes a transition to state `ff_15` when the most significant bit of `r3` is zero and the machine is in either state `ff_15` or state `ff_11`. Transitioning to state `ff_15` from state `ff_11` corresponds to entering the `while` loop for the first time. Transitioning to state `ff_15` from state `ff_15` corresponds to remaining in the `while` loop for an additional cycle.

In addition to the one hot controller, the VITO preprocessor also generates an architecture composed of combinational logic, muxes and simple D-type registers in the style described in section 7.2.2.1. For example, here is what VITO generates corresponding to `r2`:

```
assign new_r2 = s_5 ? y : s_18 ? r2>>1 : r2;
always @ (posedge sysclk) r2 = new_r2;
```

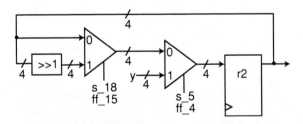

Figure 11-5. Block diagram for r2 portion of architecture.

The above Verilog is equivalent to figure 11-5, which is a kind of specialized shift register that is loadable (in state `ff_4` which includes statement `s_5`) and only shifts right (in state `ff_15` which includes statement `s_18`). Again, logic equations are given in the `.doc` file that describe the inputs to each macrocell flip flop. The following is the manually translated Verilog for the logic equations that correspond to `r2`:

```
always @ (posedge sysclk)
 begin
  r2[0] <= #0 ((r2[0]&~ff_15&~ff_4)
   |(y[0]&ff_4)|(r2[1]&ff_15&~ff_4));
  r2[1] <= #0 ((r2[1]&~ff_15&~ff_4)
   |(y[1]&ff_4)|(r2[2]&ff_15&~ff_4));
  r2[2] <= #0 ((r2[2]&~ff_15&~ff_4)
   |(y[2]&ff_4)|(r2[3]&ff_15&~ff_4));
  r2[3] <= #0 ((r2[3]&~ff_15&~ff_4)|(y[3]&ff_4));
 end
```

Although it might have appeared from figure 11-5 that there would be two macrocells of delay (for each mux), the synthesis tool merged the logic equations of the two muxes together into a single macrocell per bit slice. Except for r2[3], each bit slice is similar to the others. For example, there are three cases to consider for r2[0]. First, when ff_4 is active, two terms of the logic equation, r2[0]&~ff_15&**~ff_4** and r2[1]&ff_15&**~ff_4**, are guaranteed to be zero. This leaves only y[0]&**ff_4**, which passes through the proper bit of y into the input of the r2[0] flip flop. Second, when ff_15 is active, we know (because of the nature of one hot controllers) that ff_4 could not be active, but the synthesis tool did not know this. Therefore, the tool generates r2[1]&ff_15&~ff_4. The ~ff_4 is not necessary considering the total one hot system but is necessary to achieve the mux behavior shown in figure 11-5. Because the other two terms of the logic equation, r2[0]&**~ff_15**&~ff_4 and y[0]&**ff_4**, are guaranteed to be zero in this case (ff_15 active and ff_4 inactive), the remaining term, r2[1]&**ff_15**&~ff_4, passes through the right-shifted bit (r2[1]) into the input of the r2[0] flip flop. Third, the last possibility is that neither ff_4 nor ff_15 is active. In this case, r2[0]&**~ff_15&~ff_4** holds the former value of the r2[0] flip flop.

Of course, as mentioned earlier, it is tedious to have to manually translate the non-standard .doc file back into Verilog. The designer would probably prefer to use the structural Verilog automatically generated by PLDesigner. The instance and wire names shown in figure 11-6 and in the following may vary slightly, depending on tool- specific details:

```
mbuf B5(tmp85,pin_13);and A17(tmp90,tmp91,tmp92,pin_12);
and A16(tmp87,ff_15,tmp88,pin_46);not I19(tmp88,ff_4);
not I20(tmp91,ff_15);and A18(tmp93,ff_4,pin_25);
not I21(tmp92,ff_4);or O5(tmp86,tmp87,tmp90,tmp93);
dffarap DFF5(pin_12,tmp85,tmp86,GND,GND);
mbuf B6(tmp94,pin_13);and A20(tmp98,tmp99,tmp100,pin_46);
and A19(tmp96,ff_15,tmp97,pin_44);not I22(tmp97,ff_4);
not I24(tmp100,ff_4);and A21(tmp101,ff_4,pin_23);
not I23(tmp99,ff_15);or O6(tmp95,tmp96,tmp98,tmp101);
dffarap DFF6(pin_46,tmp94,tmp95,GND,GND);
...
```

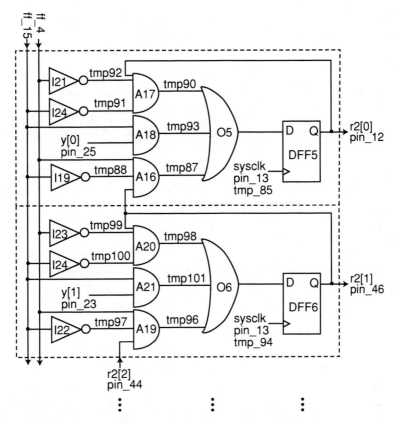

Figure 11-6. Macrocells implementing low-order bit slices of r2 .

11.6 Switch debouncing and single pulsing

To have a useful machine, it is often necessary for the machine to receive information from a person. In many situations, the most economical way to design such a machine is to use mechanical switches and buttons. In previous chapters, we have assumed the existence of ideal switches and buttons. For example, in section 2.2.1, we assumed a push button would assert its output for exactly one clock cycle when it is pushed. The problem is that mechanical switches and buttons are not perfect. They are neither synchronous nor is their output reasonable for use with a machine being clocked millions of times per second. This is because mechanical switches exhibit an undesirable property, known as *bounce*.

Ideally, when a person flips a switch on, we would hope that the output of the switch would become and remain one until the person flips the switch off. Unfortunately, real switches do not behave this way, as is illustrated by the following timing diagram:

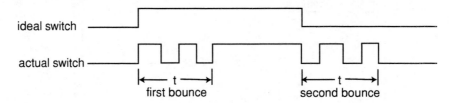

Figure 11-7. Ideal versus actual switch behavior shows need for debouncing.

Happily, real switches bounce for less than a constant time t seconds. For example, even the very awkward DIP switches[14] soldered onto the M4-128/64 demoboard bounce for less than a quarter of a second.

One solution to the bounce problem is to design a debounce machine[15] whose input is the actual switch, and whose output is the idealized pb signal needed by many of the designs in this book. Most of the time, the actual switch is quiet; thus the debounce machine continually reassigns 0 to pb. The debounce machine does something different when the actual switch makes its first transition to a one. During this first t second period when bounce occurs, we assume that the output of the actual switch will eventually stabilize to 1. Therefore, the number of clock cycles when the actual switch could be zero during this first bounce period is less than t times the clock frequency. After the first bounce period but before the second bounce period, the actual switch continually reads as a one. A second bounce period occurs when the switch is released.

The total number of clock cycles during which the actual switch reads as a zero from the time of the first transition to one until the final transition to zero is less than twice t times the clock frequency. The designer precomputes this constant, which will be loaded into a counter when the machine first detects that the actual switch has become a one. For example, with the M4-128/64, two times 0.25 seconds times 1.8432 MHz is approximately one million. Since 0.25 is an overestimation of t, the exact number of clock cycles is not too important, as long it is near one million. A convenient number around this size is $2^{20}-1$.

---

[14] People often use pencils to move these tiny switches, which aggravates the bounce problem. The constant t tends to be smaller for switches that are easier for people to manipulate, but the underlying cause of bounce is always electrical.

[15] The design here assumes that a single-pole single-throw switch is used and that the debounce machine must be completely digital. Much more economical solutions are possible that either use a few analog components, such as a capacitor and a resistor, or that use a single-pole double-throw switch. In the case of the M4-128/64 demoboard, neither alternative is possible without external components.

In addition to debouncing the switch, we need to make sure that the pb output of the debounce machine lasts for exactly one clock cycle. Otherwise, it would be as though the user is making millions of requests for computation, when in fact the user makes just one request. The following implicit style module solves both the debouncing and single pulsing aspects of this problem:

```
module debounce(sw3,pb,cnt,sysclk,reset);
   input sw3,sysclk,reset;
   output pb;
   output [19:0] cnt;
   wire sw3,sysclk,reset;
   reg pb;
   reg [19:0] cnt;
   always
    begin
     @(posedge sysclk) `ENS;
     pb <= `CLK 0;
     if (sw3 == 1)
       cnt <= `CLK 20'hfffff;
     else
       while (cnt[19:1] != 0)
         begin
           @(posedge sysclk) `ENS;
           if (sw3 == 0)
             cnt <= `CLK cnt - 1;
           if (cnt[19:1] == 0)
             pb <= `CLK 1;
         end
    end
endmodule
```

Assuming cnt is zero and the actual switch, sw3, is zero, the machine leaves cnt alone and therefore does not enter the while loop. The first time the machine detects that sw3 is one, the machine assigns the constant to cnt. Eventually, sw3 becomes zero again during the first bounce period. Since cnt now contains the constant, the machine enters the while loop. Inside the while loop, cnt is decremented only when sw3 is zero. The while loop exits when all bits of cnt other than the least significant are zero (i.e., cnt is 1). During this last clock cycle in the while loop, cnt might or might not be decremented one last time (hence the reason for ignoring the least signifcant bit). In that same clock cycle, pb is scheduled to become one for a single cycle. (pb will be scheduled to return to zero in the next clock cycle when the machine returns to the top state.) Therefore, the above code allows us to use the rather primitive DIP switch, sw3, as an ideal push button, pb.

11.7 Explicit style switch debouncer

As explained in chapter 4, a pure behavioral design can be broken down manually into an architecture and a controller. For example, the controller can be written in the explicit style, where each state transition must be given explictly, using a `case` statement:

```
`define TOP 0
`define BOT 1
module deboun_control(sw3,cnteq0_1,clrpb,
         ldpb,ldcnt,deccnt,sysclk,reset);
  input sw3,cnteq0_1,sysclk,reset;
  output clrpb,ldpb,ldcnt,deccnt;
  wire sw3,cnteq0_1,sysclk,reset;
  reg clrpb,ldpb,ldcnt,deccnt,ps,ns;
  always @(posedge sysclk or negedge reset)
    begin
      if (~reset)
        ps = 0;
      else
        ps = ns;
    end
  always @(ps or sw3 or cnteq0_1)
    begin
      clrpb = 0;ldcnt = 0;deccnt = 0;ldpb = 0;
      case (ps)
      `TOP: begin
             clrpb = 1;
             if (sw3 == 1)
               begin
                 ldcnt = 1;
                 ns = `TOP;
               end
             else
               if (cnteq0_1)
                 ns = `TOP;
               else
                 ns = `BOT;
           end
      `BOT: begin
             if (sw3 == 0)
               deccnt = 1;
             if (cnteq0_1)
               begin
                 ldpb = 1;
                 ns = `TOP;
```

Continued

```
                end
            else
                ns = `BOT;
            end
        endcase
    end
endmodule
```

The above code corresponds to what was called the pure structural stage in chapter 4, but for brevity, the above uses only behavioral statements. (The present state register and next state logic are not given in separate modules as was done in chapter 4.) Although similar in operation to the implicit style design given in section 11.6, the explicit style design is much more tedious to understand. Also, the designer must give a Verilog architecture (not shown) consisting of a counter (controlled by ldcnt and deccnt) and an enabled register (controlled by ldpb and clrpb). Finally, the designer must instantiate the controller and architecture to make a module that is identical to section 11.6:

```
module debounce(sw3,pb,cnt,sysclk,reset);
    input sw3,sysclk,reset;
    output pb;
    output [19:0] cnt;
    wire sw3,sysclk,reset;
    wire [19:0] cnt;
    wire pb,cnteq0_1,clrpb,ldpb,ldcnt,deccnt;
    deboun_arch architec(pb,cnteq0_1,cnt,
        clrpb,ldpb,ldcnt,deccnt,sysclk);
    deboun_control controller(sw3,cnteq0_1,
        clrpb,ldpb,ldcnt,deccnt,sysclk,reset);
endmodule
```

In this case, the binary encoding makes only a slight savings in macrocells (3%) compared to the one hot encoding used by VITO. As in many other designs, the majority of the macrocells are devoted to the architecture. Those macrocells must be present, regardless of whether the original Verilog was implicit or explicit style. All of the extra manual coding required for the explicit style was not worth the effort.

11.8 Putting it all together: structural synthesis

A typical design often uses a combination of the above techniques. For example, consider a machine activated by the debounced sw3 DIP switch that takes a three bit binary number from the other DIP switches ({sw2, sw1, sw0}) and does bit serial addition of this to a four-bit accumulator, r1, whose output is displayed in hexadecimal on the LEDs a1 .. g1. In order to reuse the code given above, the designer needs structural instances of vsyadd1 and debounce:

```
module mach445(sw3,sw2,sw1,sw0,sysclk,reset,
               a1,b1,c1,d1,e1,f1,g1);
   input sw3,sw2,sw1,sw0,sysclk,reset;
   output a1,b1,c1,d1,e1,f1,g1;
   wire sw3,sw2,sw1,sw0,sysclk,reset;
   reg a1,b1,c1,d1,e1,f1,g1;
   function [7:0] seven_seg;
     input [3:0] i;
     ...
   endfunction
   wire pb,ready;
   wire [3:0] r1,r2;
   reg [3:0] y;
   wire [19:0] cnt;
   vsyadd1 v1(pb,ready,r1,y,r1,r2,reset,sysclk);
   debounce deb1(sw3,pb,cnt,sysclk,reset);
   always @(sw2 or sw1 or sw0)
     y = {sw2,sw1,sw0};
   always @(r1)
     {a1,b1,c1,d1,e1,f1,g1} = ~seven_seg(r1);
endmodule
```

The vsyadd1 and debounce module definitions are given in the same file as the above module. In the above, y is simply another name for {sw2, sw1, sw0}. Note that r1 connects both to the v1.r1 output as well as the v1.x input for the instance of vsyadd1. In other words, r1 plus y will eventually replace the old value of r1.

The function seven_seg (whose case statement definition is not shown) takes a four-bit binary input, i, and outputs the seven bits required to drive one LED digit in hexadecimal. This combinational logic output is complemented to accommodate the active low requirements of the LEDs.

The .pi file must be defined using the pin numbers given in section 11.3.6. When synthesized and downloaded to the M4-128/64 demoboard, the above design will operate properly.

11.9 A bit serial PDP-8

All the designs in chapters 8 through 10 use bit parallel arithmetic to illustrate concepts about general-purpose computers. In contrast, many early general-purpose computers, including the Manchester Mark I, used bit serial arithmetic because it required less hardware. Most modern general-purpose computers are designed with bit parallel arithmetic because it is faster and easier. As a concluding synthesis example, however, let us build a bit serial PDP-8. This allows the CPU to fit within one M4-128/64 chip, and it simplifies the connections to an external memory chip, which must be wired manually to the demoboard.

The PDP-8 subset chosen for this example is the same as section 9.6 (CLA, TAD, DCA, HLT, JMP, SPA, SMA and CIA), with the addition of the SNA, SZA, CMA and IAC instructions described in appendix B. The link as well as additional instructions are not implemented in this example. This subset is sufficient for the childish division program given in section 9.7. Bit serial arithmetic is necessarily a multi-cycle approach, and so the multi-cycle PDP-8 ASM of section 8.3.1.3 is a good starting point for the design, but there are several algorithmic variations required for the CPU to fit into the M4-128/64.

First, bit serial addition loops are used for incrementing `pc` (the user interface and states F3A and E1ASKIP), incrementing `ac` (state E0IAC) and adding to `ac` (state E0TAD). Second, bit parallel comparisons, such as `ir==12'o7200` for CLA, need to be replaced with comparisons of only the appropriate bits, such as `ir[11:9]==7 & ir[8]==0 & ir[7]==1` for CLA. Third, like the original PDP-8 (but unlike chapters 8 and 9), combined instructions (e.g., CMA and IAC to form CIA) are allowed at no extra cost because the bits of the instruction register are tested individually. Fourth, memory accesses occur one bit at a time with a one-bit-wide `mb` register wired to the data in pin of the memory chip and a one-bit-wide `membus` wired to the data out pin of the memory chip. Fifth, like section 8.3.2.4 and figure 8-11, memory must be a separate actor so that it can be physically wired to the M4-128/64. Sixth, the `write` signal is active low for the memory chip used here, which is the opposite of figure 8-11. Seventh, since the number of bits in a memory chip is a power of two but the number of bits in the PDP-8's memory is a multiple of twelve, the simplest approach is to disregard four out of every sixteen bits from the one-bit-wide memory chip. In other words, `bitmem[0]` through `bitmem[11]` form the twelve-bit `m[0]`, and `bitmem[16]` through `bitmem[27]` form `m[1]`. Eighth, in addition to the memory address register, `ma`, the bit serial approach needs a bit address register, `ba`, which provides the low-order four bits of the address going to the memory chip. At any time, the bit from the memory chip currently being processed by the CPU is `bitmem[{ma,ba}]`. Ninth, `ba` also serves as a binary counter for bit serial arithmetic loops, rather than the unary `r3` counter described in section 11.5.4. Tenth, because this subset only implements the direct page zero addressing mode (and not the full set of addressing modes described in appendix B), the memory address register only needs to be seven bits wide (a reduction which saves several macrocells). Eleventh, the user interface of chapter 8 (but_DEP, but_PC, but_MA, cont and the twelve-bit switch register) has been replaced with a simpler but workable scheme using four undebounced switches and a push button, cont, that must be externally debounced. Twelfth, swin, which is the concatenation of the four switches, determines the user interface action taken when cont is pressed:

```
swin action
0000 ba ← 0
001- ba ← 0; pc ← {swin[0],pc[11:1]}
010- bitmem[{pc,ba}] ← swin[0]; Advance {pc,ba}
011- Advance {pc,ba}
1000 Execute
```

where advancing {pc,ba} means incrementing just ba, except in the case when ba==4'b1011. In that special case, pc is incremented and ba becomes zero.

11.9.1 Verilog for the bit serial CPU

In the following implicit style Verilog, the comments indicate names of states similar (but not identical) to those of figure 8-11. Many of the states, especially those for the user interface, have no direct correspondence to figure 8-11:

```
always
 begin
  @(posedge sysclk) `ENS; //INIT
   halt <= `CLK 1; write <= `CLK 1;
   forever
    begin
     @(posedge sysclk) `ENS; //F1
      ma <= `CLK pc; ba <= `CLK 0;
      c <= `CLK 1;
      if (halt)
       begin
        while (~(cont&swin[3]))
         begin
          @(posedge sysclk) `ENS; //IDLE
           halt <= `CLK 0; ma <= `CLK pc;
           mb <= `CLK swin[0]; c <= `CLK 1;
           if       (cont&(swin[3:2] == 2'b00))
            begin
             @(posedge sysclk) `ENS;
```

Continued

```
        ba <= `CLK 0;
        if (swin[1])
         begin
          @(posedge sysclk) `ENS;
            pc <= `CLK {swin[0],pc[11:1]};
         end
      end
    else if (cont&(swin[3:2] == 2'b01))
     begin
      @(posedge sysclk) `ENS;
       write <= `CLK swin[1];
      @(posedge sysclk) `ENS;
       write <= `CLK 1;
      @(posedge sysclk) `ENS;
       ba <= `CLK ba + 1;
       if (ba == 11)
        begin
         @(posedge sysclk) `ENS;
          ba <= `CLK 0;
          while (ba != 11)
           begin
            @(posedge sysclk) `ENS;
             pc <= `CLK
               {sum(pc[0],0,c),pc[11:1]};
             c <= `CLK car(pc[0],0,c);
             ba <= `CLK ba + 1;
           end
          @(posedge sysclk) `ENS;
            ba <= `CLK 0;
        end
      end
    end
  end
else
 begin
  @(posedge sysclk) `ENS; //F2
   while (ba != 11)
    begin
     @(posedge sysclk) `ENS; //F3A
      ir <= `CLK {membus,ir[11:1]};
      pc <= `CLK {sum(pc[0],0,c),pc[11:1]};
      c <= `CLK car(pc[0],0,c);
      ba <= `CLK ba + 1;
    end
```

Continued

```
@(posedge sysclk) 'ENS; //F3B
ma <= 'CLK ea(ir); ba <= 'CLK 0;
mb <= 'CLK ac[0]; c <= 'CLK ir[11];
if      (ir[11:9] == 1)
 begin
  while (ba != 11)
   begin
    @(posedge sysclk) 'ENS; //E0TAD
     ac <= 'CLK {sum(ac[0],membus,c),ac[11:1]};
     c <= 'CLK car(ac[0],membus,c);
     ba <= 'CLK ba + 1;
   end
 end
else if (ir[11:9] == 3)
 begin
  while (ba != 11)
   begin
    @(posedge sysclk) 'ENS;//E0DCA
     ac <= 'CLK {1'b0,ac[11:1]};
     write <= 'CLK 0;
    @(posedge sysclk) 'ENS;//E1ADCA
     write <= 'CLK 1;
    @(posedge sysclk) 'ENS;//E1BDCA
     ba <= 'CLK ba + 1; mb <='CLK ac[0];
   end
 end
else if (ir[11:9] == 5)
 begin
  @(posedge sysclk) 'ENS; //E0JMP
   pc <= 'CLK ma;
 end
else if (ir[11:9] == 7)
 begin
  if (ir[8])
   begin
    if (ir[1])
     begin
      @(posedge sysclk) 'ENS; //E0HLT
       halt <= 'CLK 1;
     end
    if (ir[3]^(ir[6]&ac[11]|ir[5]&(ac==0)))
     begin //SPA,SZA,SMA,SNA
      while (ba != 11)
       begin
```

Continued

```
                    @(posedge sysclk) `ENS; //E0ASKIP
                    pc <= `CLK {sum(pc[0],0,c),pc[11:1]};
                    c <= `CLK car(pc[0],0,c);
                    ba <= `CLK ba + 1;
                  end
              end
          end
        else
          begin
            if (ir[7])
              begin
                @(posedge sysclk) `ENS; //E0CLA
                ac <= `CLK 0;
              end
            if (ir[5])
              begin
                @(posedge sysclk) `ENS; //E0CMA
                ac <= `CLK ~ac;
              end
            if (ir[0])
              begin
                while (ba != 11)
                  begin
                    @(posedge sysclk) `ENS; //E0IAC
                    ac <= `CLK {sum(ac[0],0,c),ac[11:1]};
                    c <= `CLK car(ac[0],0,c);
                    ba <= `CLK ba + 1;
                  end
              end
          end
      end
    end
end
```

11.9.2   Test code

For a design as complicated as this, it is important to simulate before synthesis. Even a tiny bug could prevent the fabricated hardware from operating at all and give no trace as to the cause. In order to simulate the above, we need a non-synthesizable model of the memory chip that will be connected to the fabricated CPU:

```
module mem(mabus,babus,mbbus,membus,write);
  input mabus,babus,mbbus,write;
  output membus;
  wire [11:0] mabus;
  wire [3:0] babus;
  wire mbbus, write;
  reg membus;
  reg [11:0] m[0:127];
  reg [11:0] temp;
  always @(mabus or babus)
   begin
    temp = m[mabus]; membus = temp[babus];
   end
  always @(negedge write)
   begin
    #50 membus = mbbus; temp = m[mabus];
    temp[babus] = membus; m[mabus] = temp;
   end
endmodule
```

The above models memory as twelve-bit words but interfaces to the CPU one bit at a time. An attempt to access one of the four unused bits will result in 1'bx because of the way Verilog treats bit selects that are out of bounds. The above must be instantiated together with the CPU:

```
module  pdp8_system(swin,cont,halt,sysclk,reset);
  input swin,cont,sysclk,reset;
  output halt;
  wire cont,sysclk,reset,halt,mb, membus, write;
  wire [3:0] swin,ba;
  wire [11:0] ma;
  pdp8_cpu cpu(swin,write,membus,cont,
                  ba,ma,mb,halt,reset,sysclk);
  mem memory(ma,ba,mb,membus,write);
endmodule
```

Assuming pdp8_system is instantiated as pdp8_machine, the test code can initialize a memory location using a twelve-bit word refered to with hierarchical reference to the array pdp8_machine.memory.m[...]. In order to simulate the pushing of cont, a task is helpful:

```
task push;
  input [3:0] sw;
  begin
    swin = sw; #200 cont = 1;
    #100 cont = 0; #300;
    case (swin[3:2])
      0:   #200;
      1:   #2000;
      2,3: #100 wait(halt);
    endcase
    #300;
  end
endtask
```

The time control in the task depends upon what swin selection was requested. For example, for the test code to set the program counter to 12'o0100 and then execute a program, the task waits 200 units of $time for each bit shifted into the program counter and then waits until the CPU halts:

```
push(4'b0010);push(4'b0010);push(4'b0010);//0
push(4'b0010);push(4'b0010);push(4'b0010);//0
push(4'b0011);push(4'b0010);push(4'b0010);//1
push(4'b0010);push(4'b0010);push(4'b0010);//0
push(4'b1000);//Execute until HLT
```

11.9.3 Our old friend: division

In running this simulation with the childish division program of section 9.7, we observe that this bit serial implementation takes 558 cycles when 1 is the quotient, 827 cycles when 2 is the quotient and 1096 cycles when 3 is the quotient. Let us put this in perspective with running the childish division software on the other PDP-8 implementations discussed earlier:

```
section  kind    arithmetic  clock cycles
11.9.1   multi   serial      289+269*quotient
8.3.2.1  multi   parallel    55+55*quotient
9.6      pipe    parallel    12+10*quotient
```

Assuming the same clock period, the bit serial approach is about five times slower than the multi-cycle bit parallel approach of chapter 8, which in turn is about five times slower than the pipelined bit parallel approach of chapter 9. To execute one instruction, it takes on average about one cycle for the pipelined bit parallel machine of section 9.6, five cycles for the multi-cycle bit parallel machine of section 8.3.2.1 and twenty-seven cycles for the multi-cycle bit serial machine of section 11.9.1. In the latter case, it takes twelve cycles to fetch the instruction, twelve cycles to fetch the data and three cycles for the other typical states (i.e., F1, F2 and F3B).

11.9.4 Synthesizing and fabricating the PDP-8

This design will occupy the majority of the macrocells in the M4-128/64. After synthesis with VITO and PLSynthesizer, it is necessary to let PLDesigner choose the pins where the signals are routed. If the designer provides complete `.pi` information at first, it is likely that PLDesigner would be unable to fit this design into a single M4-128/64. Instead, the designer should only constrain critical pins. This design does not make use of any of the hardware on the demoboard, other than `sysclk` and `reset`. The only other critical pins are 18, 54, 63 and 68, which should not be used since these are tied to the DIP switches. Instead, `swin` will come from external switches. Once the design does get placed in a single chip, the pins selected by PLDesigner should be put in a `.pi` file so that future minor modifications of the design will not require physical rewiring of the memory chip to the demoboard:

```
membus:10    sysclk:13    contin:19    ma[5]:22
ma[4]:23     ma[6]:24     swin[1]:37   swin[3]:38
mb:43        reset:4      ba[3]:56     ba[2]:58
ba[1]:59     ba[0]:60     ma[0]:62     swin[0]:70
swin[2]:73   halt:74      write:84     ma[3]:93
ma[1]:96     ma[2]:98
```

Each I/O pin of the M4-128/64 is attached to a pin on one of two headers (JP4 or JP5) soldered to the demoboard. A small, low-cost static memory chip that can be used is the 2102, which is arranged as 1 x 1024 bits. Using wirewrap wire and a sixteen-pin dual in-line wirewrap socket, the memory can be attached to the demoboard as follows:

PDP-8	M4	2102	PDP-8	M4	2102
signal	header	pin	signal	header	pin
ba[0]	JP5-27	1	GND	JP4-2	9
ba[1]	JP5-25	2	Vcc	solder	10
write	JP5-12	3	mb	JP5-1	11
ba[2]	JP5-23	4	membus	JP4-27	12
ba[3]	JP5-19	5	GND	JP4-2	13
ma[0]	JP5-31	6	ma[3]	JP4-1	14
ma[1]	JP4-7	7	ma[4]	JP4-26	15
ma[2]	JP4-11	8	ma[5]	JP4-28	16

It is desirable that the ma and ba signals also be attached to external LEDs to provide feedback to the user. (The onboard LEDs cannot be used because of the place and route limitations of the M4-128/64.) The five-volt power supply (Vcc) to the memory chip must be soldered on the demoboard power connection. In addition, the following external switches must be connected: contin (externally debounced) to JP4-34, swin[1] to JP4-6, swin[3] to JP4-4, swin[0] to JP5-32 and swin[2] to JP5-26.

11.10 Conclusions

Five kinds of synthesizable Verilog were considered in this chapter: behavioral registers, behavioral combinational logic, behavioral implicit style state machines, behavioral explicit style state machines and structural instantiation. Of these, the implicit style is the best choice because it has such a close relationship to the behavioral ASMs discussed in earlier chapters. Often a designer must use some of the other kinds of Verilog, such as combinational logic, to create a complete design, but implicit style should be the first choice for synthesizing hardware.

This chapter has used the M4-128/64 CPLD with VITO, PLSynthesizer and PLDesigner. Although the details of performing synthesis using chips and software from different vendors may vary somewhat from those described here, the design flow for Verilog synthesis is similar. Simulation is a critical part of this design flow. Even though simulation takes some effort by the designer, in most cases, a bug discovered during simulation will be much less expensive than one that remains hidden until after the hardware is fabricated. Synthesis as well as place and route tools output structural Verilog netlists, which can be used with test code to verify the operation of the synthesized design.

11.11   Further reading

PALNITKAR, S., *Verilog HDL: A Guide to Digital Design and Synthesis,* Prentice Hall PTR, Upper Saddle River, NJ, 1996. Chapter 14.

11.12   Exercises

11-1. Give the synthesizable `seven_seg` function used in section 11.8.

11-2. Synthesize a 3-bit childish division machine based on the Verilog given in section 7.4.2 that will work with the hardware resources of the M4-128/64 demoboard. The code should be modified so that x is a register (rather than a bus) that is loaded with $y=\{sw2,sw1,sw0\}$ when the debounced $sw3$ generates the first pb pulse. The second pb pulse starts the computation of x/y, which will be displayed in hexadecimal on the seven-segment display. Use the `debounce` module of section 11.6 and a top-level module similar to section 11.8 with the function from problem 11-1. Use test code that verifies the design after each step in the design flow.

11-3. Synthesize a factorial machine based on problem 2-4 that will work with the hardware resources of the M4-128/64 demoboard. $\{sw2,sw1,sw0\}$ is the 3-bit value of n which is used when the debounced $sw3$ generates the pb pulse. The 13-bit factorial of n will be displayed in hexadecimal on the LEDs. Use the `debounce` module of section 11.6 and a top-level module similar to section 11.8 with the function from problem 11-1. Use test code that verifies the design after each step in the design flow.

11-4. Modify the design of section 11.9 to include the link and the CLL, CML, RAR and RAL instructions (appendix B) in a way that allows the design to fit in the M4-128/64. Make appropriate changes to other instructions. Use test code based on the machine language program in section 8.3.2.5.3. Hint: because of the restrictions on <= in VITO, you need to define a 13-bit `lac` register, rather than separate `link` and `ac` registers.

11-5. Modify the design of section 11.9 to include the ISZ instruction (appendix B) in a way that allows the design to fit in a single M4-128/64. Use appropriate test code, such as the machine language code from problem 9-2.

11-6. Give Verilog for the architecture of the `debounce` module in section 11.7.

A.   MACHINE AND ASSEMBLY LANGUAGE[1]

Most people use programs written in *high-level languages*. High-level languages are hardware-independent, complex languages that are relatively easy to use. *Hardware independent* means that programs written in high-level languages will run on nearly any general-purpose computer. Examples of high-level languages include Pascal, Verilog and C.

In contrast, low-level languages are simple in form and closer to how computers actually operate. This makes them harder for the programmer to use. Low-level languages are hardware dependent and have one statement per machine operation. Hardware dependent means that low-level languages are designed for a specific computer's hardware. Each statement is called a mnemonic. Mnemonics are easily memorized symbols that represent each fundamental computer operation in a textual form for the programmer's use. An instruction is a binary word that represents these fundamental operations in a form the computer can process.

Low level languages include machine language and assembly language. Assembly language is made up of instructions represented by mnemonics. Machine language consists of the instructions represented in binary. Assembly language has four major parts:

1. labels - symbolic names for places in memory (where variables are stored).
2. mnemonics - indications of what the computer will do.
3. operand - the data operated on by the instruction.
4. comments - a guide to the program that are ignored by the computer.

One statement in a high-level language program often corresponds to many assembly language and machine language instructions. For example, the machine language file of a program written in C and the machine language file of the same program written in assembly language are basically equivalent. But, the assembly language version is much longer than the C program. Consider the following very simple program:

```
/* Total tuition for three classes*/
int tuit,engl=74,cosc=106,math=148;
main(){tuit = engl + cosc + math;}
```

---

[1] This appendix was written by Susan Taylor McClendon and Mark G. Arnold.

Appendix A   485

This is equivalent to the following assembly language program written for the PDP-8, a simple general-purpose computer used as an example in chapters 8, 9 and 11:

```
       label   mnemonic operand comment
               *0100            /starting addr
               CLA              /put zero in AC
               TAD      ENGL    /add ENGL to 0
               TAD      COSC    /add COSC to ENGL
               TAD      MATH    /add MATH to COSC+ENGL
               DCA      TUIT    /store in TUIT, clear AC
               HLT              /halt
       ENGL,   0112             /74 dollars
       COSC,   0152             /106 dollars
       MATH,   0224             /148 dollars
       TUIT,   0000
       }
```

The *0100 indicates the starting address of the program in octal. The mnemonics indicate what each instruction does. The operand refers to a label defined later in the program. The following shows this example program translated to PDP-8 machine language code:

```
               0100/7200
               0101/1106
               0102/1107
               0103/1110
               0104/3111
               0105/7402
               0106/0112
               0107/0152
               0110/0224
               0111/0000
```

The four digits on the right of the "/" indicate a memory address in octal. The four digits on the left indicate the contents which show the octal values of the bit patterns representing the machine language equivalent of each mnemonic. Starting at address 0106_8 the contents are data values, not instructions.

TAD performs a *Two's* complement *AD*dition of the operand to the contents held in the AC. DCA, *D*eposit and *C*lear the *AC*, deposits the value held in the AC into memory and then clears the AC. CLA and HLT are *non-memory reference instructions*. The CLA instruction *CL*ears the *AC* and the HLT instruction causes the fetch/execute algorithm to stop. The machine language code for CLA is 7200_8 and for HLT is 7402_8. More details about these and other instructions of the PDP-8 are given in appendix B.

B.   PDP-8 COMMANDS[1]

The commands listed below are the Memory Reference Instructions (MRI) and the non-memory reference instructions of the PDP-8. The bits referred to below are given in little endian notation.

Memory reference instructions

1. <u>TAD</u> ($1xxx_8$) - *T*wo's complement *AD*d contents of memory address xxx_8 to the link and ac.

2. <u>DCA</u> ($3xxx_8$) - *D*eposit contents of ac at memory address xxx_8 and then *C*lear ac.

3. <u>AND</u> ($0xxx_8$) - Logical *AND* of ac with contents of memory address xxx_8.

4. <u>JMP</u> ($5xxx_8$) - *Ju*M*P* to memory address xxx_8 so that the fetch/execute cycle will process the instruction stored there instead of the next sequential instruction. The PC is simply loaded with xxx_8.

5. <u>ISZ</u> ($2xxx_8$) - *I*ncrement (add 0001_8) to contents of memory address xxx_8 and *S*kip next instruction if contents become *Z*ero.

6. <u>JMS</u> ($4xxx_8$) - *Ju*M*p* to *S*ubroutine located at memory address xxx_8. The JMS instruction saves the return address at memory addess xxx_8 and then the PC becomes $xxx_8 + 0001_8$. (The return address is the value of the PC indicating which instruction would have otherwise executed next.)

The "xxx_8" in the MRI instructions indicates a memory address used by the instruction. There are four addressing modes of the PDP-8: direct page zero, indirect page zero, direct current page and indirect current page. There is also a variation of the indirect addressing mode known as *autoincrement*.

Why do we need other addressing modes? One reason lies in the number of addresses we can represent using the page addressing bits. The page addressing bits are bits 6-0 of the ir. Only 2^7 or 128_{10} addresses (starting at address 0_{10}) can be represented by these seven bits. To represent the other 3968_{10} memory locations possible with the 12-bit address bus, the PDP-8 subdivides the 4096_{10} memory locations into 32_{10} *pages* (starting at page zero) of 128_{10} memory locations (4096_{10} DIV $128_{10} = 32_{10}$). To access a particular page, the PDP-8 uses two types of addressing modes: *direct* and *indirect*. Bit eight indicates either direct or indirect addressing mode and bit seven indicates

---

[1] This appendix was written by Susan T. McClendon and Mark G. Arnold.

Appendix B   *487*

either *page zero* or *current page*. Page zero is normally used for global variables and constants and the current page (01_8-37_8) is normally used for local data and corresponding code. The following lists the combinations of bits seven and eight for each possible addressing mode:

```
ir[8] ir[7] Addressing Mode        Effective Address
  0     0    Direct Page Zero       ir[6:0]
  0     1    Direct Current Page    {pc[11:7],ir[6:0]}
  1     0    Indirect Page Zero     m[ir[6:0]]
  1     1    Indirect Current Page  m[{pc[11:7],ir[6:0]}]
```

Direct page zero computes the Effective Address (EA) as simply the low-order seven bits of the instruction register. This is the only addressing mode used in appendix A. Direct current page computes EA as the high-order five bits of the program counter concatenated to the low-order seven bits of the instruction register. This is useful for programs that do not fit in the 128 words of page zero. Indirect page zero computes EA as the contents of memory pointed to by the low-order seven bits of the instruction register. Similarly, indirect current page computes EA as the contents of memory pointed to by the concatenation of the high-order five bits of the program counter and low-order seven bits of the instruction register. These indirect addressing modes are useful when the address of data varies during runtime, and also in conjunction with the JMP instruction to return from a subroutine (called by a JMS instruction) or from an interrupt service routine.

The indirect addressing modes are slower, but more powerful, than the direct addressing modes since the EA comes from memory. First, the machine obtains the address of the EA from the instruction register (and possibly the program counter). Next, it accesses memory to obtain the EA. Finally, it accesses memory to obtain the data.

Autoincrement occurs on the PDP-8 with indirect addressing when the address of the EA (not the EA itself) is between 0010_8 and 0017_8. In these eight cases, the EA in memory is incremented *prior* to execution of the instruction. For example, the instruction 1417_8 increments the word at m[0017_8], and then adds m[m[0017_8]] to the accumulator.

Non-memory reference instructions

Group 1 microinstructions

1. <u>CLA</u> (7200_8) - *CL*ear the *Ac*cumulator, bit 7 on. This instruction sets the `ac` to 0000_8.

2. <u>CLL</u> (7100_8) - *CL*ear the `link`, bit 6 on. This instruction sets the `link` to 0.

3. <u>CMA</u> (7040_8) - *CoM*plement the *Ac*cumulator, bit 5 on. This instruction complements (sets all 1's to 0's and 0's to 1's) the `ac`.

4. <u>CML</u> (7020_8) - *CoM*plement the `link`, bit 4 on. This instruction complements the `link`.

5. <u>RAR</u> (7010_8) - *R*otate the *Ac*cumulator and `link` *R*ight, bit 3 on. This instruction shifts bit 11 through bit 0 one position to the right. The `link` shifts to bit 11 and bit 0 shifts to the `link`. All other bits shift one position to the right.

6. <u>RTR</u> - (7012_8) - *R*otate the accumulator and `link` *T*wice *R*ight, bit 3 and 1 on. Bit 0 shifts to bit 11, the `link` shifts to bit 10 and bit 1 shifts to the `link`. All other bits shift two positions to the right.

7. <u>RAL</u> (7004_8) - *R*otate the *A*ccumulator and `link` *L*eft, bit 2 on. This instruction shifts bit 10 through 0 one position to the left. The `link` shifts to bit 0 and bit 11 shifts to the `link`. All other bits shift one position to the left.

8. <u>RTL</u> (7006_8) - *R*otate the accumulator and `link` *T*wice *L*eft, bit 2 and 1 on. Bit 11 shifts to bit 0, the `link` shifts to bit 1 and bit 10 shifts to the `link`. All other bits shift two positions to the left.

9. <u>IAC</u> - (7001_8) - *I*ncrement the *AC*cumulator, bit 0 on. Adds 1 to the contents of the `ac`. If the `ac` is 7777_8, the `link` will be complemented (as in the CML instruction). This allows the `link` and `ac` to act together as a 13- bit counter register.

10. <u>NOP</u> (7000_8) - *No OP*eration, bits 0-7 off.

The Group 1 Microinstructions can be combined together. For example CLA CLL is 7300_8.

Group 2 microinstructions

1. SMA (7500_8) - *Skip on Minus* Accumulator, bit 6 is 1_2 and bit 3 is 0_2. Normally used with signed data. Skips the next instruction if the value in the ac is negative.

2. SPA (7510_8) - *Skip on Positive* Accumulator, bit 6 is 1_2 and bit 3 is 1_2. Normally used with signed data. Skips the next instruction if the value in the ac is positive.

3. SZA (7440_8) - *Skip on Zero* Accumulator, bit 5 is 1_2 and bit 3 is 0_2. Skips the next instruction if the value in the ac is equal to zero.

4. SNA (7450_8) - *Skip on Non-zero* Accumulator, bit 5 is 1_8 and bit 3 is 1_2. Skips the next instruction if the value in the ac is not equal to zero.

5. SZL (7430_8) - *Skip on Zero* link, bit 4 is 1_2 and bit 3 is 1_2. Skips the next instruction if the link is 0_2.

6. SNL (7420_8) - *Skip on Non-zero* link, bit 4 is 1_2 and bit 3 is 0_2. Skips the next instruction if the link is not equal to zero.

7. SKP (7410_8) - *SKiP* unconditionally, bit 3 is 1_2. Skips the next instruction.

8. HLT (7402_8) - *HaLTs* the computer. Implemented by setting the HALT bit.

9. OSR (7404_8) - Inclusive *Or* of the *Switch Register* with the ac. The result is left in the ac and the original content of the ac is destroyed.

Note that all the memory reference instructions begin with 0_8 to 5_8, and that all the non-memory reference instructions (group 1 and group 2 microinstructions) begin with 7_8. The I/O (Input/Output) instructions are not given here, but they all begin with 6_8.

Interrupts are external signals that cause temporary suspension of the fetch/execute cycle. On the PDP-8, there are two instructions, ION (6001_8) and IOF (6002_8) that control whether interrupts are ignored. ION sets the interrupt enable flag, and IOF clears it. On the PDP-8, an interrupt is ignored unless the last instruction was not 6001_8 and interrupt enable flag is 1. If these conditions are met, the interrupt causes the same action as executing the instruction 4000_8 without fetching such a machine code from memory. The interrupt also causes the interrupt enable flag to become 0. At that point, the fetch/execute cycle resumes. At the end of the interrupt service routine, the programmer must put an ION instruction followed by a JMP indirect instruction.

C. COMBINATIONAL LOGIC BUILDING BLOCKS

Combinational logic (also known as *combinatorial logic*) is the term used to describe the kind of digital hardware whose output depends only on its inputs. Combinational logic is critical to the operation of all computers; however, by itself, combinational logic has no memory. Because of this inability for combinational logic to "remember" previous results, combinational logic, by itself, is insufficient to implement non-trivial algorithms. Combinational logic needs to be combined with sequential logic (see appendix D) to implement such algorithms. The early chapters of this book explain the design process by which an algorithm is transformed into a structure composed of a mixture of combinational and sequential logic. This appendix provides a review of the combinational logic building blocks used throughout this book, and the notations used to describe them.

This appendix does not focus on circuit diagrams or netlists. Instead it describes things at a higher level, known as block diagrams. As described in section 2.5, hierarchical design (the relationship between circuit diagrams and block diagrams) allows us to look at a design at several levels of detail. Designers should work at the highest possible level, which means the notation used should conceal as much of the detail as possible. Lower levels of detail (such as the gate level, circuit diagram or netlist levels) must be dealt with at some point. Modern approaches, such as Verilog synthesis tools, have largely eliminated the need for designers to manipulate these lower level details manually. The bottom-up skills traditionally taught in an introductory digital design course (dealing with optimization of gates) are precisely those manipulations that nowadays are carried out automatically. This appendix assumes the reader has had enough exposure to such details previously to believe that they can be carried out automatically. Instead, sections C.2 through C.11 focus on a top-down approach, based on combinational logic building blocks. First, section C.1 discusses how detailed we might want to be when describing these devices.

C.1 Models of reality

All scientific and engineering disciplines use simplified and idealized models of reality that are easier to describe with mathematics than the reality itself. The role of the computer scientist or engineer is to create a useful product and get it to market rapidly. Such a designer does this by applying a model of reality to a practical problem. By using a model of reality that is too complicated, the designer will be burdened with unnecessary details, and the product will be late to market. The designer needs to choose a model of reality that is appropriate, as illustrated by the following planetary analogy. A planetary model that says that the sun orbits around the earth in a perfect circle every twenty-four hours is an acceptable model of reality for everyday problems. The mathematical simplicity of a circle is compelling, but there are problems where the highly simplified model is insufficient. A more accurate model would say that the earth orbits the sun in an elliptical path as the earth itself rotates. Although not as simple as a circle, an ellipse is still fairly straightforward to describe with simple mathematics. For some problems the elliptical model would also be insufficient, and a very complex model, considering lunar interaction, etc., might be required.[1]

C.1.1 Ideal combinational logic model

Speed and cost are not the first concerns of the designer. Producing a design that implements a correct algorithm is the top priority. For this reason, it will be convenient to think of combinational logic as being instantaneous. Such idealized combinational logic cannot exist in the physical world and is analogous to saying the sun orbits around the earth. Although an idealized model may seem too simple, it is the proper model for automatic Verilog synthesis, which helps ensure that the designer gets the product to market on time. Most of this book (with the primary exception of chapter 6) assumes idealized combinational logic.

C.1.2 Worst case delay model

Just as in the planetary analogy, sometimes the problem will demand something more accurate. As illustrated in chapter 6, there are problems where the computer designer must meet certain speed and cost constraints. Rather than jumping from no detail to every detail, it would be nice to have a simple, but reasonably accurate, model, analogous to the elliptical model of planetary motion. In computer design, the worst case propagation delay model satisfies this need. Worst case propagation indicates the maximum number of gates through which a signal change must pass in the worst case. An assumption commonly used in this model is that each gate has a delay of one unit of $time in a Verilog simulation.

C.1.3 Actual gate-level delay model

The delay in a combinational logic device depends on the values being processed by that device. Sometimes, the delay may be shorter than predicted by the worst case model. As described in chapter 6, it is difficult to consider all the possible paths through

---

[1] In fact, such complex planetary models are only practical because of electronic computers that can simulate these complex interactions.

the gates that compose the device. A Verilog simulator is a tool that allows the designer to try out many combinations of values to see how long it takes for the simulated combinational device to process the information under different circumstances.

C.1.4   Physical delay model

The most accurate but cumbersome model considers all the physical and geometric factors that compose the machine. Such a model considers physical laws that govern analog electronics, including factors such as the speed of light, capacitance, inductance, etc. Although there are times when computer designers must confront these harsh realities, the goal of a good top-down technique is to insulate the designer from physical reality to as great an extent as possible. This book never considers this level of detail.

C.2   Bus

The fundamental building block of all combinational logic is the bus. A bus is a device that transmits information from a source to a destination. The symbol for a bus is a line with a slash drawn through it. Next to the slash is a number, which indicates the number of bits that the bus transmits at any instant.

C.2.1   Unidirectional

A bus is either unidirectional or bidirectional. Most of the buses used in this book are unidirectional. A unidirectional bus is drawn as a line with an arrow pointing in one direction. The arrow indicates the direction in which information flows through the bus. For example, the following is a four-bit unidirectional bus:

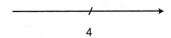

4

Figure C-1. Symbol for a four-bit unidirectional bus.

Being four bits wide, this bus can only transmit numbers that range from 0 to 15.

All but a few experimental computers have been built with electronics. Although other technologies besides electronics (such as the relatively new field of photonics) can implement these abstractions, modern synthesis tools and the entire computer industry are oriented toward the following electronic approach. In this electronic approach, the abstraction of an n-bit-wide bus is physically implemented as n "wires" running conceptually in parallel to each other. For example, the above four-bit bus would actually be four "wires":

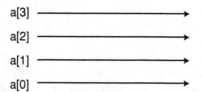

Figure C-2. Implementation of a four-bit bus.

Each "wire" transmits one bit of binary information from the source. For example, to send the number a=15 from the source to the destination, all four transmit a one:

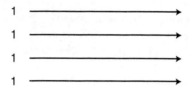

Figure C-3. Transmitting 15 on a four-bit bus.

On the other hand, to send the number a=7, the most significant "wire" instead transmits a zero:

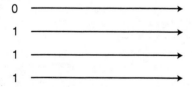

Figure C-4. Transmitting 7 on a four-bit bus.

C.2.2 Place and route

Whether or not the physical implementations of these four "wires" (composed perhaps of a connection between insulated wire dangling in the air, copper plating on a circuit board and traces within an integrated circuit) actually run geometrically parallel to each other is irrelevant. For example the following:

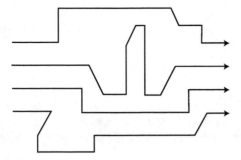

Figure C-5. One possible routing of a four-bit bus.

is equivalent to figure C-4. Such geometric details need not concern us because tools (known as place and route) determine this automatically.

C.2.3 Active high versus active low

At any instant, each "wire" will be at one of two voltages, known as *high* and *low*. The physical values of these voltages seldom concern the designer.[2]

Voltage by itself is not information. The goal of this book is to describe how to design machines that process binary information. Additional abstractions are necessary to relate physical voltages to the binary information being processed by an algorithm. On each wire, one of two possible abstractions is chosen (perhaps by the synthesis tool rather than the human designer) to forever describe how a voltage on that wire translates into binary information. These two abstractions are *active high* and *active low*.

In the active high abstraction, the high voltage means 1 and the low voltage means 0. In the active low abstraction, the opposite holds.[3] The easiest approach is to assume all wires are active high, which is the approach used in this book in order to avoid confusion. Beware that with actual physical chips it is common that some wires will be

---

[2] The numeric values of these voltages vary depending on the technology used. Typically, the lower the voltage, the faster the machine. For the rugged TTL logic families commonly used in educational labs, high is five volts, and low is zero volts. Faster, more modern but less rugged chips based on CMOS use lower voltages, such as 3.3 volts. Slow vacuum tube machines of the 1950s used around +50 volts and -25 volts.

[3] When all the signals are active high, the system is know as positive logic. When all the signals are active low, the system is known as negative logic. When the system is a mixture of both, it is known as mixed logic (not to be confused with the very different concept in Verilog of mixed behavioral structural design, as described in chapter 4).

active low while other are active high. If you use test equipment to observe the operation of an actual physical chip, you must understand the active low abstraction; however, during the design process, you can ignore this confusing issue.

C.2.4   Speed and cost

In this book, we assume an ideal bus, even if other combinational devices in the design are not ideal. Such an ideal bus, which would transmit a signal change from the source to the destination in zero seconds, cannot exist in the physical world.

Unlike the other devices described later in this appendix, the speed of a unidirectional bus does not depend on gate-level propagation delay. Therefore, in the worst case model, the speed of a bus is also instantaneous. As described in chapter 6, Verilog allows a designer to simulate propagation delay on other devices, but buses in Verilog typically have no delay.

The physical speed of a unidirectional bus can be determined by dividing its geometric length by a constant which describes how fast a change in a signal travels along the bus. For most electronic buses, this constant is approximately the speed of light, which is roughly one third of a meter (one foot) per nanosecond.

The reason unidirectional buses have been preferred is that they are extremely cheap and fast compared to the other combinational devices described later that are built out of gates. Bidirectional buses (appendix E) have gate-level propagation delay, like any other combinational device.

Cost is usually related to how much area a device takes on a chip. The area of a bus is at least the width of a wire times its geometric length times the number of bits in the bus. The area of a bus may be larger when problems occur in place and route.

As technology has improved, the relative speed and cost advantage of buses have diminished. In modern "deep submicron" silicon fabrication, interconnection delay, particularly between chips, is a significant factor.

C.2.5   Broadcasting with a bus

Another advantage of unidirectional buses is that they allow broadcasting of information from a single source to a reasonably large number of destinations[4] at little additional cost. It is a common misconception among novice hardware designers that to send the same information to two places in a design requires some special device:

---

[4] How many destinations is determined by the fanout of the logic family.

Figure C-6. Unnecessary device.

This misconception is understandable since to make b and c synonymous with a in software requires two explicit (and possibly expensive) steps:

```
b=a;

c=a;
```

Accomplishing the same thing in hardware is essentially free. You simply run the bus two different places, and refer to the same physical bus by different names at these new locations:

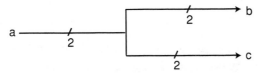

Figure C-7. Transmitting on one bus to multiple destinations for free.

One of many geometric arrangements of "wires" that can accomplish this is:

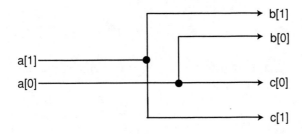

Figure C-8. Implementation of figure C-7.

Note there is no connection between a[1] and a[0], although there are connections between a[0], b[0] and c[0] and between a[1], b[1] and c[1]. Within the time it takes light to travel the physical distance of the bus, the voltages at b[1] and c[1] will be the same as a[1], which means that bit of information has been transmitted to those two locations.

It is common for a designer to use different names for the same bus, but doing so can be confusing. It would be better whenever possible to use the same name at both the source and destinations:

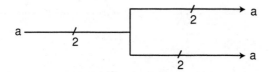

Figure C-9. Using the same name at every node.

since this more accurately reflects physical reality. Nevertheless, there will be times when it is advantageous to re-label the same bus with different names. A rose by any other name is just as sweet, and a bus by any other name is just as cheap.

C.2.6 Subbus

There are certain operations in the binary number system that are trivial to implement. For example, unsigned division by two, b=a/2 (also known as shift right, b=a>>1) appears to require some special device:

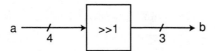

Figure C-10. Combinational device to divide by two (three-bit output).

but in fact can be implemented at no cost simply by rearranging how a subset of the wires of the bus a is connected to b:

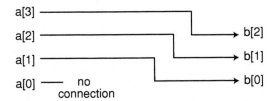

Figure C-11. Implementation of figure C-10.

This subset bus is known as a subbus. A designer can select any bits of the source bus to form a subbus. The notation we use for this is the concatenation syntax of Verilog, which separates the name of the individual wires with commas inside { }. For example, the bus b can be described as {a[3],a[2],a[1]}. Since subbuses that take a continuous group of bits from the source are common, there is another notation, known as bit select, that can be used: a[3:1] means the same as {a[3],a[2],a[1]}.

If the destination was also supposed to be four bits:

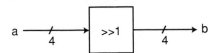

Figure C-12. Combinational device to divide by two (four-bit output).

b[3] would have to be tied to a constant 0:

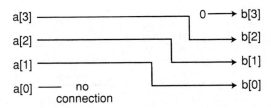

Figure C-13. Implementation of figure C-12.

In concatenation notation, this[5] would be {0,a[3],a[2],a[1]} or simply {0,a[3:1]}

---

[5] In correct Verilog notation, the constant 0 would have to be described as 1'b0.

C.3 Adder

Many algorithms, even those that are not primarily mathematical, often need to do addition of binary numbers. One way to accomplish this is to provide a combinational logic unit that performs this computation:

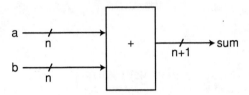

Figure C-14. Combinational device to add two n-bit values (n+1 bit output).

The block diagram symbol for an adder is simply a rectangle with a "+" or the word "adder" inside it. The number of bits in the output bus is one more than the number of bits in the larger of the input buses to allow for the largest possible sum. Note that a, b and sum are typically unsigned. (The low-order n bits of sum are also valid when they are signed twos complement; however there are complications with signed values beyond the scope of the discussion here.)

C.3.1 Carry out

It is common for the extra bit in the sum to be broken into a separate carry out (cout) signal, with wordsum being a subbus:

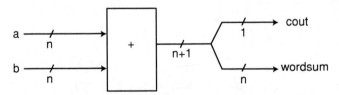

Figure C-15. Treating the high-order bit as carry out.

where sum={cout,wordsum}. The above is often drawn as:

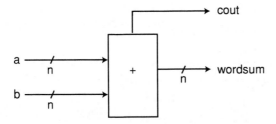

Figure C-16. Alternate symbol for figure C-15.

where the n-bit `wordsum` is a valid result that fits within the same sized word as a and b only when `cout` is 0. When `cout` is 1, an *overflow* error is said to have occurred. The physical implementation of this approach is identical to the earlier view of the adder. The advantage of this view is that all buses are the same width, which often simplifies the design of a larger system. The disadvantage is that `cout` must be observed while the system is in operation to detect the possibility of an error. Sometimes, however, the designer has a priori knowledge that `wordsum` is small, and so `cout` can be ignored.

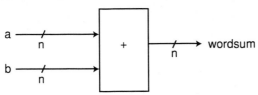

Figure C-17. Adder without carry out.

C.3.2   Speed and cost

There are many ways that adders can be implemented physically. One of the most common techniques is the ripple carry approach, which requires a circuit composed of about $3*n$ OR gates (each having two inputs) and $6*n$ AND gates (each having two inputs). Another way to state this is that it requires n full-adder modules (as described in sections 2.5 and 3.10.6). The worst case propagation delay for a ripple carry adder is proportional to n (as illustrated in section 6.3).

Faster techniques exist that require more gates. Some commonly used techniques include carry lookahead and carry skip.

C.4 Multiplexer

The multiplexer (commonly referred to as a mux) is the most important combinational logic building block next to the bus. Its purpose is to select one of its inputs to be its output and to ignore its other inputs. This "out of many, choose one" behavior is symbolized in this book as a triangle, whose tip is the one chosen output:

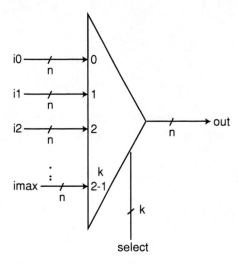

Figure C-18. Symbol for multiplexer.

Some people draw muxes as a rectangle with the word "mux" written inside. The mux has a select input, which is k bits wide. The mux also has (at most) 2^k other buses that are data inputs (i0, i1, i2, ... imax), each n bits wide. The mux has one data output which is n bits wide. If any input bus has fewer bits, assume zeros are concatenated on the left.

C.4.1 Speed and cost

There are several ways a mux can be implemented physically. In the most common approach, the mux shown above would be implemented using n OR gates (each having 2^k inputs), $n*2^k$ AND gates (each having k inputs) and k inverters. This approach needs only three stages of propagation delay. Sometimes it is possible to reduce this down to two stages (by eliminating the need for the inverters[6]) and so muxes implemented this way are quite fast.

C.5 Other arithmetic units

Although addition is the arithmetic operation for which we typically use a combinational building block, other operations can be implemented similarly. This section describes several arithmetic operations for which it is reasonable to fabricate specialized combinational logic.

C.5.1 Incrementor

One of the most common operations is adding one to a number:

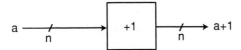

Figure C-19. Symbol for incrementor.

Although conceptually this could be implemented as:

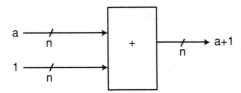

Figure C-20. Inefficient implementation of incrementor.

it is better to specify an incrementor if that is all the problem needs. Using a general adder is inefficient both in terms of speed and cost.

---

[6] The need for inverters can be eliminated in so-called *dual rail* designs, where every signal is provided in both active high and active low form. The reason the inverters are not needed is because certain devices, such as flip flops, naturally provide both active high and active low versions of the same signal at no extra cost.

C.5.1.1 Speed and cost

Although ripple carry addition of two arbitrary numbers requires n stages of worst case propagation delay, incrementation can be done in only two stages of propagation delay using n-1 OR gates (each with two inputs), 2*n-2 AND gates (each with two inputs), n-1 AND gates (of various sizes) and some inverters.

C.5.2 Ones complementor

The ones complement, -a-1 (also known as bitwise not, ~a):

Figure C-21. Symbol for ones complementor.

is often part of a larger computation.

C.5.2.1 Speed and cost

The ones complement only takes n inverters, and one unit of propagation delay.

C.5.3 Twos complementor

Forming the negative of a signed number is necessary in many algorithms:

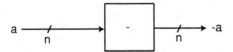

Figure C-22. Symbol for twos complementor.

This can be implemented as

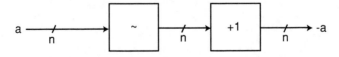

Figure C-23. Possible implementation of twos complementor.

C.5.4 Subtractor

The building block for a combinational logic subtractor is analogous to addition.

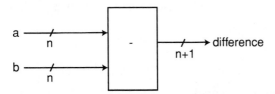

Figure C-24. Symbol for subtractor.

Although a and b are unsigned, `difference` is a signed twos complement value. The additional bit in the output indicates whether the difference is positive or negative.

One approach to implement a subtractor would be to use an adder and a twos complementor. A more efficient but less common approach would be to derive specialized logic for subtraction ("full subtractors").

C.5.5 Shifters

Multiplication and division by constant powers of two can be accomplished at essentially no cost through subbusing and concatenation. For example, multiplication by 4 (shifting left two places):

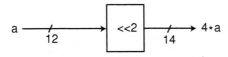

Figure C-25. Symbol for shifter.

simply concatenates a to two bits that are zero on the right. The reason this does not cost anything is because the power of two is a constant.

Sometimes the shifter has another input, known as the shift in (si), that allows the designer to specify what the least significant bits are:

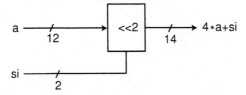

Figure C-26. Symbol for shifter with shift input.

Again, such a device is essentially free because it is implemented as the concatenation of a to si.

A barrel shifter allows a variable number of places for the shift. The number of places to shift is given by a k-bit shift count (sc) bus:

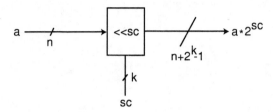

Figure C-27. Symbol for barrel shifter with shift count input.

This can be implemented in two (or three) levels of worst case propagation delay as constant shifters and a mux:

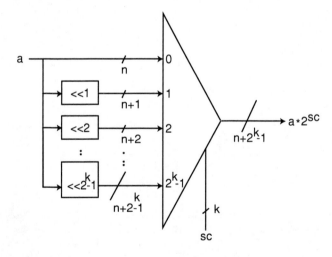

Figure C-28. Possible implementation of barrel shifter.

An alternative implementation which is slower but less costly uses k muxes, each with two inputs.

A similar right shifter can be implemented for division by variable powers of two. Barrel shifters can be arranged to allow for both multiplication or division by arbitrary powers of two, and to allow for arbitrary shift input (rather than concatenation with zeros).

C.5.6 Multiplier

The left barrel shifter in the last section only allows multiplication by a power of two. Many algorithms need multiplication by variables that are not powers of two. The multiplier is a fairly costly hardware device that allows two arbitrary numbers to be multiplied:

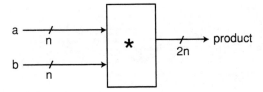

Figure C-29. Symbol for multiplier.

Note that the product has twice as many bits as the input buses. We will normally assume that a, b and product are unsigned. It takes a physically different device to multiply signed numbers.

C.5.6.1 Speed and cost

There are many ways to do multiplication; however the most commonly used techniques require on the order of n^2 gates, and at least twice the worst case propagation delay of addition. Because the cost and speed of a combinational multiplier is so high compared to the devices discussed in this appendix, a slower but cheaper approach involving sequential logic (appendix D) and ASM charts (chapter 2) is often used to generate a product.

C.5.7 Division

Division (by non-powers of two) is even more costly than multiplication when implemented as a combinational logic building block. Division is seldom implemented as combinational logic. Most of this book uses an example of one simple way that division can be implemented using sequential logic and ASM charts.

C.6 Arithmetic logic unit

In many problems, the same building block needs to compute different mathematical functions under different circumstances. A single unit that can handle most of the functions needed for a system is known as an Arithmetic Logic Unit (ALU). A designer can choose to put whatever functionality in an ALU as is appropriate for a particular problem, however it may be convenient[7] to use an ALU that has already been designed, such as the 74xx181.

Regardless of what details are inside the ALU that a designer chooses, the basic principle of how a combinational logic ALU operates is the same. There is a k-bit bus, aluctrl, that customizes the ALU for the particular function that needs to be computed.

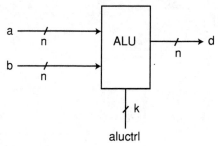

Figure C-30. Symbol for Arithmetic Logic Unit (ALU).

Conceptually, an ALU could be implemented as a mux which selects from the various combinational functions which that particular ALU is capable of performing:

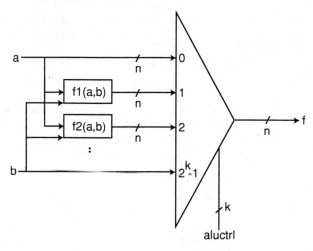

Figure C-31. Possible implementation of ALU.

---

[7] Especially in an educational lab setting.

Note that ALUs typically allow the passing through of a or b unmodified, and so an ALU can also serve the role of a mux. In physical reality, the implementation of an ALU might be quite different than a mux, but conceptually an ALU is equivalent to the above.

C.6.1   Logical operations

As its name implies, an ALU can perform both arithmetic and logical operations. Logical operations are those in which each bit of f depends only on the corresponding bit in a and also possibly on the corresponding bit in b. For example, consider the operation of a four-bit ALU doing the bitwise 'AND operation, f=a&b:

```
            a       0011
            b       1001

            f       0001
```

In other words, f[3]=a[3]&b[3], f[2]=a[2]&b[2], f[1]=a[1]&b[1] and f[0]=a[0]&b[0]. These bitwise operations are the only dependencies in computing 'AND. For example, in computing f[1], there is no dependence on a[0] or b[0]. Breaking a bitwise operation such as a&b apart into separate single-bit logic equations as shown above is known as *bit blasting*.[8] Bit blasting is one of the many trivial but tedious details of hardware design that designers seldom need be concerned with because Verilog synthesis tools do such things automatically. Of course, the designer needs to understand that an n-bit-wide operation like a&b eventually becomes n separate AND gates operating independently. From this knowledge, it is easy to understand the worst case propagation delay of a&b is only one unit of gate delay, regardless of how many bits are in a and b.

Mathematically, there are only sixteen primitive bitwise logical operations involving no more than two variables. All other formulas involving no more than two variables and involving only combinations of these sixteen primitive operations can be simplified by the laws of Boolean algebra to one of these sixteen operations. The following is a table of the sixteen primitive logical operations:

---

[8] This quite descriptive term was coined by Synopsys, the pioneering vendor of Verilog synthesis tools in the early 1990s.

mnemonic	aluctrl	alternative aluctrl	operation
'NOT	000010		f = ~a;
'NOR	000110		f = ~(a|b);
'ANDNA	001010		f = (~a)&b;
'ZERO	001110		f = 0;
'NAND	010010		f = ~(a&b);
'NOTB	010110		f = ~b;
'XOR	011010		f = a^b;
'ANDNB	011110		f = a&(~b);
'ORNA	100010		f = (~a)|b;
'EQU	100110		f = ~(a^b);
'PASSB	101010		f = b;
'AND	101110	101101	f = a&b;
'ONES	110010		f = -1;
'ORNB	110110		f = a|(~b);
'OR	111010	000100	f = a|b;
'PASS	111110	000000	f = a;

The mnemonic column gives arbitrary names to these sixteen operations which will be used throughout this book.

When designing an ALU for a particular problem, it may not be necessary to include all of the sixteen mathematically possible operations inside the ALU. Omitting some of these operations may economize the total area required for the ALU and also may reduce the number of bits required for aluctrl.

The 74xx181 is an ALU that implements all sixteen of the possible logical operations. Since it also implements many arithmetic operations, it needs a six-bit aluctrl. In the above table, the aluctrl and alternative aluctrl columns indicate the six-bit pattern that must be input to the 74xx181 in order to perform the desired operation. For certain operations, such as 'AND, 'OR and 'PASS, more than one bit pattern can be used to produce the desired result.

C.6.2 Arithmetic operations

In contrast to logical operations, arithmetic operations are those where a change in one bit position of a or b potentially affects several bit positions of f. For example, consider addition ('PLUS) with the same ALU as the last example using the same values for a and b:

a	0011	
b	<u>1001</u>	
f	1100	

The fact that a[0] and b[0] are both one in this example ultimately affects f[0], f[1] and f[2]. This ripple effect is why addition has a worst case propagation delay proportional to n.

The following table shows some of the most useful arithmetic operations available in the 74xx181 ALU:

'INCREMENT	000001		f = a+1
'DECREMENT	111100		f = a-1
'PLUS	100100		f = a+b
'DIFFERENCE	011001		f = a-b
'DOUB	110000		f = 2*a
'DOUBINCR	110001		f = 2*a+1

Because the 74xx181 is a low-cost ALU which is readily available for educational laboratory experiments, it does not implement multiplication or division. Section 2.3.1 shows how to use one of these ALUs to implement division using a slow but simple algorithm.

C.6.3   Status

An ALU commonly outputs extra information besides just the n-bit-wide result, f. For example, the 74xx181 has two status outputs that provide information about the computation currently being performed by the ALU. The first of these, cout, comes from the ALU's internal adder. It may be used to detect overflow ($a+b>=2^n$) when 'PLUS is being performed, and to detect a<b when 'DIFFERENCE is being performed.

The second status output, zero, detects whether f==0. It may be used to detect whether a==b when 'DIFFERENCE is being performed.

C.7   Comparator

There are six mathematical relational operators (==, !=, <, >, <= and >=). The vast majority of useful algorithms use one or more of these to make decisions that determine how the algorithm proceeds. Although the status outputs of an ALU may be able to answer such questions, it is often not efficient to use an ALU to do so. Instead, a specialized combinational building block, known as a comparator, is used instead. A comparator has two n-bit-wide input buses, a and b. At most, a comparator has three bits of output:

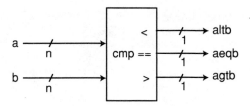

Figure C-32. Symbol for comparator.

From these three outputs, the other three conditions can be derived, for example ageb=agtb|aeqb. Many problems only need an equality test:

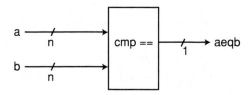

Figure C-33. Symbol for equality comparator.

C.7.1   Speed and cost

A comparator that only provides the equality output is much cheaper than one that also provides inequalities, such as altb. Such an equality only comparator needs 2*n inverters, 2*n AND gates (each having two inputs), n OR gates (each having two inputs) and one AND gate (having n inputs). The propagation delay for an equality only comparator is four units of gate delay under these assumptions. The cost is even lower when one input is a constant.

A comparator that also provides inequality outputs will have a worst case propagation delay proportional to n. It will also use considerably more area than an equality only comparator.

C.8 Demux

The demultiplexor (demux) is a specialized combinational building block which is not used in the early chapters of this book. It plays an important role in implementing concepts found in later chapters.

As the name implies, the demux is the opposite of a mux. The symbol used in this book for a demux reflects this. It is a triangle where the only input to the demux is connected to the tip of the triangle. The many outputs of the demux are drawn on the opposite side:

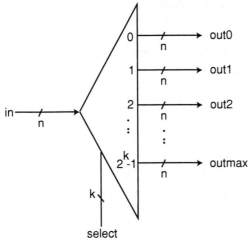

Figure C-34. Symbol for demux.

Like the mux, the demux has a k-bit input bus known as `select`. Some people may draw the demux as a box. All but one of the n-bit output buses will be zero. The selected output bus will pass through unchanged the value on the input bus.

C.8.1 Speed and cost

Demuxes are simply a large collection of AND gates that operate independently. The demux shown above requires $n*2^k$ AND gates (each having $k+1$ inputs) and k inverters. Such an implementation would have a worst case propagation delay of only two gates. Sometimes, the inverters can be eliminated, in which case the propagation delay is only one gate.

C.8.2   Misuse of demuxes

Novice hardware designers often use demuxes where they are unnecessary. For example, suppose part of the time a machine needs to increment a number, a, and part of the time the machine needs to multiply it by two, but the machine never needs to increment and double the number at the same time. Novice designers often put a demux in such a design:

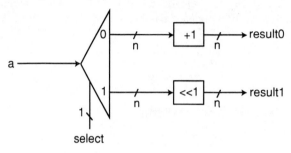

Figure C-35. Misuse of demux.

In the above, result0 is a+1 if select==0 but is 0 otherwise. Also, result1 is a*2 if select==1 but is 0 otherwise. Since it was assumed that a+1 and 2*a do not need to be simultaneously available, the above might work, but it would be considered a bad design for three reasons. First, the demux is an unnecessary and expensive (both in terms of speed and cost). Second, if the problem changes so that result1 is supposed to be 2*a simultaneously with result0 being a+1, the above design would be completely wrong. Third, even if 2*a and a+1 are never needed simultaneously, the designer is burdened with providing the proper select.

It is understandable that novice designers make this mistake. In software, the programmer only specifies the operation required (either a+1 or 2*a) based on select. But in hardware, as was discussed in section C.2.5, it is easier to route a bus to every place where it is needed. The cost of doing this is usually quite low, and certainly less than using a demux:

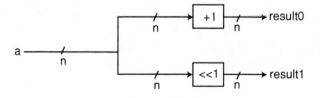

Figure C-36. Proper design omits demux.

There is no harm in `result1` being `2*a` simultaneously with `result0` being `a+1`. **In hardware**, it is often **more economical to compute everything you might need and ignore those results that are not pertinent** under particular circumstances.

Therefore, demuxes are not needed in the early chapters of this book. Demuxes are important in more advanced design topics. Demuxes are important in the design of memory systems (section 8.2.2.3.1) and in the implementation of one-hot controllers (chapter 7).

C.9 Decoders

A decoder is a specialized combinational device that converts from a binary code to some other code. The most common decoder converts from binary to what is called a unary code. The following table lists these codes for the numbers between 0 and 7

value	binary	unary out
0	000	00000001
1	001	00000010
2	010	00000100
3	011	00001000
4	100	00010000
5	101	00100000
6	110	01000000
7	111	10000000

Such a decoder can be thought of in two ways. First, it can be thought of as a building block that simply takes k bits of binary input, and produces 2^k bits of unary output:

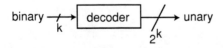

Figure C-37. Symbol for binary to unary decoder.

Second, a binary to unary decoder can be thought of as a building block composed of 2^k comparators. The output of each comparator provides one of the bits of the unary code. The second input of each comparator is connected to a binary constant. Each comparator is comparing against a different k-bit binary constant (from 0 to 2^k-1):

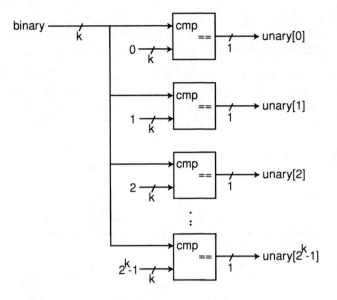

Figure C-38. Possible implementation of decoder.

Finally, an alternative way of looking at a binary to unary decoder is as a demux whose one-bit-wide input bus is tied to 1:

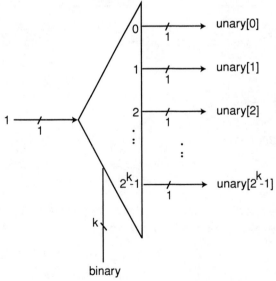

Figure C-39. Alternate implementation of decoder.

C.9.1 Speed and cost

The speed and cost of a binary to unary decoder is similar to a demux.

C.9.2 Other kinds of decoders

Decoders exist that involve other codes besides unary, such as those for seven segment displays. Such decoders are more specialized, and not widely used in computer design.

C.10 Encoders

Sometimes, a designer needs to convert from a unary code to binary. A combinational building block that performs this conversion is known as an encoder. If the designer could be sure the input were always a proper unary code (with exactly one bit that is one), the encoder could be implemented simply with k OR gates. But there are only 2^k valid unary codes out of the very large number (two *raised to the* k) of bit patterns that might appear on the input.

Instead, designers use a priority encoder.

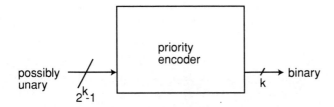

Figure C-40. Symbol for priority encoder.

It outputs the binary code corresponding to the bit position of the least significant leading zero.

000	00
001	01
010	10
011	10
100	11
101	11
110	11
111	11

The priority encoder is useful for counting how many leading bits of a number are zero. This is a computation that is necessary to implement floating-point arithmetic. Priority encoders are also often used so that a general -purpose computer can select which one of several external interrupts has the highest priority.

C.11   Programmable devices

Almost any imaginable mathematical function can be realized as a combinational building block if it involves a small enough number of bits of input. With Verilog synthesis tools available since the mid 1990s, functions involving around sixteen or fewer bits of input are routinely converted into combinational logic without the designer having to worry about their technological or gate-level implementation. The synthesis tool produces a file that can be downloaded into one of many kinds of programmable devices. The process of transferring the design into a programmable device is known as *programming* it. Such programming is a mechanical process, which does not require human intervention or creativity. The term *burning* is sometimes used to mean the same thing as programming. This use of the term programming should not be confused with its use in software (chapter 8), where the term programming means the same thing as design, which, of course, requires lots of creativity.

There are many kinds of programmable devices available, including Programmable Logic Arrays (PLAs), simple and Complex Programmable Logic Devices (CPLDs), Field Programmable Gate Arrays (FPGAs) and Read Only Memories (ROMs). CPLDs and FPGAs also have provision for sequential logic (see appendix D), but ROMs are pure combinational logic. The combinational logic implemented by all ROMs and by many FPGAs are based on truth tables without the need for expressing logic equations. In contrast, PLAs and CPLDs are based on logic equations (sum of products) rather than truth tables. Synthesis tools automatically produce truth tables or logic equations, depending on the target technology the designer selects.

C.11.1   Read only memory

Automatic synthesis of combinational logic for functions involving more than about sixteen bits depends on the complexity of the function. A simple function like addition can be implemented for an arbitrarily large number of input bits with combinational logic because the function decomposes into smaller combinational logic units, e.g., full adders in the case of addition. The synthesis tool is well aware of the properties of commonly used functions like addition. The decomposition of more complicated functions (whose properties are not built into the synthesis tool) is often less obvious. Synthesis tools explore many possible implementations for the combinational logic requested by the designer, however; as the number of input bits increases, the number of design alternatives grows exponentially. It becomes difficult for the synthesis tool to derive the logic equations needed for technologies such as CPLDs. ROMs tend to be the most practical approach for complex functions as the number of input bits grow. This is because with ROMs, all the designer has to do is tabulate the desired behavior, rather than find logic equations that produce that behavior. This avoids using a synthesis tool that has to explore exponential possibilities.

There are several ways of describing a ROM. The usual viewpoint of the designer is that the ROM is a black box, specialized for computing some particular function:

Figure C-41. Symbol for a Read Only Memory (ROM).

The number of input bits, k, and output bits, n, need not be the same, although often $k==n$. The input bus to the ROM is known as the address. The output bus of the ROM is known as the contents. This address and contents terminology is borrowed from memory systems (section 8.2.2.3.1). However, such a ROM is not truly a "memory" because once a value is burned into a ROM, it cannot be changed.[9]

Normally, the designer will indicate more than just the word "ROM" inside the box, since the ROM could be programmed to implement any function. For instance, the designer might need a "square root ROM," or something like that. The designer is then responsible for providing a table of the contents that need to be burned into that particular ROM.

Another viewpoint of the ROM is to describe it in terms of the combinational logic that it implements. A ROM is simply a mux whose data inputs are connected internally to the constants (c0, c1, c2, ... cmax) that the designer has burned into the ROM.

---

[9] So-called Electrically Erasable Programmable ROMs (EEPROMs) are not truly ROMs when the system in which they are used controls their erasure, such as when they are used as memory in general-purpose computers. In the terminology of chapter 8, such EEPROMs are non-volatile memory with slow access time. EEPROMs are however truly ROMs when their erasure is activated only by a separate development system under the control of the designer and not the hardware being designed. Put another way, if the designer does not use an EEPROM's erasure property in the design of the system itself, then the EEPROM is acting like a ROM, which is to say, the EEPROM implements some combinational logic function.

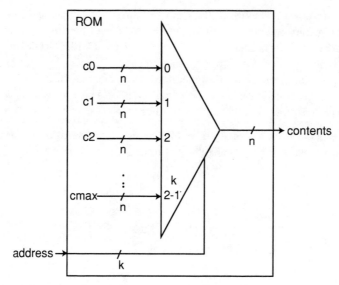

Figure C-42. Possible implementation of a ROM.

C.11.2   Complex PLDs

An alternative to using a ROM is to use a CPLD. The internal structure of a CPLD is too complex to describe here. A designer is seldom concerned with such details. Instead, a synthesis tool takes care of the details for the logic equations required inside the CPLD.

C.12   Conclusions

Combinational logic is an important building block for computer design. The distinguishing characteristic of combinational logic is that it lacks memory: its output is a function of its input. Common combinational building blocks include muxes, demuxes, incrementors, shifters, adders, ALUs, comparators, encoders and decoders. Some devices, such as buses and fixed position shifters, can be implemented at zero cost. Others, such as multipliers, are quite expensive. Ideally, we model combinational logic as a pure mathematical function having no propagation delay, but in reality, different approaches to implementing combinational building blocks have different propagation delays and costs.

Read Only Memories (ROMs) are not actually memory because they do not have the ability to forget. ROMs are simply a different, more convenient, approach for implementing combinational logic. The use of ROMs as well as the use of programmable logic with Verilog synthesis tools has made the design of specialized combinational logic relatively easy.

C.13   Further reading

GAJSKI, DANIEL D., *Principles of Digital Design*, Prentice Hall, Upper Saddle River, NJ, 1997. Chapter 5.

PROSSER, FRANKLIN P. and DAVID E. WINKEL, *The Art of Digital Design: An Introduction to Top Down Design*, 2nd ed., Prentice Hall PTR, Englewood Cliffs, NJ, 1987. Chapter 3.

C.14   Exercises

Using the combinational logic building block devices listed in each of the following problems, give a block diagram that implements the more complex combinational building block described by the data output(s). The buses in these problems should be interpreted as unsigned binary integers.

C-1.   Control Inputs: CTRL (3 bits)
Data Inputs: A (32 bits), B (32 bits), C (32 bits), D (32 bits), E (32 bits)
Data Output: F (32 bits)
Devices: one 32-bit adder, one 32-bit 2-input mux, one 32-bit 4 input mux

CTRL	Data outputs
000	F=A
001	F=B
010	F=C
011	F=D
100	F=A+E
101	F=B+E
110	F=C+E
111	F=D+E

C-2. Control Inputs: CTRL (3 bits)
Data Inputs: A (32 bits), B (32 bits), C (32 bits), D (32 bits), E (32 bits)
Data Output: F (32 bits)
Devices: one 32-bit adder, four 32-bit 2-input muxes

CTRL	Data outputs
000	F=A
001	F=B
011	F=D
100	F=A+E
101	F=B+E
110	F=C+E
111	F=D+E

C-3. Control Inputs: CTRL (3 bits)
Data Inputs: A (32 bits), B (32 bits), C (32 bits), D (32 bits), E (32 bits)
Data Output: F (32 bits)
Devices: one 32-bit 8 input mux, four 32-bit incrementors.

CTRL	Data outputs
000	F=A
001	F=B
010	F=C
011	F=D
100	F=A+1
101	F=B+1
110	F=C+1
111	F=D+1

C-4. Control Inputs: none
Data Inputs: A (8 bits), B (8 bits), C (8 bits)
Data Output: A+B+C+2 (10 bits)
Devices: one 8-bit adder, one 9-bit adder

C-5. Control Inputs: none
Data Inputs: A (32 bits), B (32 bits)
Data Output: max(A,B), min(A,B)
Devices: one 32-bit comparator, two 32-bit 2-input muxes

C-6. Control Inputs: none
Data Inputs: array of four unsorted 32-bit integers
Data Output: same integers in sorted order
Devices: five of the devices from problem C-5

Hint: This is hierarchical design. Do not draw any muxes or comparators.

C-7. Control Inputs: CTRL (3 bits)
Data Inputs: A (32 bits), B (32 bits), C (32 bits)
Data Outputs: D (32 bits), E (32 bits)
Devices: three 32-bit adders, one 32-bit 2 input mux, one 32-bit 8-input mux

CTRL	Data outputs	
000	D=A;	E=A
001	D=B;	E=B
010	D=C;	E=0
011	D=C;	E=0
100	D=A+C;	E=A+C
101	D=B+C;	E=B+C
110	D=A+B;	E=0
111	D=A+B;	E=0

C-8. Control Inputs: ALUCTRL (6 bits), CTRL(1 bit),
Data Inputs: H (8 bits), L (8 bits), M(8 bits)
Data Outputs: F (8 bits), G (8 bits)
Devices: two 8-bit integer ALUs (74LS181),
                one 8-bit 2 input mux

ALUCTRL	CTRL	Data	Output
100100	0	F=H+L;	G=H+2*L
100100	1	F=H+M;	G=H+L+M
101101	0	F=H&L;	G=H&L
101101	1	F=H&M;	G=H&L&M
000100	0	F=H|L;	G=H|L
000100	1	F=H|M;	G=H|L|M

Hint: It is a theorem of Boolean algebra that L&L==L.

C-9. Control Inputs: CTRL(2 bits)
Data Inputs: X (4 bits), Y (4 bits), Z (8 bits),
Data Output: W (9 bits)
Devices: one 8-bit adder, one 4-bit multiplier, any number of 8-bit 2-input
          muxes

CTRL	Data output
00	W=X
01	W=X+Z
10	W=X*Y
11	W=X*Y+Z

C-10.   Control Inputs: CTRL(2 bits)
Data Inputs: X (4 bits), Y (4 bits), Z (8 bits)
Data Output: W (9 bits) Same as in problem C-9.
Devices: any number of adders of any width (specify), any number of shifters, any number of 2-input muxes of any width (specify)

Hint: With these building blocks, you need to implement the 4-bit multiplier using the shift and add algorithm for multiplication, which is analogous to the pencil and paper algorithm for decimal multiplication:

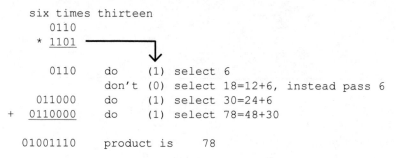

```
six times thirteen
    0110
  * 1101

    0110     do     (1) select 6
             don't  (0) select 18=12+6, instead pass 6
  011000     do     (1) select 30=24+6
+ 0110000    do     (1) select 78=48+30

 01001110    product is    78
```

D. SEQUENTIAL LOGIC BUILDING BLOCKS

Although combinational logic (appendix C) is useful for implementing mathematical functions in hardware, combinational logic has no memory of values that were computed previously. Most practical algorithms make use of old values to compute new ones. Therefore, combinational logic by itself is insufficient to implement interesting algorithms. In addition to combinational logic building blocks, interesting machines must include sequential logic building blocks, commonly referred to as registers, that allow the hardware to remember old values. This appendix reviews several important synchronous sequential logic building blocks.

D.1 System clock

The term synchronous means that all of the sequential building blocks are connected to a single signal, known as the system clock or `sysclk` for short. The place where the system clock is connected is shown as a wedge in the lower-left corner of each synchronous building block:

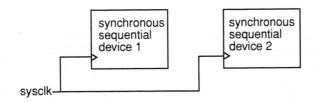

Figure D-1. Universal connection to system clock signal shown.

This connection need not be drawn because it is understood that all synchronous sequential devices connect to this same signal. For example, the following is understood to mean the same as the above:

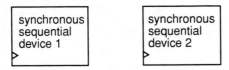

Figure D-2. Universal connection to system clock signal assumed.

D.2   Timing Diagrams

To describe the behavior of sequential logic, it is often helpful to use timing diagrams. A timing diagram plots values against time. For single-bit signals, like `sysclk`, this plot appears similar to some kind of graph you might have drawn in an algebra class.

The concept of a timing diagram originated with the display produced by an oscilloscope. Computers operate at speeds too fast to be observed by the unaided eye. When testing (or repairing) an actual physical computer, a computer designer needs some kind of test equipment to observe signals inside the computer down to a time resolution of about a nanosecond. The oscilloscope is a kind of test equipment that plots voltage versus time on a phosphor screen. The earliest electronic computer designers half a century ago used primitive oscilloscopes, and modern versions of oscilloscopes are still used by computer designers today.[1]   For example, if you were to connect an oscilloscope to the `sysclk` signal, you might see:

Figure D-3. An analog waveform for the system clock signal.

which shows how the analog voltage (vertical axis) on the `sysclk` wire varies with time (horizontal axis). Physical properties, like capacitance and inductance, affect the ragged shape of the analog voltage shown on the oscilloscope. Computer designers are not concerned with analog voltages, and so this rather messy physical reality is abstracted to an idealized square wave:

sysclk

Figure D-4. A digital abstraction of the system clock signal.

Such a square wave is not physically possible; however, as explained in section C.1.1, computer designers often use models of reality that are physically unrealistic because such simplified models emphasize only those things which are algorithmically important.

---

[1] Computer designers now often use more sophisticated kinds of test equipment.

In the case of the `sysclk` signal, the only thing that is important is that it subdivides time into equal-sized intervals, known as clock periods or clock cycles:

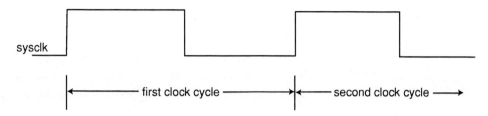

Figure D-5. The system clock divides time into cycles.

Each clock period begins and ends on the rising edge of `sysclk`. (Some kinds of sequential logic use the falling edge; however in this book all synchronous sequential building blocks use the rising edge.)

D.3   Synchronous Logic

Synchronous logic is a restriction on physical reality where changes in the values shown in a timing diagram occur only at the exact instant of the rising edge of `sysclk`. For example, in the following timing diagram, one bit of data is being manipulated by an algorithm:

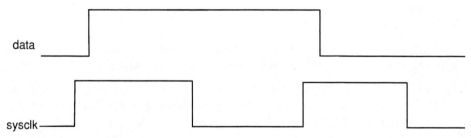

Figure D-6. An ideal synchronous timing diagram.

The above is a valid synchronous timing diagram because the changes in data occur only at the rising edge of `sysclk`. In physical reality, the changes in data occur slightly later than the actual instant of the rising edge due to propagation delay of the circuits used to generate the `data` signal:

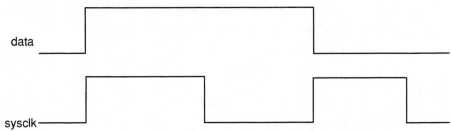

Figure D-7. A realistic synchronous timing diagram with propagation delay.

but as discussed in section C.1.1, we normally ignore propagation delay. At the beginning of the design process, the primary concern of the designer is getting the algorithm right. Worrying about physical reality is a distraction from the designer's most important mission—ensuring that the algorithm is correct.

The following diagram is not synchronous. It is known as *asynchronous* because the data pulse might occur at any time with respect to `sysclk`:

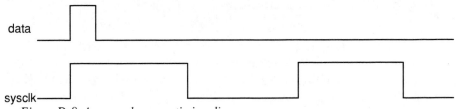

Figure D-8. An asynchronous timing diagram.

With only one exception that happens when a machine is first turned on (described in sections 4.4.5 and 7.1.6), we will not use such asynchronous logic.

Synchronous design is safe and easy. Asynchronous design is hard and dangerous. Commercial synthesis tools concentrate on synchronous design. Therefore, synchronous design is widely used in industry.

D.4 Bus timing diagrams

Digital computers represent values other than zero and one using a group of bits on a bus with the binary number system. The physical reality is that each wire in a bus represents a separate bit of information. But from an algorithmic viewpoint, the designer wants to look at the bus as containing a single binary value. Suppose the value of a variable, v, goes through the sequence 0, 1, 2, 3, 0, 1, 2, 3, 0 .... This could be shown on a timing diagram as two separate bits that change synchronously with sysclk:

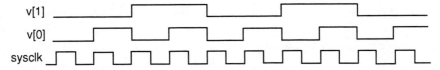

Figure D-9. Timing diagram showing individual bits of a bus.

However dealing with separate bits is quite tedious. Instead timing diagrams usually show the numeric value of the complete bus during each clock cycle:

Figure D-10. Timing diagram showing numeric values on a bus in decimal.

In timing diagrams, the notation shown in figure D-11:

Figure D-11. Notation used for bus timing diagrams.

shows the instant in time (a particular rising edge of sysclk) when the numeric value of the bus changes. It is only necessary for one bit of the bus to change for the numeric value of the bus to be completely different.

D.5   The D-type register

The simplest sequential building block is the D-type (or delay type) register:

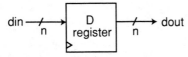

Figure D-12. Symbol for D-type register.

Some people refer to `din` as the D input and `dout` as the Q output. When n=1, this device is referred to as a D-flip flop. In fact, an n-bit D-type register is usually built from n D-type flip flops.

In the D-type register, `dout` is simply a delayed version of `din`. Put another way, `dout` in the present clock cycle is the same as `din` in the previous clock cycle. Suppose that `din` just happens to be going through the binary sequence:

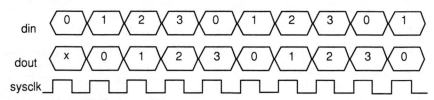

Figure D-13. Example timing diagram for D-type register.

`dout` will also go through the same sequence, but it will lag by one clock cycle. In the above, x means unknown (see section 3.5.3 for details on how `'bx` is used in Verilog simulation), because there is not enough information to predict what is in the register at the beginning.

As another example, consider what happens when `din` is somewhat more random:

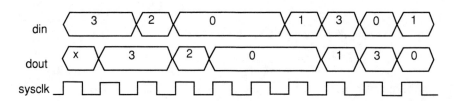

Figure D-14. Another timing diagram for D-type register.

The D-type register is not used by itself in computer design very often. The two most common uses of the D-type register are synchronizers and the present state register for controllers (see sections 2.4.1 and 7.1.1).

All of the more useful registers described below can be constructed from a simple D-type register combined with combinational logic, but it is often not helpful to think of things that way. It is usually better to think in terms of one of the more sophisticated building blocks described below. On the other hand, D-type registers together with specialized logic are often included in designs created by synthesis tools (see section D.11).

The reason the D-type register by itself is often inadequate for many problems is that the D-type register only remembers the old value for one clock cycle. Most algorithms have variables that must remain unchanged for multiple clock cycles. This requires a more sophisticated kind of register, discussed in the next section.

D.6   Enabled D-type register

Algorithms are the starting point for the hardware design approach described throughout this book. Algorithms are composed of steps that manipulate variables at certain moments in time. The rising clock edge determines when those moments occur. The vast majority of algorithms manipulate their variables in complicated ways, so that the variables do not change at **every** rising clock edge. For this reason, we need a kind of register building block that can hold its former contents for multiple clock cycles as well as being able to load itself with new contents. The building block that has such a capability is known as an enabled D-type register, or just simply an enabled register.

In order to allow the designer to choose between these two different actions, the enabled register has a command input. This command input is sometimes known as the load signal or the enable signal. In this book, this command input is typically abbreviated with a name like ld.

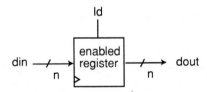

Figure D-15. Symbol for enabled D-type register.

The following action table describes what the enabled register does based on the ld input:

ld	action
0	hold
1	load

An action table is not a truth table, because unlike a truth table, an action table includes the concept of time.

For example, suppose the following din and ld signals are provided to an enabled register:

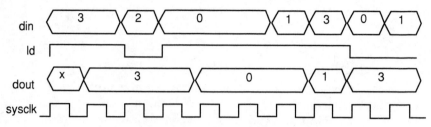

Figure D-16. Example timing diagram for enabled D-type register.

In this example, ld happens to be 0 at certain times when dout happens to be 3. This means in the clock cycle after ld is 0, dout will continue to hold the value 3, regardless of what din happens to be. On the other hand, when ld is 1, the enabled register acts just like a simple D-type register.

The enabled register can be implemented as a mux connected to a simple D-type register:

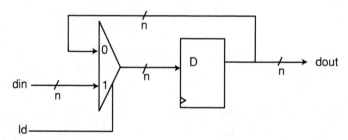

Figure D-17. Implementation of enabled D-type register using simple D-type and mux.

When ld is 0, the mux passes through the old value of dout to be reloaded into the simple D-type register. When ld is 1, the mux passes through the new din value to be loaded into the simple D-type register.

Other arrangements of hardware not based on the simple D-type register can also implement an enabled register. Therefore, in the top-down approach, designers typically specify an enabled (or loadable) register without concern for how it is implemented.

In the TTL logic family, the 74xx377 (for n=8) and 74xx378 (for n=6) chips implement the same actions as the above.[2]

D.7 Up counter register

When combined with combinational logic, the enabled D-type register is sufficient to implement any algorithm, however, certain register operations occur so frequently that these operations deserve special implementation as sequential building blocks in their own right. The distinguishing characteristic of these operations is that the register has within itself all the information necessary to perform the operation.

Perhaps the most important of these special operations is counting. Most algorithms include steps that involve counting. In fact, the very first practical machine ever built with digital electronics (by Wynn-Williams in 1932) was a counter used to count alpha particles for a physics experiment conducted by Lord Rutherford. Since that time, billions of counters have been fabricated.

There are many variations on how to build a counter. In this book, we will concentrate only on the two most important kinds of counters: the synchronous loadable binary up counter (described in this section), and the synchronous loadable binary up/down counter (described in section D.8). We will refer to these more simply as the up counter and the up/down counter, respectively. When the word *counter* is used by itself in this book, it means the synchronous loadable binary up counter.

The up counter has three command inputs. The ld command signal is the same as it is in an enabled register. The clr command signal causes the counter to become zero at the next rising edge of the clock. The count command signal (sometimes referred to as the inc command signal) causes the counter to increment at the next rising edge of the clock.

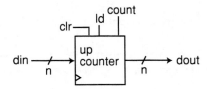

Figure D-18. Symbol for up counter register.

---

[2] Except ld is active low, which has the apparent effect of reversing the 0 and 1 inside the mux of the TTL chip.

The behavior of the up counter is summarized by the following action table:

ld	clr	count	action
0	0	0	hold
0	0	1	increment
0	1	0	clear
0	1	1	clear
1	0	0	load
1	0	1	load
1	1	0	load
1	1	1	load

Note that the ld signal has a higher priority than clr and count. Also clr has a higher priority than count. An up counter can be constructed from a simple D-type register, three muxes and an incrementor:

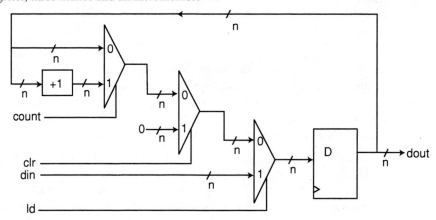

Figure D-19. Implementation of up counter register using simple D-type register and combinational logic.

Recall that the combinational logic incrementor (section C.5.1) is considerably faster than an adder. Even so, there are other more efficient ways of constructing a counter than the technique shown above. For example, in the TTL logic family, the 74xx163 chip provides for n=4 the same actions[3] as the above using fewer gates and less propagation delay.

---

[3] Except that clr and ld are active low.

D.8 Up/down counter

Some algorithms involve both incrementing and decrementing the same variable. For such algorithms, the use of an up/down counter may be appropriate. The up/down counter has three command inputs. The `ld` command signal is the same as in the earlier registers. The `count` command signal causes the counter to increment or decrement at the next rising edge of the clock, depending on the `up` command signal. If `up` is 1 when `count` is 1, the counter increments. If `up` is 0 when `count` is 1, the counter decrements.

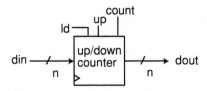

Figure D-20. Symbol for up/down counter register.

The behavior of the up/down counter is summarized by the following action table:

ld	count	up	action
0	0	0	hold
0	0	1	hold
0	1	0	decrement
0	1	1	increment
1	0	0	load
1	0	1	load
1	1	0	load
1	1	1	load

An up/down counter can be constructed from a simple D-type register, two muxes, a combinational logic incrementor and a combinational logic decrementor:

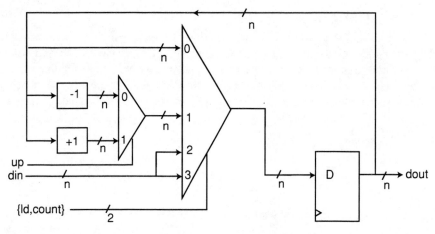

Figure D-21. Implementation of up/down counter register.

There are other ways of constructing this than the technique shown above. For example, in the TTL logic family, the 74xx669 chip provides for n=4 the same actions as the above using fewer gates and less propagation delay.

D.9   Shift register

Like counting, multiplication and division by two, as well as the related operations of rotation, can be implemented within a specialized device. Shift registers are sequential building blocks that implement these operations internally. There are many kinds of shift registers. The kind used in this book is a synchronous parallel loadable left/right shift register, with left and right shift (serial) inputs. This device is referred to simply as a shift register in this book.

The shift register has a `clr` signal (similar to the up counter) and a two-bit `shiftctrl` signal. The action table for this shift register is:

clr	shiftctrl	action
0	00	hold
0	01	right
0	10	left
0	11	load
1	00	zero
1	01	zero
1	10	zero
1	11	zero

In addition to the n-bit-wide `din` bus that all synchronous registers have, the shift register has two inputs, `rsi` and `lsi`, each one bit wide, that only play a role when the shift register is shifting:

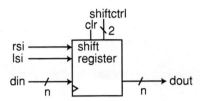

Figure D-22. Symbol for shift register.

The one-bit input `rsi` is ignored except when the register is shifting right (`shiftctrl=01`), in which case `rsi` determines the value of the most significant bit of `dout` for the next clock cycle. Similarly, `lsi` is ignored except when the register is shifting left (`shiftctrl=10`), in which case `lsi` determines the value of the least significant bit of `dout` for the next clock cycle.

This can be implemented using two combinational logic shifters, two muxes and a simple D-type register.

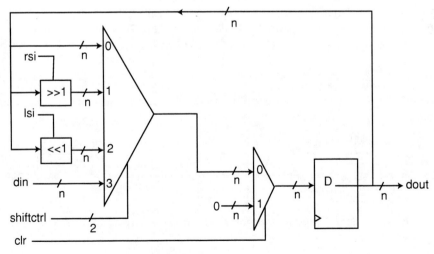

Figure D-23. Implementation of shift register.

Recall that the combinational logic shifters do not cost anything. There are other ways of constructing this than the technique shown above. For example, in the TTL logic family, the 74xx194 chip provides for $n=4$ the same actions[4] as the above using fewer gates and less propagation delay.

D.10   Unused inputs

Sometimes a designer needs more capability than an enabled register, but not as much as is offered by one of the other register building blocks described above. For example, a designer may need a register that omits any one of the three command inputs of an up counter:

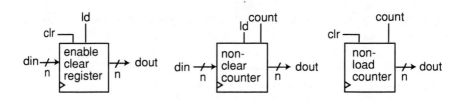

Figure D-24. Symbols for other registers.

The register on the left omits the count signal and is therefore not truly a counter. The register on the left is known as a enabled clearable register. The register in the middle is a counter that does not ever need to be cleared but that instead is loaded with `din`. The register on the right is a counter that never has to be loaded and therefore does not need a `din` bus.

All three of these are specializations of the up counter described in section D.7. They can be implemented by tying one of the three command inputs of an up counter to 0:

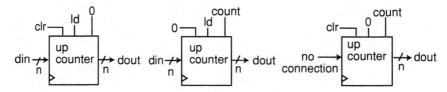

Figure D-25. Implementations for these registers using a loadable clearable up counter.

---

[4] Except that `clr` is active low.

It is important to understand the distinction between a block diagram and a circuit diagram. A circuit diagram is a detailed description used by people (or more likely automated manufacturing equipment) that put together a machine with no understanding of how the machine was designed. A block diagram is an abstract description used by designers as they think through various design alternatives.

The guiding philosophy for drawing block diagrams is how well the diagram describes the thoughts of the designer. A block diagram should be as simple as possible. Even if the circuit diagram will eventually use a counter with one of its command inputs tied to zero, it is easier for designers to communicate with each other by simply omitting that detail. Designers understand that one way of implementing the following:

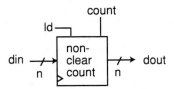

Figure D-26. Symbol for a non-clearable up counter.

is as:

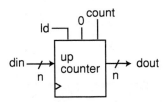

Figure D-27. Possible implementation using a clearable up counter.

although there is probably a more efficient way. Rather than overspecifying a block diagram with details, the designer only shows what is essential to the problem being solved. This is the same philosophical reason why we omit drawing the connection to sysclk, ground and Vcc: we know they have to be there,[5] and so why clutter the block diagram with a detail that adds nothing to our understanding?

In a similar way, it is common to use a shift register that never needs to be cleared:

---

[5] Vcc and ground supply power to a chip. The chip will not operate without these connections. Likewise, synchronous devices will not operate without a connection to sysclk.

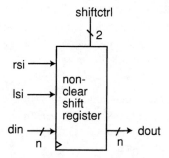

Figure D-28. Symbol for a non-clearable shift register.

This can be implemented as:

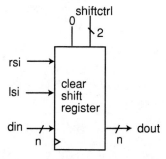

Figure D-29. Possible implementation using a clearable shift register.

D.11 Highly specialized registers

The registers shown above are but a small sample of the ones that are theoretically possible. A designer can create a specialized building block just for a particular problem if the value of dout in the next clock cycle can usually be computed as a combinational function of the current dout. As with the registers shown above, such specialized building blocks are typically implemented with a simple D-type register combined with muxes and other combinational logic. With the introduction of Verilog synthesis tools in the mid 1990s, designers may start conceptualizing a problem in terms of the building blocks given in earlier sections, only to have the synthesis tool convert those building blocks into some more efficient specialized one which is specific to their particular algorithm.

From a theoretical viewpoint, every computer can be thought of as a single very big register, whose value is meaningless to the human mind. In essence, this theoretical approach treats this one register as the concatenation of every piece of information the computer needs to remember. Mathematicians like to conceptualize things this way, but such an approach is an oversimplification that does not help a practical designer.

The building blocks given earlier are at the right level of abstraction for practical use. They are available as isolated chips (74xx377, 74xx378, 74xx163, 74xx669 and 74xx194) suitable for laboratory experiments which build the confidence of novice designers. They are commonly used by synthesis tools, even though synthesis tools may sometimes do something more sophisticated. In order to understand the more sophisticated things that synthesis tools do, one must already be familiar with the building blocks given in the earlier sections of this appendix.

D.12 Further Reading

GAJSKI, DANIEL D., *Principles of Digital Design*, Prentice Hall, Upper Saddle River, NJ, 1997. Chapter 6.

PROSSER, FRANKLIN P. and DAVID E. WINKEL, *The Art of Digital Design: An Introduction to Top Down Design*, 2nd ed., Prentice Hall PTR, Englewood Cliffs, NJ, 1987. Chapter 4.

D.13 Exercises

D-1. Complete the following timing diagram to show dout, given an enabled register with a 4-bit din, and a control input ld:

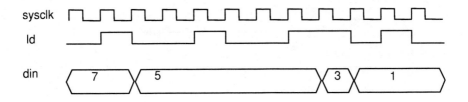

D-2. Complete the following timing diagram to show dout, given a shift register with a 4-bit din, a 2-bit control input shf and inputs clr, rsi and lsi:

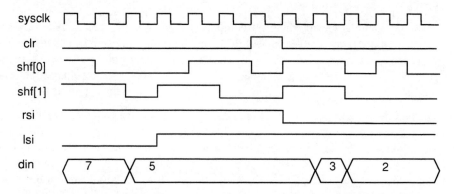

D-3. Complete the following timing diagram to show dout, given an up counter register with a 4-bit din, and control inputs clr, ld and inc:

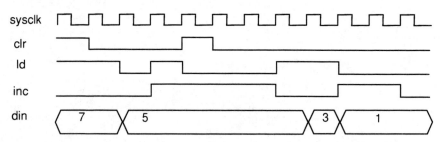

D-4. Complete the following timing diagram to show dout, given an up/down counter register with a 4-bit din, and control inputs count, ld and up:

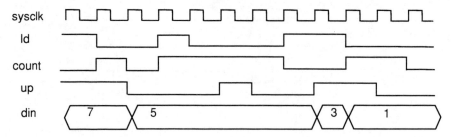

E. TRI-STATE DEVICES

A *tri-state device* is a special kind of combinational building block that has the ability to disconnect its output logically from the bus to which that output is physically connected. For simplicity, the combinational devices defined in appendix C and used throughout most of this book do not have tri-state capabilities, although many actual chips do. This appendix describes what tri-state devices are, and shows two common uses for them.

E.1 Switches

As explained in section C.2.1, a bus is composed of several wires that run in parallel to each other. The bit transmitted on each wire of the bus originates at the output of some gate (such as an AND gate), and is received at the input(s) of other gate(s). Although computer designers normally prefer to abstract away the electronic details of how a gate operates, some understanding of how a non-tri-state device operates is necessary to understand the extra feature provided by a tri-state device.

Each non-tri-state gate is actually composed of several simpler switching devices, such as transistors. Although the details in the operation of these switching devices depend upon the technology family used (CMOS, TTL, etc.), the effect they have on the gate's output is partly analogous to the effect that a wall switch has on the voltage across the filament of a light bulb. When the wall switch is open, the light is turned off because the voltage at point a is independent of the voltage at point b:

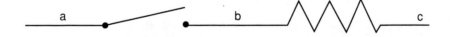

Figure E-1. An open switch causes a light to be off.

Saying that the switch is open is the same as saying a is disconnected from b. Since the filament of an ordinary light bulb is really just a wire that is a poor conductor (a resistor), the voltage at b will be the same as at c. For this reason, the filament is cool, and the light does not shine. On the other hand, when the switch is closed, the light is turned on because the voltage at a is identical to the voltage at b.

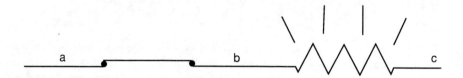

Figure E-2. A closed switch causes the light to be on.

Saying that the switch is closed is the same as saying a is connected to b.

E.1.1 Use of switches in non-tri-state gates

Non-tri-state gates are more complicated than light switches in two ways. First, the gate has to compute the desired output bit (which may require switching devices not described here). Second, the gate has to connect the output wire to the proper voltage.

In most technologies, connecting the output wire to the proper voltage requires two switches: the top switch connects the output wire to the voltage[1] for the bit 1, and the bottom switch connects the output wire to the voltage[2] for the bit 0. For example, suppose the gate needs to output the bit 0. To do this, the "1" switch is open and the "0" switch is closed:

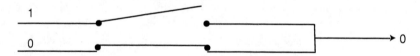

Figure E-3. A gate producing 0 as output.

The only other possibility for a non-tri-state gate is that the gate needs to output the bit 1. To do this, the "1" switch is closed and the "0" switch is open:

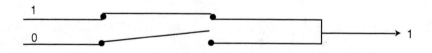

Figure E-4. A gate producing 1 as output.

---

[1] For active high TTL, 5 volts.
[2] For active high TTL, 0 volts.

A non-tri-state gate is always in one of these two configuration ("1" switch open and "0" switch closed or vice versa).

E.1.2 Use of switches in tri-state gates

The electronic distinction between a non-tri-state gate and a tri-state gate is that a tri-state gate allows a third configuration[3] (both the "1" switch and the "0" switch open):

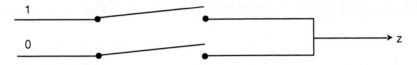

Figure E-5. A gate producing z as output.

The output wire is logically disconnected from the part of the gate the computes an answer. The voltage on the output wire will not be determined by this gate (but could be determined by some other gate). To denote this situation symbolically, we say that the output bit is z (1'bz in proper Verilog notation), which stands for high impedance.

E.2 Single bit tri-state gate in structural Verilog

A tri-state gate has two inputs, `enable` and `in`, and one output, `out`:

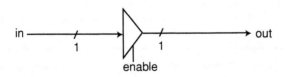

Figure E-6. A tri-state gate.

The behavior of this gate can be described by the following truth table:

enable	in	out
0	0	z
0	1	z
1	0	0
1	1	1

---

[3] Hence the name tri-state.

In other words, the tri-state driver gate is really nothing more than an electronically controlled switch. When `enable` is 1, the switch is closed:

in ——————————→ out

Figure E-7. Effect of tri-state gate when enable is 1.

When `enable` is 0, the switch is open:

in ——————————→ out

Figure E-8. Effect of tri-state gate when enable is 0.

There is a Verilog built-in gate, known as `bufif1`, that implements this. For example, the following instance:

```
wire out,in,enable;
bufif1 b1(out,in,enable);
```

is equivalent to the single-bit tri-state gate shown above.

As described in section 6.3.4, Verilog allows you to indicate the propagation delay of a built-in gate, such as the `bufif1`:

```
wire out,in,enable;
bufif1 #10 b1(out,in,enable);
```

Also, Verilog allows you to indicate different delays for the (rising) time required to change to a one and the (falling) time required to change to a zero. For built-in gates such as `bufif1`, there is a third separate time that may be of interest in some designs, the *turn off* delay, which is how long it takes when the output changes to `1'bz`. For example:

```
wire out,in,enable;
bufif1 #(10,20,30) b1(out,in,enable);
```

takes 10 units of $time if out becomes one, 20 units of $time if out becomes zero and 30 units of $time if out becomes 1'bz.

Also Verilog provides other forms of tri-state gates, such as bufif0, which has an active low enable signal. For example, the following:

```
wire out,in,enable,enable_low;
not i1(enable_low,enable);
bufif0 #(10,20,30) b1(out,in,enable_low);
```

is functionally identical to the above, as explained by the following:

enable_low	in	out
0	0	0
0	1	1
1	0	z
1	1	z

E.3 Bus drivers

It is the computer designer's job to avoid details such as the ones given above. In order to work with tri-state devices without having to get down to the gate level, we need a more abstract model of what a tri-state device does. Such an abstract model will allow us to use bus-width tri-state devices without having to be concerned with how the tri-state gates switch on and off for each individual bit. This abstract model describes the actions of the tri-state device in terms of a non-tri-state device connected to a special device, known as a tri-state bus driver:

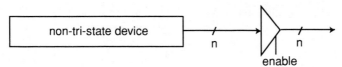

Figure E-9. Tri-state bus driver.

The symbol for a tri-state bus driver looks like a mux, except there is only one input bus (which is n bits wide). Since a mux always has at least two input busses, there should be no reason to confuse these two devices, both of which are symbolically represented as triangles.

Physically, the tri-state bus driver is composed of n independent tri-state driver gates, each one of which is physically a `bufif1` instance. Like all other gate-level features of Verilog, working with `bufif1` gates is not easy, and so it is better to think of an n-bit-wide tri-state bus driver like any other bus-width building block device, using the combinational logic modeling technique described in section 3.7.2.1:

```
module tristate_buffer(out,in,en);
  parameter SIZE=1;
  output out;
  input in,en;
  reg [SIZE-1:0] out;
  wire [SIZE-1:0] in;
  wire en;
  always @(in or en)
    begin
       if (en === 1)
          out = in;
       else if (en === 0)
          out = 'bz;
       else
          out = 'bx;
    end
endmodule
```

The `'bz` provides as many `1'bz` values as is required by `SIZE`.

E.4   Uses of tri-state
There are two main uses of tri-state devices: replacement for muxes and bidirectional buses.

E.4.1   Tri-state buffers as a mux replacement
The first primary use of tri-state bus drivers is to create a structure that is a replacement for a mux. For example, the following:

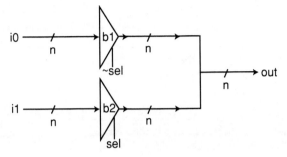

Figure E-10. Using tri-state bus drivers to form a mux.

serves the same role as a two-input mux. The above can be described in Verilog by using two instances of the `tristate_buffer` defined in the last section:

```
module silly_mux(out,i0,i1,sel);
   parameter SIZE=1;
   output out;
   input i0, i1, sel;
   wire [SIZE-1:0] out,i0,i1;
   wire sel;
   wire nsel;

   not n1(nsel, sel);
   tristate_buffer #SIZE b1(out, i0, nsel);
   tristate_buffer #SIZE b2(out, i1, sel);
endmodule
```

E.4.1.1   How Verilog processes four-valued logic

Section 3.5.3 describes the four-valued logic (0, 1, 1'bz, 1'bx) used for each bit of Verilog `wire`s and `reg`s. The need for the binary values 0 and 1 is obvious. The value 1'bx is often the result of a misconnection of gates. In this appendix, the reason for the fourth value, 1'bz, should now become clear.

If it were not for the high-impedance value, 1'bz, it would never make sense for the outputs of two devices to be tied together, such as shown above in the diagram of section E.4.1. Because of 1'bz, smoke does not come out of the chip when the two tri-state buffers are wired together.

There is an algorithm built into Verilog that models the physical behavior of a `wire`, based upon the `output` port(s) of instantiated modules to which that `wire` is connected. When there is only one output port connected to a `wire`, the value of the `wire` in question reflects the value of that single-output port. When that single-output port changes, the `wire` connected to it is instantaneously and automatically changed. This is the situation that occurs throughout most of the structural examples this book.

The situation is more complicated when there are two or more `output` ports connected to the same wire. In this example, the `output` ports of `b1` and `b2` both drive the same wire. In hierarchical naming (section 3.10.8) the `output` ports are `b1.out` and `b2.out`, and the wire they both drive is simply `out`. The following table describes what Verilog computes automatically as a particular bit of the wire `out`, given the corresponding bits of `b1.out` and `b2.out`:

	b2.out	0	1	z	x
b1.out					
0		0	x	**0**	x
1		x	1	**1**	x
z		**0**	**1**	**z**	x
x		x	x	x	x

If we guarantee either that every bit of either `b1.out` is `1'bz` or that every bit of `b2.out` is `1'bz`, we can be certain that no bit of `out` will be `1'bx` (see bold above). This is precisely what the two tri-state drivers do for us. When `sel` is 1, every bit of `b1.out` is tri-stated, but when `sel` is 0, every bit of `b2.out` is tri-stated.

E.4.1.2   The *tri* declaration

Verilog provides an alternative to declaring a `wire` when tri-state drivers are in use, known as `tri`. The following would also have been legal inside the declaration of `silly_mux`:

```
wire [SIZE-1:0] i0,i1;
tri [SIZE-1:0] out;
```

The `wire` and `tri` declarations do the same thing, and so which one to use is a matter of personal taste.

E.4.2 Bidirectional buses

Although most of this book assumes that a `wire` is essentially free in the fabricated hardware, in fact a wire does cost something. The cost is fairly reasonable when the wire remains hidden inside a physical chip, but the cost is quite high when that wire must be routed outside the chip.

Each bit of a Verilog wire that must be routed outside a chip requires a physical pin. One of the most severe limitations in hardware design is the number of pins available for a chip. Therefore, hardware designers often wish to make maximum utilization of the pins that are available. A *bidirectional bus* is one which sends information both ways:

Figure E-11. One bidirectional bus.

Routing a bidirectional bus off chip requires half the number of pins that routing two unidirectional buses requires:

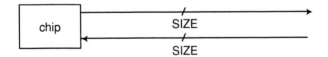

Figure E-12. Two unidirectional buses.

Bidirectional buses are especially important in the design of memory systems (section 8.2.2.1).

E.4.2.1 The *inout* declaration

In order for a Verilog module to use a bidirectional bus, the port for the bidirectional bus must be declared as `inout`. An `inout` port must be declared as a `wire` (or `tri`). Therefore, an `inout` port cannot be directly given its value by the behavioral code of the module in which it is declared. The `inout` port means the wire connecting the instantiated and instantiating modules is physically tied together, without an intervening buffer.[4]

---

[4] Although it is not necessary, an `input` port could have an intervening buffer into the chip, and an `output` port could have an intervening buffer out of the chip.

The algorithm Verilog uses to determine the value on an `inout` port combines the value outside the module together with the value inside the module, according to the table given in section E.4.1.1 (except the names will be at a different point in the hierarchy than `b1.out` and `b2.out`). The distinction between an `input` port and an `inout` port is not visible within the instantiated module (containing the `inout` declaration). This distinction is only visible within some other instantiating module (which instantiates the module having the `inout` port).

E.4.2.2 A read/write register

To illustrate how a bidirectional bus can reduce the number of pins on a chip, consider a register whose values can be read and written using a single bidirectional bus:

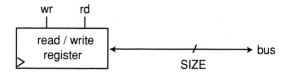

Figure E-13. A read/write register with a bidirectional bus.

If this device were fabricated on a single chip, it would require 5+SIZE pins (including the clock and power). In comparison, the enabled register using unidirectional buses (described in sections D.6 and 4.2.1.1) would require 4+2*SIZE pins, which is almost twice as many.

In order for the bidirectional bus to do double duty, there must be two command inputs: `rd` and `wr`. When `rd` is one, this device drives the bus (provides output) to show the current contents of the register. When `wr` is one, this device leaves the bus alone (`'bz`) and instead the bus provides the input which the register will load at the next rising edge of the clock. Here is the internal structure of this register:

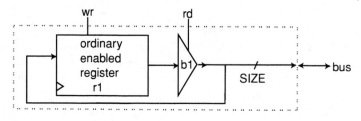

Figure E-14. Implementation of figure E-13.

and here is how this can be described in Verilog:

```
module rw_register(bus,rd,wr,clk);
  parameter SIZE = 1;
  inout bus;
  input rd,wr,clk;
  wire [SIZE-1:0] bus;
  wire [SIZE-1:0] do;
  wire rd,wr;

  enabled_register #SIZE r1(bus,do,wr,clk);
  tristate_buffer #SIZE b1(bus,do,rd);
endmodule
```

Here is an example of using two instances of the read/write register defined above:

```
reg r1rd,r1wr,r2rd,r2wr;
wire [3:0] bus1;

rw_register #4 r1(bus1, r1rd, r1wr, sysclk);
rw_register #4 r2(bus1, r2rd, r2wr, sysclk);
```

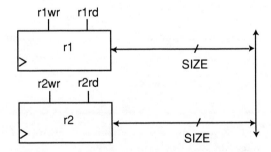

Figure E-15. Instantiation of two read/write registers.

Unlike the `silly_mux` example, there is nothing in the above to guarantee that `bus1` will avoid becoming `'bx`. Instead, it is the responsibility of the designer to ensure that `r1rd` and `r2rd` are never simultaneously one.

For example, to implement the register transfer r1 ← r2 requires generating the commands:

```
            r1rd = 0;
            r1wr = 1;
            r2rd = 1;
            r2wr = 0;
```

E.5 Further Reading

PALNITKAR, S., *Verilog HDL: A Guide to Digital Design and Synthesis,* Prentice Hall
PTR, Upper Saddle River, NJ, 1996. Chapter 5.

E.6 Exercises

E-1. Revise the architecture of the two-state division machine (whose Verilog code is
given in section 4.2.3) so as to eliminate the instance of `mux2` and instead use two
instances of the `tri-state_buffer` defined in section E.3. Use the test code given
in section 4.1.1.1.

E-2. Define a behavioral Verilog module (`bi_mem`) for an asynchronous bidirectional
memory (section 8.2.2.1) consisting of 4096 words, each 12-bits wide. The ports are a
12-bit `addr` bus, a 12-bit `data` bus and the commands `write` and `enable`. The
following table describes the actions of this memory:

```
    enable    write        action
      0         -       data = 12'bz
      1         0       data = m[addr]
      1         1       m[addr] ← data
```

E-3. Show how the ASM of section 8.4.6 and the architecture of section 8.4.4 need to
be modified to work with the memory defined in problem E-2.

E-4. Define the Verilog corresponding to problem E-3.

E-5. One of the reasons why the tri-state approach is attractive for memory system
design is that it allows multiple memory modules to be connected together to form a
larger memory without the need for a mux. Using eight instances of the memory defined in problem E-2 together with a decoder having a 3-bit input, give a block diagram
that implements a memory of 32,768 12-bit words.

E-6. Give the structural Verilog for problem E-5.

E-7. Pins are so limited in many memory packaging technologies that industry has resorted to several contorted techniques to minimize pin count. One common approach with dynamic memories is to transmit the address in two parts: the *row* address and the *column* address. Internally, this makes for a square geometric arrangement consisting of rows and columns of identical memory cells. Externally, this cuts the number of address pins in half by (approximately) doubling the time to access a word. To distinguish the use of the half-size `addr` bus, there are two input signals: `ras` and `cas`. If `ras` is asserted when `cas` is not, the `addr` bus indicates the row. When both are asserted, the `addr` bus indicates the column. For simplicity, assume `ras` and `cas` signals are synchronous to the `sysclk` input which is provided to this chip. (Often such chips are asynchronous with much stricter timing constraints than are shown below). The following timing diagram illustrates reading a word from such a memory:

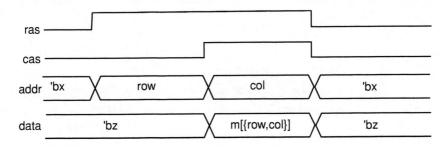

Writing to such a memory is similar, except a `write` signal is asserted and the new content is provided to the chip on the data bus during the entire time. Define a structural Verilog module for a memory containing 16 twelve-bit words using twenty instances of the `rw_register` defined in section E.4.2.2 together with additional combinational logic.

F. TOOLS AND RESOURCES

There are several Verilog example files used in this book. There are also several design automation software packages (tools) used with these files. In addition, there are several information resources that may be helpful on the Internet. This appendix briefly describes how to obtain and use these tools and resources. The details are subject to change, and the respective Web sites should contain the most up-to-date information.

F.1 Prentice Hall

Selected Verilog examples can be downloaded from the Prentice Hall Web site, `www.phptr.com`.

F.2 VeriWell Simulator

Most of the examples in this book have been tested using the Verilog simulator from Wellspring Solutions, Inc, known as VeriWell. At the time of this writing, this excellent software package is available at no charge by downloading it from `www.wellspring.com`. The downloaded version has limits on the size of Verilog source files that it accepts and does not provide graphical (timing diagram) output. The downloaded version is available for MSDOS (command-line), Windows 95/NT (GUI), Macintosh (GUI) and several UNIX (command-line) dialects. For the command-line versions, simply type:

```
veriwell file1.v file2.v ...
```

which will produce the output of `$display` commands both on the screen and in a file known as `veriwell.log`. For the GUI versions, you need to create a "project file" by selecting Project (Alt P) New (Alt N) and choose a name (`.prj`) for the project file. Then select Project (Alt P) Add file (Alt F) to specify the `.v` file name(s). To run the simulator, select Project (Alt P) rUn (Alt U).

Most of the designs in this book are able to simulate on the free version. Wellspring Solutions sells a hardware key that removes the limitations of the free version and also sells a separate package for graphical output:

Wellspring Solutions
7 Tudor Drive, Suite 300
Salem, NH 03079
(603) 898-1100

F.3   M4-128/64 demoboard

Documentation for the M4-128/64 CPLD chip used in chapter 11 can be downloaded from `www.vantis.com`. The demoboard, power supply, download cable and MACHPRO software can be obtained from:

Vantis
Box 3755
Sunnyvale, CA 94088

F.4   Wirewrap supplies

To build the CPU described in chapter 11 requires wirewrap wire, a wirewrap tool (such as an "all in one" tool that strips, wraps and unwraps) and a wirewrap socket. It also requires a memory ("RAM") chip, such as the 2102. Some of these may be available at local electronics stores, but there are several mail-order companies, such as Jamesco (`www.jamesco.com`), that carry a complete selection of such supplies.

F.5   VerilogEASY

The synthesis package used in chapter 11, known as VerilogEASY, is sold by MINC, Inc. (`www.minc.com`). VerilogEASY comes in several versions, each targeting different vendors' programmable logic. A limited version of VerilogEASY that targets the M4-128/64, but that is restricted on the number of inputs and outputs, will be available to readers of this book in the last quarter of 1998. There is no charge for downloading this limited version, but MINC requires that people downloading their software register at their Web site. VerilogEASY accepts the common synthesizable subset of Verilog other than implicit style. VerilogEASY produces two output files: `.src` (in the proprietary DSL language) and `.v` (structural Verilog netlist). To fabricate working hardware requires other tools, described in sections F.3 and F.6. MINC also sells a full version of VerilogEASY and an even more powerful synthesis tool, known as PLSynthesizer:

MINC, Inc.
6755 Earl Drive
Colorado Springs, CO 80918-1039
(719) 590-1155
info@minc.com

F.6 PLDesigner

The place and route tool used in this book for CPLDs, such as the Vantis M4-128/64, is known as PLDesigner. It runs on Windows 95/NT. It is not possible to fabricate a design for the M4-128/64 without using this tool. PLDesigner can be purchased from MINC. The following directions apply to PLDesigner: At the PLDesigner menu, choose File (Alt F) Open (Alt O), and enter the name of the `.src` file created by VerilogEASY. Do a File (Alt F) eXit (Alt X). Select Device (Alt D) Parameters (P) and choose the M4-128/64 (MACH445) and say OK. Select Settings (Alt S) Options (Alt O) and be sure Timing Models are set only to generic Verilog. Select Project (alt P) Build all (Alt B) to create the JEDEC file (`.j1`). To create the back annotated Verilog, select Project (Alt P) Generate Timing Model (Alt T), which will put the `.v` file in a `model` subdirectory (since a similarly named `.v` file (the input to VerilogEASY) will already exist).

F.7 VITO

The Verilog Implicit To One hot (VITO) preprocessor is a freely available synthesis preprocessor written by James D. Shuler and Mark G. Arnold. It may be downloaded from the Prenticite Hall Web site. It can also be downloaded from `www.cs.brockport.edu/~jshuler` or `plum.uwyo.edu/~vito`. UNIX and MSDOS versions are available at those Web sites. The theory of how this tool operates is discussed in chapter 7. It is a command-line program, and the following is a typical use:

```
vito -t implicit.v >explicit.v
```

where `implicit.v` is the name of a file consisting of one or more modules that have implicit style state machines. The `-t` option generates comments that explain the transformation. The output of VITO is redirected to another file (`explicit.v`), which would then be used as the input to VerilogEASY (or another synthesis tool). The designer is free to choose other file names.

Each module must include a `sysclk` and `reset` port. The names of these are also given in a special file known as `vito.rc`. For the M4-128/64 (with its active low `reset`), the `vito.rc` file should contain:

```
        vito.out
        vito.stmt
        vito.arch

    s_
    sT_
    tmp_
    join_
    ff_
    qual_
    reset
    sysclk
    @(posedge sysclk)
    new_
    @(posedge sysclk or negedge reset)
    ~reset
        vito.tail
```

The other names in this file are the prefixes of wire names that VITO will generate, and the temporary files VITO uses. This file is based on position, and so extra blank lines are not allowed.

F.8   Open Verilog International (OVI)

The independent organization that developed the Verilog standard (IEEE 1364) is known as Open Verilog International (`www.ovi.org`). OVI is co-sponsor of the International Verilog Conference (`www.hdlcon.org`) held each spring at the Santa Clara Convention Center. OVI sells a language reference manual, which is the authoritative source for questions of Verilog syntax and semantics:

> Open Verilog International
> 15466 Los Gatos Blvd.
> Suite 109-071
> Los Gatos, CA 95032
> (408) 353-8899
> `ovi@netcom.com`

F.9   Other Verilog and programmable logic vendors

Here is a partial list of other Verilog and vendors' Web sites: `www.altera.com`, `www.avanticorp.com`, `www.cadence.com`, `www.fintronic.com`, `www.sunburst-design.com`, `www.synopsys.com`, `www.simucad.com`, `www.synplicity.com`, `www.veribest.com`, `www.xilinx.com` and `www.eg.bucknell.edu/~cs320/1995-fall/verilog-manual.html`.

F.10   PDP-8

Additional resources relating to the PDP-8 can be found at `strawberry.uwyo.edu`, `www.in.net/~bstern/PDP8/pdp8.html`, `www.faqs.org/faqs/dec-fac` and `www.cs.uiowa.edu/~jones/pdp8`. A portion of this information is also available at the Prentice Hall website.

F.11   ARM

Additional resources relating to the ARM can be found at `www.arm.com`.

G. ARM INSTRUCTIONS[1]

1. Efficient instruction set

The foundation of all processor architectures is the instruction set. When designing it there are two contradictory aims: high code density and easy instruction decoding. The ARM instruction set strikes an optimal balance between these. The instructions are powerful so programs in ARM are short, saving memory and speeding execution because of reduced bus bandwidth requirements. Yet because the instructions decode so easily the ARM processor is small and cheap, consumes very little power and runs at high speed.

In general RISC processors code less densely than CISCs because of their very nature - yet ARM code is generally as dense as code for 32-bit CISC processors, which is significantly better than other 32-bit RISC processors. With the 16-bit Thumb extension, ARM code density is the best in the business.

Features of the ARM fundamental to easy decoding include:

- A small number of highly flexible instruction types
- Consistent instruction data formats

Features implemented for high code density include:

- Barrel shifter to perform arbitrary shifts within the same cycle, at no speed penalty
- Conditional execution on every instruction to eliminate many branches
- Load and store multiple instructions for rapid context switching and memory transfer

2. Instruction set summary

The ARM instruction set is a good target for compilers of many different high-level languages. Where required, though, assembly code programming in ARM is straightforward and enjoyable. The instructions are flexible and orthogonal, the memory model is flat and there are no complicated instruction interdependencies as there are for some RISC processors. Because a whole line of C code can often be performed within one or two instructions, the instructions correspond closely to natural program steps. See the ARM Code Examples which demonstrate the true power and magic of ARM machine code.

---

[1] ARM documentation is copyright 1997 Advanced RISC Machines, Ltd. and is reprinted by permission.

The ARM7 instruction set comprises 10 basic instruction types

- Two of these make use of the on-chip arithmetic logic unit, barrel shifter and multiplier to perform high-speed operations on the data in the 16 visible 32-bit registers.

- three classes of instruction control the transfer of data between main memory and the register bank, one optimized for flexibility of addressing, another for rapid context switching and the third for managing semaphores.
- two instructions control the flow and privilege level of execution.
- three types of instruction are dedicated to the control of external coprocessors which allow the functionality of the instruction set to be extended in an open and uniform way.

The ARM7 instruction set is summarised in the table below:

Instruction Type	Instruction bit format, 31...0
Data Processing	cond |0|0|I| Opcode |S| Rn | Rd | Operand 2
Multiply	cond |0|0|0|0|0|0|A|S| Rd | Rn | Rs |1|0|0|1| Rm
Single Data Swap	cond |0|0|0|1|0|B|0|0| Rn | Rd |0|0|0|0|1|0|0|1| Rm
Single Data Transfer	cond |0|1|I|P|U|B|W|L| Rn | Rd | Offset
Undefined	cond |0|1|1| xxxx |1| xxxx
Block Data Transfer	cond |1|0|0|P|U|S|W|L| Rn | Register List
Branch	cond |1|0|1|L| offset
Copro Data Transfer	cond |1|1|0|P|U|N|W|L| Rn |CRd| CP# | Offset
Copro Data Operation	cond |1|1|1|0| CP OpcP |CRn|CRd| CP# | CP |0|CRm
Copro Regester Transfer	cond |1|1|1|0|CP OpcP|L|CRn|CRd| CP# | CP |1|CRm
Software Interrupt	cond |1|1|1|1| Ignored by processor

Register Model

The processor has a total of 37 registers made up of 31 general 32 bit registers and 6 status registers. At any one time 16 general registers (R0 to R15) and one or two status registers are visible to the programmer. The visible registers depend on the processor mode and the other registers (the *banked registers*) are switched in to support rapid interrupt response and context switching.

Below is a list of the visible registers for each of the processor modes. The banked registers are in italics.

User32	FIQ32	Svc32	Abort32	IRQ32	Undef32
R0	R0	R0	R0	R0	R0
R1	R1	R1	R1	R1	R1
R2	R2	R2	R2	R2	R2
R3	R3	R3	R3	R3	R3
R4	R4	R4	R4	R4	R4
R5	R5	R5	R5	R5	R5
R6	R6	R6	R6	R6	R6
R7	R7	R7	R7	R7	R7
R8	*R8_fiq*	R8	R8	R8	R8
R9	*R9_fiq*	R9	R9	R9	R9
R10	*R10_fiq*	R10	R10	R10	R10
R11	*R11_fiq*	R11	R11	R11	R11
R12	*R12_fiq*	R12	R12	R12	R12
R13	*R13_fiq*	*R13_svc*	*R13_abt*	*R13_irq*	*R13_und*
R14	*R14_fiq*	*R14_svc*	*R14_abt*	*R14_irq*	*R14_und*
R15 (PC)	R15 (PC)	R15 (PC)	R15 (PC)	R15 (PC)	R15 (PC)
CPSR	CPSR	CPSR	CPSR	CPSR	CPSR
_	*SPSR_fiq*	*SPSR_svc*	*SPSR_abt*	*SPSR_irq*	*SPSR_und*

More complete documentation can be found at www.arm.com.

H.   ANOTHER VIEW ON NON-BLOCKING ASSIGNMENT

There are other forms of the non-blocking assignment besides the RTN form (`<= @(posedge sysclk)`) used earlier in this book:

```
var <= expression;

var <= #delay expression;
```

The semantics of `<=` and `<=#delay` are more subtle than the extra `always` analogy explained in section 3.8.2, which only applies to `<= @(posedge sysclk)`. In general, non-blocking expressions are evaluated immediately and put into a simulator queue to be stored after all blocking assignments at the $time given by the specified `#delay`. Clifford Cummings of Sunburst Design, Inc. (`www.sunburst-design.com`) gave a very informative presentation on the use of such non-blocking assignment statements at the 1998 International Verilog Conference. He suggested that the blocking assignment, `=`, should be primarily limited to modeling combinational logic, as in:

```
always @(a or b)
   sum = a + b;
```

H.1   Sequential logic

According to Cumming's guidelines, sequential logic, such as a simple D type register, should use the non-blocking assignment:

```
always @(posedge sysclk)
   dout <= din;
```

The $<=$ without time control above has the same meaning as $<=$ #0 (sections 11.3.3 and 11.5.6). It causes the simulator to put the assignment into a special "non-blocking event queue" that stores a new value after all = and =#0 have finished but before $time advances. The $<=$ without time control is useful for situations where an explicit style module uses the same regs on opposite sides (example in bold) of different assignments that model distinct sequential devices, as in:

```
        always @ (posedge sysclk)
          r2 <= r1;
        always @ (posedge sysclk)
          r1 <= r2;
```

Assuming r1 and r2 were initialized (not shown), the above exchanges r1 and r2, regardless of the order in which the simulator schedules the two always blocks. Using = instead of $<=$ above would have the incorrect effect of duplicating the value of one of the registers (seemingly chosen at random) into both of them. To use = in this situation requires intervening combinational logic:

```
        always @ (posedge sysclk)
          r2 = new_r1;
        always @ (r1)
          new_r1 = r1;   //identity
```

with similar code for r2, which is hard for designers to remember when the combinational logic is simply the identity function. This problem does not occur when the interacting always blocks are in separate modules because the port(s) act like the intervening combinational logic.

Most existing explicit style designs, including examples in this book (sections 3.7.2.2, 7.2.2.1, 11.3 and 11.7), use = properly (with intervening combinational logic or ports) rather than $<=$. Probably many designers stumble onto correct sequential logic using = without understanding why it is correct. Even more alarming, some incorrect sequential logic using = may appear to be correct because of the arbitrary order in which the

Verilog simulator schedules the assignments. Cummings suggested designers use only <= for sequential logic to guarantee correct operation without making the designer remember the intervening combinational logic or ports.

H.2 $strobe

Cummings also suggested using the $strobe system task, which works like $display, but shows the result of non-blocking assignment at the same $time the assignment is made. For example, instead of the $display code with delay used in many examples in this book, as typified by section 3.8.2.3.2:

```
always @ (posedge sysclk) #20

   $display("%d a=%d b=%d ", $time, a, b);
```

Cummings would recommend:

```
always ? posedge sysclk)

   $strobe("%d a=%d b=%d ", $time, a, b);
```

which has the advantage that the values that will take effect during a particular clock cycle will be displayed at the actual $time of the rising edge. With $display, there must be at least #1 delay (#20 in this example) beyond @ (posedge sysclk) to view the values changed by non-blocking assignments.

H.3 Inertial versus transport delay

An interesting contrast between blocking and non-blocking assignment that Cummings illustrated is the difference between *transport delay*, which retains all the values signal has, regardless of how briefly they exist:

```
always @ (signal)

     dela <= #3 signal;
```

and *inertial* delay, which filters out certain values of `signal` when values change more rapidly than a specified amount of `$time` (3 in this example):

```
always @ (signal)
   delb = #3 signal;
always @ (signal)
   #3 delc = signal;


$time  0  1  2  3  4  5  6  7  8  9
signal 1  1  1  1  2  3  3  3  3  3 ...
dela   x  x  x  1  1  1  1  2  3  3 ...
delb   x  x  x  1  1  1  1  2  2  2 ...
delc   x  x  x  1  1  1  1  3  3  3 ...
```

Of course, transport delay is more realistic of how physical signals behave.

H.4   Sequence preservation

Because of their queued implementation, non-blocking assignments at the same `$time` occur in the sequence the `<=`s executed. For example, a`<=1` followed by a`<=2` stores 2 into a. As emphasized in section 9.6, hardware registers (described in implicit style code with `<= @ (posedge sysclk)`) cannot store multiple values during a single clock cycle. Even though Verilog allows it, it is inappropriate for implicit style code to have more than one `<= @ (posedge sysclk)` to a given `reg` during a particular clock cycle. On the other hand, Cummings pointed out that it is useful in explicit style code for a plain `<=` to give a default values to the output of a state machine, which a later `<=` can modify at the same `$time`.

H.5   Further reading

CUMMINGS, CLIFFORD E., "Verilog Nonblocking Assignments Demystified", *7th International Verilog HDL Conference,* Santa Clara, CA, Mar 16-19, 1998, pp. 67-69.

I.   GLOSSARY

The following include terms used in computer design. Terms marked with * are unique to this book. Synonyms for terms not used in this book are also given. In addition, the following includes Verilog features (courier font), some of which are not described elsewhere in this book. See the references given at the end of chapter 3 for details about Verilog features not described in this book.

Access time: The *propagation delay* of a *memory*.

Active high: A *pin* of a physical chip where 1 is represented as a high voltage.

Active low: A *pin* of a physical chip where 1 is represented as a low voltage.

*Actor: A machine or person that interacts with the machine being designed.

Address bus: A *bus* used to indicate which word of a *memory* is selected.

Algorithmic State Machine, *see* ASM

Architecture: 1. The hardware of a machine that manipulates data, as opposed to the *controller*. Is present in *mixed(1)* and *pure structural* designs. Also known as a datapath. 2. The *programmer's model* and *instruction set* of a *general-purpose computer*. See also *computer architecture*. 3. A feature of *VHDL* that provides greater abstraction of *instantiation* than *Verilog* does.

ALU (Arithmetic Logic Unit): *Combinational* logic capable of computing several different functions of its input based on a command signal. Typically, the functions include arithmetic operations, such as addition, and bitwise (logical) operations, such as AND.

ASM (Algorithmic State Machine): A graphic notation for finite state machines consisting of *rectangles(1)*, *diamonds* (or equivalently *hexagons*), and possibly (for *Mealy machines*) *ovals*. A *pure behavioral* ASM is equivalent to *implicit style* Verilog with *non-blocking assignment*. Moore *mixed(1)* ASMs can be implemented as implicit style Mealy Verilog.

Asynchronous: Logic which has memory but which does not use the *system clock*.

Backannotation: Recording the *propagation delay* in a *netlist* after *synthesis*.

Behavioral: Code which describes what a machine does, rather than how to build it. *see also pure behavioral*.

Big endian notation: The most significant bit (byte, word, etc.) is labeled as 0.

Blocking procedural assignment: A Verilog statement (=) that evaluates an expression now, causes the process to delay for a specified time and then stores the result.

Bottom testing loop: A loop that is guaranteed to execute at least once. Such loops are difficult to code in *implicit style* Verilog. For *simulation*, use `enter_new_state` task with `!==` to the bottom state. For synthesis, use `disable` inside `forever`.

Bus: A groups of wires that transmit information.

Bus driver: A *tri-state* device that passes through its input when it is enabled, but outputs `'bz` when it is not.

Cache: A *memory* that allows faster access to words used most frequently.

casex: A variation of `case` that treats `1'bz` or `1'bx` as a don't care. For example, the 13 bits of output for the truth table given in section 2.4.1 could be coded as:

```
always @(ps or pb or r1gey)
  begin
    casex ({ps,pb,r1gey})
      3'b00x: t=13'b0110001010101;
      3'b01x: t=13'b1110001010101;
      3'b100: t=13'b0101110110010;
      3'b110: t=13'b1101110110010;
      3'b1x1: t=13'b1101110110010;
    endcase
    {ns,ldr1,clrr2,incr2,ldr3,muxctrl,aluctrl,ready}=t;
  end
```

casez: Like `casex`, except it only uses `1'bz`.

*Central ALU: An *architecture(1)* that uses a single *ALU* for all computation. The associated *pure behavioral ASM* is usually restricted to one register transfer (*RTN* or *non-blocking assignment*) per *state*, and so algorithms designed for a central ALU architecture are usually slower than those designed for *methodical* architectures.

Central Processing Unit: see *CPU*.

CPU (Central Processing Unit): The main element of a *general-purpose computer*, besides *memory*.

Combinational: Logic which has no memory. In Verilog, *ideal* combinational logic (including a *bus* or *tri-state* device) is modeled with @ followed by a *sensitivity list* or by a *continuous assignment*.

Combinatorial: *see* Combinational

Command signal: 1. An internal signal output from a *controller* that tells the *architecture(1)* what to do. Found only at the *mixed(1)* and *pure structural* stages. 2. An external signal output from a controller to another *actor*.

Computer architecture: 1. *see Programmer's model* and *instruction set architecture*. 2. A generic term for a field of study that encompasses the computer design topics in this book along with more abstract modeling concerns not discussed here, such as networked general-purpose computers, disk drives and associated software operating system issues.

Concatenation: The joining together of bits, indicated by { } in Verilog.

Conditional: *Non-blocking assignments (RTN)* and/or *command signals* that occur in a particular *state* only under certain conditions. See *Mealy* and *oval*.

Continuous assignment: A shorthand for instantiating a hidden module that defines behavioral *combinational logic*. Allows assignment to a Verilog `wire`. Eliminates the need to declare *ports* and *sensitivity lists*.

Controller: The hardware of a machine that keeps track of what step of the algorithm is currently active. Described as an *ASM* at the *mixed(1)* stage, but as a *present state* and *next state* logic at the *pure structural* stage.

CPLD (Complex Programmable Logic Device): A fixed set of AND/OR gates optionally attached to flip flops with a programmable interconnection network allowing the downloading of arbitrary *netlist*s.

Data bus: A *bus* used to transmit words to and from a *memory*.

Datapath, *see architecture(1)*

defparam: An alternative way of instantiating a different constant for a `parameter`.

Dependent: Two or more computations where the evaluation of some parts depends on the result of other parts. It is hard to design *pipeline* and *superscalar* computers when computations are dependent.

Diamond: The *ASM* symbol for a decision, usually equivalent to an `if` or `while` in *implicit style* Verilog.

Digital: Pertaining to discrete information, e.g., bits. See also *special-purpose computer*.

$dumpfile: System task, whose argument is a quoted file name, for VCD. For a complete dump, also need $dumpvars and $dumpon. Other tasks exist for more limited dumps.

Enabled register: A *synchronous* register that has the ability to hold as well as load data.

*External status: Information from outside the machine used to make decisions.

`enter_new_state`: A user defined task only for *simulation* at the *pure behavioral* and *mixed(1)* stages. Establishes default *command signals*. Helps with *Mealy* machines and *bottom testing loops*.

Explicit style: A finite state machine described in terms of *next state* transitions. Does not have multiple `@(posedge sysclk)`s inside the `always`. Roughly equivalent to the *pure structural* stage, except instead of separate modules for the *present state* register and the *next state* logic, both are often coded in the same module, and the *architecture* is modeled in a behavioral style, often called *RTL*. Explicit style requires the designer to think in terms of *goto*s. Contrast *implicit style*.

Falling delay: The propagation delay it takes for an output to change to 0.

`$fclose`: System task, whose argument is a file handle, that closes the associated file. See also `$fopen`.

`$fdisplay`: A variation of `$display` that outputs to a file. See also `$fopen`, `$fclose`.

Field Programmable Gate Array: *see FPGA*

Finite state machine, *see ASM*.

Flip Flop: A *sequential(1)* logic device that stores one bit of information. Used to build *register*s and *controllers*.

`$fopen`: System function, whose argument is a quoted file name, that returns an integer file handle used by `$fdisplay`, `$fstrobe` or `$fwrite`:

```
integer handle;
initial
 begin
   handle = $fopen("example.txt");
   $fdisplay(handle,"Example of file output");
   $fclose(handle);
 end
```

`fork`: An alternative to `begin` that allows parallel execution of each statement. For example, the following stores into b at `$time` 2 but stores into d at `$time` 3:

```
initial
  fork
    #1  a=10;
    #2  b=20;
  join
initial
  begin
    #1  c=10;
    #2  d=20;
  end
```

Four-valued logic: A *simulation* feature of Verilog that models each bit as being one of four possible values: 0, 1, *high impedance* ($1'bz$) or *unknown value* ($1'bx$).

FPGA (Field Programmable Gate Array): A fixed set of lookup (truth) tables optionally attached to flip flops with a programmable interconnection network allowing the downloading of arbitrary *netlist*s.

$fstrobe: A variation of $strobe that outputs to a file. See also $fopen, $fclose.

Full case: A synthesis directive that causes a case statement to act as though all possible binary patterns are listed. May cause synthesis to disagree with simulation.

$fwrite: A variation of $write that outputs to a file. See also $fopen, $fclose.

General-purpose computer: A machine that fetches machine language instructions from memory and executes them. The machine language describes the algorithm desired by the user, as opposed to a *special-purpose computer*. Also known as a stored program computer.

Glitch: *see Hazard.*

Goto: A high-level language statement not found in Verilog. Similar to state transitions in *explicit style* Verilog. Equivalent to assembly language jump or branch instructions. Gotos are useful for implementing *bottom testing loops*. The closest statement in Verilog is disable, which has drawbacks when used for this purpose. Avoidance of gotos is part of *structured programming*, and is possible with *implicit style* Verilog.

Handshaking: The synchronization required when two *actors* of different speed transfer data.

Hazard: The momentary spurious incorrect result produced by *combinational* logic of non-zero *propagation delay*.

Hexagon: Equivalent to *diamond* in *ASM* notation.

Hierarchical design: Instantiation of one *module* inside another.

Hierarchical names: A path for *test code* to access the internal variables of a *module* without a *port*. The instance names of the module(s) in the path are separated with periods.

High impedance: A value (1'bz) that models a wire that is disconnected and that is also produced by a *tri-state* gate. Used in a special way by casex and casez.

Ideal: A model of a device that ignores most physical details, such as *propagation delay* and voltage.

Implicit style: A finite state machine described in terms of multiple @ (posedge sysclk)s inside an always. Does not give *next state* transitions. Roughly equivalent to a *pure behavioral* stage *ASM*. Contrast *explicit style*.

Independent: Two or more computations that do not depend on each other. It is easier to design *pipeline* special-purpose computers when computations are independent.

inout declaration: A Verilog feature that allows the same *port* to be used for information transfer both into and out of the *module*. Corresponds to *tri-state* gates.

input: A Verilog feature that only allows a *port* to be used for information transfer into a *module*.

Instance: A copy of a *module* used in a particular place of a *structural* design.

Instantiation: The act of making an *instance*.

Instruction set: The set of machine language operations implemented by a particular *general-purpose computer*.

Instruction-Set Architecture (ISA): See *programmer's model* and *instruction set*.

*****Internal status**: Information generated by the *architecture(1)* and sent to the *controller* at the *mixed(1)* stage so that the controller can make decisions based on the data in the architecture.

Latch: An *asynchronous* data storage device. Synthesis tools produce unwanted latches when a case statement is used that is not a *full* case.

Little endian notation: The least significant bit (byte, word, etc.) is labeled as 0.

Macro: A string of source code that the simulator or synthesis tool substitutes prior to parsing.

Macrocells: The basic unit of a *CPLD*, consisting of a fixed set of AND/OR gates and an optional flip flop.

Mealy: A finite state machine that, unlike a *Moore* machine, produces command signals that are a function of both the *present state* and the *status* inputs. Such a command is indicated by an *oval* in *ASM* notation.

*Methodical: An *architecture(1)* where each register has an associated ALU or other combinational logic so that all register transfers may proceed in parallel. Typically allows for faster algorithms to be implemented than the central ALU approach.

Memory: Equivalent to a collection of *register*s. Has the ability to read (remember old data) and write (forget old data and remember new data instead). Often referred to as RAM. Although ROM is often used instead of RAM in parts of a general-purpose computer, because ROM cannot forget, it is not memory.

Mixed: 1. The stage of the design where the *controller* is specified as an *ASM* using *command* and *status* signals (rather than RTN and mathematical conditions), but the *architecture(1)* is specified as a structure. 2. Any such mixture of *behavioral* and *structural* constructs in a Verilog *module*. 3. A kind of digital logic *netlist* where 1 is sometimes represented as a high voltage (*active high*) and sometimes represented as a low voltage (*active low*).

module: The basic construct of Verilog which is instantiated to create *hierarchical* and *structural* designs.

Moore machine: A finite state machine that, unlike a *Mealy* machine, produces command signals that are a function of only the *present state*. All commands in a Moore *ASM* are given in *rectangles*.

Multi-cycle: A machine that requires several fast clock cycles to produce one result. Compare with *single cycle* and *pipeline*.

Multi-port memory: Allow simultaneous access to multiple words within one clock cycle.

Netlist: A structural design described at the level of connections between one-bit `wires` and gates.

Next state: Combinational logic that computes what the next step is in the algorithm based on the *present state* and *status* inputs to the *controller*.

Node collapsing: An optimization technique used by *place and route* tools.

Non-blocking assignment: A Verilog statement (`<=`) that evaluates an expression now but that schedules the storage of the result to occur later. Several non-blocking assignments can execute in parallel without delay. There are several forms, but the one used most in this book (`<= @(posedge sysclk)`) is equivalent to the *RTN* ← used in the *pure behavioral* stage for *ASM*s.

notif0: A variation of `bufif0` that complements its output.

notif1: A variation of `bufif1` that complements its output.

One hot: An approach for the *controller* that uses one flip flop for each *state*.

output: A Verilog feature that only allows a *port* to be used for information transfer out of a *module*.

Oval: The *ASM* symbol for a *Mealy* command.

Parallel: 1. Two or more *independent* computations that occur at the same physical time. 2. Two or more computations (*dependent* or *independent*) that occur at the same simulation `$time`. In Verilog, `$time` is a separate issue from *sequence*. 3. When one assumes physical time and *sequence* are the same, the opposite of *sequential*.

Parallel case: A synthesis directive that allows parallel evaluation of the conditions given in a `case` statement.

parameter: A constant within an *instantiation* of a *module* that can be different in each *instance*.

Pin: The physical connection of an integrated circuit to a printed circuit board.

Pipeline: A machine that requires, on average, slightly more than one fast clock cycle to produce one result, provided that each result is *independent* of other results. Compare with *single cycle* and *multi-cycle*.

Place and route: A post synthesis tool that maps the synthesized design into the limited resources of a particular technology, such as a CPLD or FPGA.

PLI (Programming Language Interface): A way to interface Verilog simulations to C software, and thus extend the capabilities of Verilog.

Port: The aspect of a *module* that allows structural instantiation.

posedge: The rising edge of a signal, such as `sysclk`

Present state: The register that indicates what is the step in the algorithm which is currently active.

Programmable logic: Integrated circuits manufactured with a fixed set of devices that can be reconfigured by downloading a *netlist*. See *CPLD* and *FPGA*.

Programming: 1. The act of downloading a synthesized *netlist* into *programmable logic* or a truth table into a *ROM*. 2. The act of designing software for a *general-purpose computer*.

Programming Language Interface: See *PLI*.

Programmer's model: The *registers* of a *general-purpose computer* visible to the machine language programmer.

Propagation delay: The time required for combinational logic to stabilize on the correct result after its inputs change.

***Pure behavioral:** The stage where the design is thought of only as an algorithm using *RTN*. Equivalent to *implicit style* Verilog.

***Pure structural:** The stage where the *controller* and the *architecture(1)* are both structural.

RAM: see *memory*.

$readmemb: System task, whose arguments are the quoted name of a text file and an array. Reads words represented as a pattern of '0','1','x' and/or 'z' from the text file into the array.

$readmemh: System task, similar to $readmemb, except for hexadecimal.

Rectangle: 1. The *ASM* symbol for a *Moore* command. 2. The block diagram symbol for most devices.

Reduction: The unary application of a bitwise operator which acts as though the operator was inserted between each bit of the word. For example, if a is three bits, &a is a[2]&[1]&a[0].

reg: The declaration used when a value is generated by *behavioral* Verilog code.

Register: A *sequential(1)* device that can load, and for some register types otherwise manipulate a value. The value in a *synchronous* register changes at the next rising edge of the clock. Contrast with *combinational*.

repeat: A Verilog loop that repeats a known number of times. Very different than the bottom testing loop.

Reset: The only *asynchronous* signal used in this book, which clears the *present state*.

Resource sharing: A *synthesis* optimization where the same hardware unit is used for multiple computations.

Rising delay: The propagation delay it takes for an output to change to 1.

ROM (Read Only Memory): A tabular replacement for combinational logic. Not an actual *memory* because it does not have the ability to forget.

RTL: 1. "Register Transfer Logic." In the pre-Verilog literature, the term RTL meant the logic equations generated by the controller to implement register transfers (section 4.4.1). Today, RTL most commonly means *explicit style* behavioral Verilog.

Some vendors (notably Synopsys) also use RTL to describe *implicit style* design and *RTN*. This book avoids the use of the term RTL, in favor of the more precise terms: implicit, explicit and RTN. 2. "Rotate Two Left", a PDP-8 instruction.

RTN (Register Transfer Notation): An ← inside a *rectangle* or *oval* of an *ASM* that evaluates an expression during the current clock cycle, but that schedules the change of the left-hand register to occur at the next rising edge of the clock. Similar to the Verilog *non-blocking assignment* (<= @ (posedge sysclk)).

SDF (Standard Delay File): A way to backannotate delay information into a netlist after place and route.

Sensitivity list: The list of input variables of *combinational* logic. The variables in the sensitivity list occur inside @ separated by or. Failure to list all variables can cause unwanted *latches*.

Sequence: The order in which Verilog statements execute in simulation. Statements in a particular always or initial block execute sequentially, regardless of *time control*.

Sequential: 1. A device that has memory, such as a *controller* or a *register*, as opposed to *combinational* logic. 2. Two or more *dependent* computations that occur in a particular *sequence*, even if they occur at the same $time in a Verilog simulation. 3. When one assumes physical time and *sequence* are the same, the opposite of *parallel*.

Simulation: The interpretation of Verilog source code to produce textual output and timing diagrams.

Single-cycle: A machine that requires one slow clock cycle to produce one result. Compare with *multi-cycle* and *pipeline*.

Special-purpose computer: A machine that is customized to implement only one algorithm, as opposed to a *general-purpose computer*. Special-purpose computers are often referred to simply as digital logic.

Standard Delay File: See SDF.

strength: Additional information about a wire that models its electrical properties.

State: A step that is active in an algorithm during a particular clock cycle.

Status: See *internal status* and *external status*.

Structural: An interconnection of wires and gates (or *combinational* and *register* devices) that forms a machine. Represented by a block diagram, circuit diagram, *instances* of *modules* or a *netlist*.

Structured programming: Describing an algorithm with high-level software control statements, such as `if`, `case` and `while`, but avoiding *goto*.

Subbus: A *concatenation* of a subset of bits from a *bus*.

Superscalar: A *general-purpose computer* that is able to execute more than one instruction per clock cycle.

Synchronous: A device that makes changes only at an edge of a clock signal, typically the rising edge.

Synthesis: The automatic translation of Verilog source code into a *netlist*.

System clock: The single clock used in a completely *synchronous* design.

*Test code: Non-*synthesizable* Verilog code used only in simulation to verify the operation of other, possibly synthesizable, Verilog code that models hardware. Test code (sometimes called a testbench) gives an abstract model of the environment in which the hardware will operate.

Testbench: *see test code*

`time`: A declaration for a variable that stores the result of `$time`. Often, `integer` is used instead.

`$time`: The current simulation time step.

Time control: Verilog features that cause `$time` to advance in *simulation*: `@`, `#` and `wait`

`triand`: Similar to `wand`.

`trior`: Similar to `wor`.

`trireg`: A variation of `wire` that holds the last value when all outputs are `1'bz`.

Tri-state: A device that has the ability to disconnect itself electronically from a bus. Verilog models this using the high impedance value (`'bz`).

Turn off delay: The propagation delay it takes for an output to change to `1'bz`, associated with a *tri-state* device.

Unary code: A bit pattern where exactly one of the bits is a one. See *one hot*.

Unconditional: *Non-blocking assignments (RTN)* and/or *command signals* that are synonymous with a machine being in a particular *state*, regardless of any other conditions that may hold. See *Moore* and *rectangle*.

Unknown value: A value (`1'bx`) that models an uncertain condition for one bit, such as two fighting output `wires`, typically indicative of an error in a *structural* design. Also used in chapter 6 to model abstract *hazards*.

Value Change Dump (VCD): A standardized text file format created by `$dumpfile` and related tasks that record values of simulation variables and that is used by post-simulation analysis tools.

Verilog: The hardware description language used in this book. (See chapter 3.)

VHDL: The other hardware description language, which is more complicated for the beginner than Verilog.

wand: A variation of `wire` that produces 0 instead of `1'bx` when two outputs fight, which implements & without using a gate (wired AND).

wor: A variation of `wire` that produces 1 instead of `1'bx` when two outputs fight, which implements | without using a gate (wired OR).

wire: The declaration used when a value is generated by *structural* Verilog code.

Wirewrap: A technique of connecting chips together involving using a tool that wraps wire around posts connected to the *pins* of the chips. A convenient way to fabricate prototypes in an educational or hobby setting.

Worst case: A model of a device that only considers the longest *propagation delay* possible.

~&: The bitwise NAND operator, `a~&b  ===  ~(a&b)`, which is not found in C.

~|: The bitwise NOR operator, `a~|b  ===  ~(a|b)`, which is not found in C.

~^: The bitwise coincidence operator, `a~^b  ===  ~(a^b)  ===  a^(~b)  ===  a^~b`, which is not found in C.

J. LIMITATIONS ON MEALY WITH IMPLICIT STYLE

Chapters 5, 9 and 10 and sections 7.4 and 11.6 discuss ASM charts and Verilog simulation for Mealy state machines, in which an operation (inside an oval of an ASM chart) is conditional. Mealy state machines are more problematic than the Moore state machines used in the rest of this book. There are limitations on using the Mealy approach with implicit style Verilog.

There are three consequences of using an oval in an ASM chart. The first consequence, which can be described with implicit style (pure behavioral) Verilog, is to allow computations dependent on a decision to be initiated in parallel to the decision. For example, the decoding and execution of a TAD instruction in chapter 9 illustrate a decision (ir2[11:9] == 1) and a computation (ac+mb2) that occur in parallel:

```
if (ir2[ 11:9]  == 1)
   ac <= @(posedge sysclk) ac + mb2;
```

As in most of the examples in this book, the statement that carries out the computation is a non-blocking assignment, so the effect will not be observable until the next rising edge of the clock. When viewed by itself, the architecture is a Moore machine (it has registers that only change at the rising edge). Since the output (ac) of the complete machine (the controller together with the architecture) only changes at the clock edge, the complete machine is Moore. Only the controller is Mealy. For this reason, implicit style Verilog with non-blocking assignment can model such situations.

The second consequence of using an oval in an ASM chart arises only at the mixed stage, such as figure 5-2. Depending on how complicated the architecture is, there may be hazards created between the controller and architecture during simulation that an implicit style Verilog description of the controller will not process properly. The 1994 paper mentioned below describes a 'bx handshaking technique with an exit_current_state task that overcomes this problem for Verilog simulation. This technique is an extension to the enter_new_state method given in this book.

The third consequence of using an oval in an ASM chart arises only when a decision involves an input to a machine, and RTN is not used to produce the corresponding output. Figure 5-7 is an illustration of such a situation. For such ASMs, the machine cannot be modeled just with implicit style Verilog (`@ (posedge sysclk)` inside `always`) because the output of the machine is supposed to follow the input. In other words, if the input makes multiple changes during one clock cycle, the output should make corresponding changes during that clock cycle. The implicit style cannot model this, since the behavioral block will execute only once. Since figure 5-7 is simple combinational logic (single-state ASM), the designer uses the appropriate sensitivity list instead of `@ (posedge sysclk)`. In general, Mealy machines often have multiple states, but there is no implicit style notation to describe this reexecution of the behavioral code that must take place in each Mealy state. It is necessary to use explicit style Verilog instead. The 1998 paper gives more information about this.

It is possible to use a hybrid implicit/explicit style to cope with a machine that has Mealy external outputs, such as the ASM in section 5.2.4 (figure 5-6). This hybrid approach is synthesizable. The following shows in bold the distinctions between the simulation only technique of section 5.3 and the hybrid implicit/explicit approach:

```
reg s;
always  //implicit block
  begin
      s <= @(posedge sysclk) 0;
    @(posedge sysclk) #1;
      r1 <= @(posedge sysclk) x;
      //ready = 1;
      if (pb)
        begin
          r2 <= @(posedge sysclk) 0;
          while (r1 >= y)
            begin
                s <= @(posedge sysclk) 1;
              @(posedge sysclk) #1;
                r1 <= @(posedge sysclk) r1 - y;
                if (r1 >= y)
                  r2 <= @(posedge sysclk) r2 + 1;
              //else
              //   ready = 1;
            end
        end
```

Continued

```
always @(s or r1 or y)   //explicit block
 begin
   if (s==0) ready = 1;
   else if (r1 >= y) ready = 0;
   else ready = 1;
 end
```

The hybrid approach eliminates the `enter_new_state` task with an internal state variable, `s`. A separate `always` block implements the combinational logic that computes the Mealy output based on the internal state variable and the condition(s) specified in the Mealy decision (`r1 >= y`). Every usage of a Mealy command, including unconditional ones such as in the in the top state of this example (`s==0`), must be generated by an `always` block dedicated to that command signal.

J.1 Further Reading

Arnold, Mark G., Neal J. Sample and James Shuler, "Guidelines for Safe Simulation and Synthesis of Implicit Style Verilog," *7th International Verilog HDL Conference,* Santa Clara, CA, March 16-19, 1998, pp. 67-69.

Index

with 'ENS, 461-462

Decision:
    in ASM charts, 12
    time control within, 191
    translated as one bit wide demux,
        250
Declaration, *see also* variable:
    `event`, 212
    `function`, 114
    `inout`, 551
    `input`, 118
    `integer`, 67
    `output`, 119
    `real`, 115
    `reg`, 67, 119
    `task`, 110
    `tri`, 550
    `triand`, 578
    `trior`, 578
    `trireg`, 578
    variable, 67
    `wand`, 579
    `wire`, 67, 118-119, 494, 579:
    `wor`, 579
Decoder, 515
Decoding instructions, 297
'DECREMENT, 511
default, 69, 116, 459
`define`, 73
`defparam`, 570
Delay:
    inertial, 566
    line, 288
    propagation, *see* propagation delay
    minimum/typical/maximum, 207
    rising/falling, 207
    transport, 566
Delayed assignment, 12
Demoboard (Vantis), *see* M4-128/64
Demultiplexer, *see* demux
Demux (demultiplexer), 513:
    in memory, 284
    misuse of, 514
    translated from a decision, 250
`depend` function, 417

Dependency:
    data, 359
    examples, 404
    software, 26
Design:
    automation, 22, 438
    flow, synthesis, 439
    hierarchical, 52
Deterministic access time, 282
Devices, programmable, 518
Diagram:
    block and circuit, 54, 539
    bus timing, 528
    timing, 526
Diamond, 7,12, 13
'DIFFERENCE, 511
Digital:
    building blocks, 6, 491, 525
    design, 3
    electronics, 533
Digital Equipment Corp., *see* DEC
DIMMs (Dual In-line Memory Modules),
    289
Direct, 202:
    addressing mode, 487
        current page, 488
        page zero, 292, 475
Directive, synthesis, 459
`disable`:
    inside `forever` with bottom test-
        ing loop, 273
    statement, 213, 273
Discrete electronic devices, 120
Division:
    childish (*see also*: Childish division):
        algorithm, 23
        with conditional instructions, 428
    combinational, 507
    mixed two-state example, 271
    pure behavioral two state example,
        270
    machine:
        architecture, 154
        controller, 157

switch debouncer, 472
versus implicit, 99
Expression, 69
External:
command output, 18
data:
input in ASM, 16
output, 18
External status, 14, 571
input, 16
Extra state for interface, 312

F

Factory, analogy to pipeline, 227
Ferranti, 279
Fetch state, 390
Fetch/execute, 1:
ASM for, 294, 304
behavioral, 290
mixed, 324
registers needed for, 292
Field Programmable Gate Array
(FPGA), 443
Fighting outputs, 78
Filling pipeline, 231
Finite state machine, 8, *see also* ASM:
ARM 388, 400,
logic equation, 167-168
Mealy, 182, 184-185
Moore, 26, 30, 34, 39, 220, 224,
232
netlist, 169
PDP-8 294, 302, 333
Verilog:
behavioral, 138-149, 186-188
explicit, 472
implicit, 258-265, 270-271, 464,
471
mixed, 158-159
structural, 162-165
Flattened netlist, 54
Flip flop:
D type, 249, 530
macrocell, 442

one hot, 249
Flushing pipeline, 231
Font, 5
`for`, 69, 80, 85
`forever`, 72, 106
`fork`, 571
Forrester, Jay W., 288
Forwarding data, 360
Four-state division machine, 134
Four-valued logic, 77, 549
FPGA (Field Programmable Gate Array),
443
Friendly user, 24, 141
Full:
adder, 54
`case`, 459
`function`, 109;
`car`, 458
combinational logic, 115
`condx`, 390
`depend`, 417
`dp`, 391
`state_gen`, 163
syntax, 114

G

Gajski, Daniel D., 59, 247, 521, 541
Gate level modeling, advanced, 207
Gate:
instantiation in Verilog, 75
non-tristate, 544
tristate, 545
General purpose computer, 1, 561:
benchmarks, 320, 351, 371, 431-
432, 481
bit serial, 476
history, 277
PDP-8, 485
pipelined, 354
RISC and CISC, 475
structure, 279
superscalar, 411
Glitch, 205
Goto, arbitrary, 194

input:
    argument of `task`, 110
    port, 15, 118-119, 442
    unused, 155-157, 538
Instance:
    behavioral, 118
    structural, 118
Instantiation, 54:
    by name, 447
    by position, 447
    module, 117
    multiple gates, 75
Instruction:
    concatenation, 305
    data processing, 382
    register (`ir`), 293, 308
    set, 291
        architecture, 573
`integer`, 67, 71:
    declaration, 67:
        in text code, 93
Intel, 288, 314, 378
Interconnection errors four-valued
    logic, 77
Interface:
    extra states for, 312
    push button, 25, 317, 469-471
Interleaved memory, 403
Internal status, 43, 573
International Verilog Conference, 559
Interrupt, 376, 490
'INVALID, 416
IOF (interrupt off), 490
ION (interrupt on), 490
Iowa State University, 3, 277
`ir`, *see* instruction register'
ISZ, 487,376,484

J

Jacquard, 277
Jamesco, 557
Java, 1, 6, 381
JEDEC format, 442
.j1, 442

JMP instruction, 309, 362, 487
JMS, 487
Joslin, R. D., 582

K

Kilburn, Tom, 279-280, 287, 336

L

Latch, 459, 573
Lavington, S., 352
LDR, 436
Lee, James M., 130
Line, 338
Little endian notation, 67
Logic:
    combinational, 52, 89, 454, 491, 543
    equation approach, 167
    four valued, 77
    gates in Verilog, 74
    sequential, building blocks, 525
    synchronous, 90, 445, 527
    `wire`, 495
Logical operators (&&, ||, !), 13, 70, 509:
    ALU, 509
Loop:
    bottom testing, 189:
        `disable` inside `forever`, 273
    `while`, implicit style, 463
Lovelace, Lady Augusta Ada, 277

M

M4-128/64, 558:
    CPLD, 442
    demoboard, 443, 470, 557
`ma` (memory address register), 219:
    multi-cycle, 224
    pipelining of, 229
    single cycle, 219
Machine language:
    cache test, 339, 349
    childish division program, 318, 369,
        424, 426, 429
    program, 298, 485
    R15 test, 422